Energy and Environment

Research for Innovative Transports Set

coordinated by
Bernard Jacob

Volume 1

Energy and Environment

Edited by
Michel André
Zissis Samaras

iSTE

WILEY

First published 2016 in Great Britain and the United States by ISTE Ltd and John Wiley & Sons, Inc.

ISTE Ltd
27-37 St George's Road
London SW19 4EU
UK

www.iste.co.uk

John Wiley & Sons, Inc.
111 River Street
Hoboken, NJ 07030
USA

www.wiley.com

Library of Congress Control Number: 2016936175

British Library Cataloguing-in-Publication Data
A CIP record for this book is available from the British Library
ISBN 978-1-78630-026-3

Contents

Acknowledgments

The European Commission (DG-MOVE and DG-RTD), the Conference of European Road Directors (CEDR), the European Road Transport Research Advisory Council (ERTRAC), the European Rail Research Advisory Council (ERRAC) and the European Technology Platform WATERBORNE-TP are acknowledged for their support and active contribution to the Programme Committee of TRA2014, in charge of reviewing and selecting the papers presented at the conference, which forms the main input of this volume.

The French Institute of Science and Technology for Transport, Development and Networks (IFSTTAR) is acknowledged for having organized the TRA2014, in which 600 high-quality papers were presented successfully.

Yves Amsler, Gianfranco Burzio, Panos Papaioannou and Mark Robinson, the coordinators of the topics on Mobility and Safety, all the other members of the Programme Committee, the reviewers who actively contributed to the review and selection of the papers, and the authors who wrote them are acknowledged for their great job that produced the material for this volume.

Joëlle Labarrère, Secretary of the Programme Committee of TRA2014, is acknowledged for her valuable help to the editors and for her support to this volume production.

Preface

The transport sector is very much concerned about environmental adaptation and mitigation issues. Most of these are related to the objective of curbing GHG emission by 20% by 2020, alternative energy and energy savings, sustainable mobility and infrastructures, safety and security, etc. These objectives require the implementation of advanced research work to develop new policies, and to adjust education and industrial innovations.

The theme and slogan of the Transport Research Arena held in Paris (TRA2014) were respectively: "Transport Solutions: From Research to Deployment" and "Innovate Mobility, Mobilise Innovation". Top researchers and engineers, as well as private and public policy and decision-makers, were mobilized to identify and take the relevant steps to implement innovative solutions in transport. All surface modes were included, including walking and cycling, as well as cross modal aspects.

Policies, technologies and behaviors must be continually adapted to new constraints, such as climate change, the diminishing supply of fossil fuels, the economic crisis, the increased demand for mobility, safety and security, i.e. all the societal issues of the 21st Century. Transport infrastructures and materials, modal share, co-modality, urban planning, public transportation and mobility, safety and security, freight, logistics, ITS, energy and environment issues are the subject of extensive studies, research work and industrial innovations that are reported in this series of books.

This book is a part of a set of six volumes called the *Research for Innovative Transports* set. This collection presents an update of the latest academic and applied research, case studies, best practices and user perspectives on transport carried out in Europe and worldwide. The presentations made during TRA2014 reflect on them. The TRAs are supported by the European Commission (DG-MOVE and DG-RTD), the Conference of European Road Directors (CEDR) and the modal European platforms, ERRAC (rail), ERTRAC (road), WATERBORNE, and ALICE (freight), and also by the European Construction Technology Platform (ECTP) and the European Transport Research Alliance (ETRA).

The volumes are made up of a selection of the best papers presented at the TRA2014. All papers were peer reviewed before being accepted at the conference, and they were then selected by the editors for the purpose of the present collection. Each volume contains complementary academic and applied inputs provided by highly qualified researchers, experts and professionals from all around the world.

Each volume of the series covers a strategic theme of TRA2014.

Volume 1, *Energy and Environment*, presents recent research work around the triptych "transports, energy and environment" that demonstrate that vehicle technologies and fuels can still improve, but it is necessary to prepare their implementation (electromobility), think about new services and involve enterprises. Mitigation strategies and policies are examined under different prospective scenarios, to develop and promote alternative fuels and technologies, multi-modality and services, and optimized transport chains while preserving climate and the environment. Evaluation and certification methodologies are key elements for assessing air pollution, noise and vibration from road, rail and maritime transports, and their impacts on the environment. Different depollution technologies and mitigation strategies are also presented.

Volume 2, *Towards Innovative Freight and Logistics*, analyzes how to optimize freight movements and logistics; it introduces new vehicle concepts, points out the governance and organization issues, and proposes an assessment framework.

Volumes 3 and 4 are complementary books covering the topic of traffic management and safety.

Volume 3, *Traffic Management*, starts with a survey of data collection processes and policies and then shows how traffic modeling and simulation may resolve major problems. Traffic management, monitoring and routing tools and experience are reported and the role of traffic information is highlighted. Impact assessments are presented.

Volume 4, *Traffic Safety*, describes the main road safety policies, accident analysis and modeling. Special focus is placed on the safety of vulnerable road users. The roles of infrastructure and ITS in safety are analyzed. Finally railway safety is focused upon.

Volume 5, *Materials and Infrastructures*, is split into two sub-volumes, investigating geotechnical issues and pavement materials' characterization, innovative materials, technologies and processes and introducing new techniques and approaches for auscultation and monitoring. Solutions to increase the durability of infrastructures and to improve maintenance and repair are presented, for recycling as well as for ensuring the sustainability of the infrastructures. Specific railways and inland navigation issues are addressed. A focus is put on climate resilient roads.

Volume 6, *Urban Mobility and Public Transport*, highlights possible innovations in order to improve transports and the quality of life in urban areas. Buses and two-wheelers could be a viable alternative in cities if they are safe and reliable. New methodologies are needed to assess urban mobility through new survey protocols, a better knowledge of user behavior or taking into account the value of travel for public transport. The interactions between urban transport and land planning are a key issue. However, these interactions have to be better assessed in order to propose scenarios for new policies.

Bernard JACOB, Chair of the TRA2014 Programme Committee

Jean-Bernard KOVARIK, Chair of the TRA2014 Management Committee

March 2016

Introduction

Toward Cleaner, Efficient and Sustainable Transports: Context and Recent Research Works

I.1. Introduction

Transport systems are facing an impossible dilemma today: on one hand they must satisfy an increasing demand of mobility for a growing world population and an intensification of the goods exchanges, while on the other hand, they are also supposed to decrease their energy requirements and shift to non-fossil fuels (rarefaction and climatic impacts), while preserving or even improving the environment, decreasing the impacts of noise and air pollution on living beings, fauna and flora, to be precise. Besides that, transports have a unique opportunity to evolve in a changing world, with new services (vehicle sharing or in self-service), technologies like intelligent transportation systems (ITS), communication, etc., and also requirements including fast delivery, reliability, improved accessibility, etc.

In this book, recent research and application works – that were presented during the 5th Conference of Transport Research Arena, Paris, France, held on 14–17 April 2014 – are reported around the triptych: “transports, energy and environment”.

Successively, works will be reported on the progress and potential of electromobility and the conditions of its implementation (Part 1), the recent developments of vehicle and engine technologies for optimizing their operation while decreasing their energy needs and their environmental impacts (Part 2). Renewable and alternative energies are studied from both their technological and implementation points of view in Part 3.

Introduction written by Michel ANDRÉ and Zissis SAMARAS.

The next three parts adopt rather an environmental perspective, with respect to climate change and of the mitigation of transportation-related greenhouse gases (Part 4) and the issues of air and noise pollution due to transports in Parts 5 and 6.

As a prelude to this scientific and technical reporting, we propose a brief contextual overview, regarding the energy, environment and transport sectors, and their connections taking into account appropriate and coherent policies, towards the development of sustainable transport systems. This overview will be followed by a brief summary of the research works reported in the following chapters.

I.2. Context

I.2.1. *Consistent and coordinated energy and environmental policies are needed in Europe*

One of the major European concerns is increased oil and gas prices due to geo-political instability that may endanger EU economic performance [BER 14]. This requires the development of sound energy policy which should take into account environmental issues such as climate change, air quality, noise and other related impacts. As underlined in the Green Paper on a European strategy for sustainable, competitive and secure energy by the European Commission in 2006, Europe has to put energy policy high on the EU agenda if it is to achieve its economic, social and environmental objectives. The EU must exploit its position as the world's second largest energy market and as a world leader in demand management and the promotion of renewable energy sources. Improving energy efficiency and focusing on regionally sourced renewables will also benefit energy security by leading to lower imports of fuel, making countries less reliant on foreign supplies. Moreover, it will result in a more diverse energy mix and improve the resilience of national or regional energy systems.

I.2.2. *Co-benefits between energy and environment can be achieved*

It is important to note that significant co-benefits can be produced for health, quality of life, or even ecosystems and the economy by linking energy policies with those for climate change and environmental protection (on that subject, see for instance [WES 13, MCC 13, RAO 13, RAF 13, ALG 13]). Switching to clean energy would pay for itself, almost immediately, because actions to reduce greenhouse gas emissions reduce co-emitted air pollutants too, thus bringing co-benefits for air quality and human health, as well as for natural environment and ecosystems.

Overall, it seems that there is a strong incentive for countries to start cutting back on fossil fuels as soon as possible, as the co-benefits of avoided air pollution mortality alone can justify substantial reductions in greenhouse gases (GHG) emissions. Air quality and health co-benefits, especially as they are mainly local and near-term, can therefore provide strong additional motivation for transitioning to a low-carbon future.

I.2.3. *Energy and environmental policies face a multiplicity of actors*

Energy and environmental policies face difficulties due to multiple layers of regulation, levels of governance, multiplicity of actors and, finally, complex interactions between multiple pollutants, pollutions and environmental impacts. Climate change policy is affected by multiple decision makers at local, sub-national, national and transnational levels [BON 12]. Usually, these decision makers are not fully coordinated with respect to goals and methods, and the existence of several layers of governance may encourage strategic behavior from powerful local actors trying to enhance their own positions. Multi-level climate change governance is also related to governance of local environment and quality of life since production processes often involve emitting several pollutants and harmful effects simultaneously. Environmental policies aiming at reducing CO_2 emissions might, therefore, create trade-offs. For example, a common strategy to reduce CO_2 emissions is switching the fuel mix from oil towards bio-fuels. However, while net CO_2 emissions do fall with biofuels, a darker side also exists with an increase in the levels of nitrogen oxides.

I.2.4. *Transport: a vital sector with increasing energy and environmental concerns*

The transport sector is a vital component for fulfilling the three freedoms of the European single market: the free movement of persons, services and goods [KAL 15]. The demand for transport has been rising for many years at both national and European levels; and it accounts for the majority of all liquid fuel consumption. Therefore, air pollutants, GHG emissions, noise and other environmental damages caused by the transport sector are expected to increase, further contributing to pollution, adverse health effects and climate change. In this framework, the push to move toward energy efficiency and alternative fuels in the transport sector is expected to continue as well. The “EU-2020 strategy” includes the flagship initiative “resource efficient Europe” [EUR 11a] in which the European Commission presents proposals aiming at cleaner, more efficient and more sustainable transport through the adoption of measures such as research and innovation, setting current standards and developing the necessary infrastructure support as well as regulatory measures such as pricing. The White Paper on Transport [EUR 11b] takes into account major policy initiatives for a competitive and resource efficient transport system under sustainable developments. Furthermore, the European Road Transport Research Advisory Group (ERTRAC [ERT 10]) sets out the following ambition: "Towards a 50% more efficient road transport system by 2030".

In that context of energy and environmental high-level requirements, a cleaner, more efficient and sustainable transport system should develop in the frame of coherent energy and environmental policies, but transport, land use and urban planning policies should also be oriented accordingly. In all, this means a highly integrated approach for successful development and the implementation of improved or alternative transport technologies and energies. Besides that, transports have today a unique opportunity to evolve in a changing world, with the development of new services (vehicle sharing or in self-service), technologies (ITS, communication, etc.) and also with changing requirements (fast delivery, reliability, improved accessibility, etc.). These opportunities can and should be seized for a possible renewal of our transportation systems and of our mobility practices.

I.2.5. *Research and application works for cleaner, efficient and sustainable transport systems*

Evidently, research, innovation and application works should be the main driving forces of such an integrated development for clean and sustainable transports, considering technical but also organizational, social and human aspects of such evolution. These works are the basis of the related-knowledge production and of the technological development and demonstration, and should contribute to the understanding of organizational and governance interactions, their optimizations, the elaboration of pertinent and acceptable policies, and finally to the real-world implementation and assessment of these innovative transport technologies and services.

I.3. Recent research results of the "transports, energy and environment" triptych: a summary

The 5th Conference of Transport Research Arena (Transport Solutions from Research to Deployment) held in Paris, France during 14–17 April 2014, was the opportunity to explore recent research works around mobility of people and goods with respect to energy, environment, safety, security and economic issues, considering all surface transport modes, including crossmodal issues.

The present volume compiles a selection of scientific and application papers related to recent and most significant research works around that triptych: "transports, energy and environment".

The works in this book demonstrate that vehicle technologies and fuel can still improve (electromobility, hybrid, hydrogen/ fuel-cell, CNG, etc.), but it is necessary to prepare their implementation, to think of them as elements of new services, and to involve all actors early in the design of transports and services, in particular – besides the vehicles, fuels, infrastructures and big companies – a high number of small and medium-sized enterprises that operate the transport or that could be the new actors of innovation. Other works have already pointed out the importance of considering human aspects, for a better appropriation and to improve behaviour (eco-mobility).

Prevision, prospective and orientation in the development of transports are key elements for reaching increasingly high and mandatory expectations in regards to climate and environment. Conditions for the transport and energy mutations should be carefully analyzed, as a strong influence of the economical context (of which fuel and mobility prices) was demonstrated and could even defeat the expected effects, while incentives or bonus/credits were envisaged to promote new technologies.

Mitigation strategies are thus proposed here as combination of measures and policies under different prospective scenarios, to promote alternative fuels, new technologies and services, promote multi-modality, decrease GHG and environmental impacts, while creating "value chains". For a better understanding and analysis of transport's economic impact, research has worked on better and agreed evaluation tools, such as carbon footprint that is particularly challenging for the complex freight transport chains and the high diversity of shippers, as also demonstrated.

With respect to air pollution, research has progressed around non-exhaust pollution, re-suspension of particles, non-regulated compounds with high concerns, and estimation tools. But a robust assessment of mitigation measures still depends on input data at the very local scale of which vehicle–fleet composition and traffic conditions remain to be key factors. Some works examine depollution technologies and fuels to decrease particulates and NOx as well.

Finally, this book proposes a compilation of non-conventional research works regarding noise and vibration from transports, addressing the acoustical impacts on marine life by shipping, on terrestrial organisms by vibrations and noise caused by trains, and on the opposite, how additional warning sounds could be managed for electric vehicles for safety while simultaneously optimizing their noise level, detectability and warning efficacy. Reduction technologies, assessment, (virtual) certification and perception (human and animal) are the main aspects of these works, which enrich the environmental noise problem by diverse application studies.

I.4. Bibliography

[ÅGR 13] ÅGREN C., ENERGY, Climate and air quality policy synergies. http://www.airclim.org/acidnews/energy-climate-and-air-quality-policy-synergies, 2013.

[BER 14] BERK M.M., BOLLEN J.C., EERENS H.C. *et al.*, Sustainable energy: trade-offs and synergies between energy security, competitiveness, and environment, MNP report 500116001/2006, Netherlands Environmental Assessment Agency, Bilthoven, The Netherlands, p. 15, 2014.

[BON 12] BONILLA J., CORIA J., SYNERGIES T., Trade-offs between Climate and Local Air Pollution Policies in Sweden, Department of Economics, University of Gothenburg, Sweden, p. 34, 2012.

[ERT 10] ERTRAC, Towards a 50% more efficient road transport system by 2030, available at: http://www.ertrac.org/uploads/documentsearch/id22/ERTRAC_SRA_2010_Technical_document.pdf, 2010.

[EUR 06] EUROPEAN COMMISSION, Green Paper on a European Strategy for sustainable, competitive, and secure energy. COM(2006)105 final, available at: http://europa.eu/documents/comm/green_papers/pdf/com2006_105_en.pdf, 2006.

[EUR 11a] EUROPEAN COMMISSION, A resource-efficient Europe – Flagship initiative under the Europe 2020 Strategy, available at: http://ec.europa.eu/resource-efficient-europe/pdf/resource_efficient_europe_en.pdf, 2011.

[EUR 11b] EUROPEAN COMMISSION, Roadmap to a Single European Transport Area – Towards a competitive and resource efficient transport system, White Paper, available at: http://eur-lex.europa.eu/legal-content/EN/TXT/PDF/?uri=CELEX:52011DC0144&from=EN, 2011.

[KAL 14] KALLAS S., Transport: driving Europe's economy forward, Transport Business Summit, Brussels, 27 March 2014.

[MCC 13] MCCOLLUM D.C., KREY V., RIAHI K. *et al.*, "Climate policies can help resolve energy security and air pollution challenges", *Climatic Change*, vol. 119, pp. 479–494, July 2013.

[RAF 13] RAFAJ P., SCHOEPP W., RUSS P. *et al.*, "Co-benefits of post-2012 global climate mitigation politics", *Mitigation and Adaptation Strategies for Global Change*, vol. 18, pp. 801–824, August, 2013.

[RAO 13] RAO S., PACHAURI, S., DENTENER F. *et al.*, "Better air for better health: Forging synergies in policies for energy access, climate change and air pollution", *Global Environmental Change*, vol. 23, pp. 1122–1130, October 2013.

[WES 13] WEST J.J., SMITH S.J., R.A. *et al.*, "Co-benefits of mitigating global greenhouse gas emissions for future air quality and human health", *Nature Climate Change*, pp. 885–889, September 2013.

PART 1

Electromobility and its Implementation

Introduction to Part 1

This first part covers a wide range of subtopics that assess different ways of implementing electromobility. Today, drive electrification represents a potent concept for achieving sustainable mobility. Electromobility is now the key technology for the replacement of fossil energy sources in the long-term, creating new markets and new technologies as the old energy order comes to an end and a new one begins.

One paper reviews the results of the first two years' work of the Green eMotion project, which aims at developing and demonstrating a commonly accepted and user-friendly framework that combines interoperable and scalable technical solutions with a sustainable business platform. Two papers address the issue of electromobility services in Italy demonstrating the effectiveness of and encouraging the deployment of intelligent vehicle systems, designed specifically for electric vehicles on European roads. Another paper outlines the development of the training programmes for safe eco-driving of clean vehicles, and describes the framework for evaluating their impact. The chapter concludes with some major results on improving the capacity of European SMEs to develop and implement products in the low-carbon transport sectors.

Introduction written by Zissis SAMARAS.

1

Toward a Europe-Wide Interoperable Electromobility System

In March 2011, the European Commission kicked off a four-year Europe-wide initiative called Green eMotion. The aim of Green eMotion is to define and demonstrate an interoperable and user-friendly electromobility system based on existing installations and the experience thus gained. One of the goals of Green eMotion is to create an ICT system concept for the Europe-wide electromobility services marketplace (including definition of the requisite interfaces) and to demonstrate it. Almost all the 10 Green eMotion demo regions in Europe are using this business-to-business (B2B) marketplace, linking their IT systems to the marketplace via open standard interfaces. The test phase, which began in September 2012, comprises functions used for locating charging points and clarifying charging contracts with third-party providers (roaming). In addition to this key topic of ICT, Green eMotion is also working towards important developments in other areas:

1) improvements in standardization;

2) technical reports and policy recommendations;

3) extensive evaluation of static and dynamic electromobility data;

4) demonstration of a new approach to the sharing of electric vehicles in urban areas.

1.1. Background

Climate change and the reduction of greenhouse gas (GHG) emissions are one of the major challenges of our time. A growing world population is leading to higher demand for mobility and energy, aggravating the problem and pushing emission levels upwards. The transport sector faces serious environmental, economic and societal challenges, which in the short- and medium-term will necessitate the transformation of the system.

Chapter written by Gabriele GIUSTINIANI, Luca PERSIA, Heike BARLAG and Norbert VIERHEILIG.

There is a need to mitigate the negative impacts on the environment and transport that will seriously impede the goal of reducing GHG emissions by 20% before 2020, as well as the longer-term climate goals. Until now transport has been the only sector in which GHG emissions have increased since 1990.

Meanwhile, the proportion of European citizens living in urban areas is increasing steadily. It is expected that by 2050 some 84% of the population will live in cities. Urban transport accounts for 40% of CO_2 emissions and 70% of emissions of other pollutants from road transport. The congestion in urban agglomerations must be brought to tolerable levels, while a growing demand for mobility has to be satisfied in the near future.

For this reason, the decarbonization of road transport, starting from urban areas, is one of the top priorities of policymakers in Europe. Battery-powered electric vehicles are seen as one of the most promising long-term sustainable options. For a number of reasons, such as improved battery performance and changes in emission performance standards, sometime in the next ten years electric and plug-in hybrid vehicles (EV and PHEV)[1] are likely to begin to gradually replace vehicles using internal combustion engines. They offer advantages in energy efficiency and a number of environmental aspects, such as their potential contribution to reducing CO_2 and local emissions and other negative external effects of mobility (e.g. noise), but some conditions are still keeping EVs from a full rollout on the market.

In some niche markets, EVs are already performing well, and EVs are gradually being introduced to more markets thanks to a large number of national and local demonstration initiatives.

The extent to which EV and PHEV will contribute to reducing GHG emissions also depends on the energy mix used to recharge the battery. Theoretically, the driver can choose whether to run his EV using conventional electrical energy sources or strictly from renewable electrical energy sources. However, EVs also have the potential to contribute to the integration of renewable energy sources (RES) into the electricity grid as they provide storage and demand response options.

The implementation of alternative mobility concepts requires consumer acceptance, meaning that EV must provide car owners with the same kind of ease-of-use, safety and reliability at reasonable prices. A comprehensive and standardized recharging infrastructure is one of the aspects that will favor consumer acceptance of EVs.

1 For simplicity, the abbreviation EV will be used for both electric vehicles and plug-in hybrid vehicles.

1.2. The Green eMotion project partnership and objectives

To enable the mass market for electromobility in Europe, important players from industry, the energy sector, municipalities and universities and research institutions have joined forces in the Green eMotion project (henceforth, GeM), which was kicked off in March 2011. GeM is a four-year EU funded project with 43 partners, coordinated by Siemens AG. An overview of the partnership is provided in Figure 1.1; for a more detailed description, see http://www.greenemotion-project.eu/.

The GeM partnership is convinced that transport, as a network industry, must be approached as a system integrating the ICT elements, services, interoperable equipment (vehicles and recharging infrastructure), and grid infrastructure as well as policies, incentives, and urban mobility concepts to ensure smooth and efficient interaction and, ultimately, a self-explanatory overall market system.

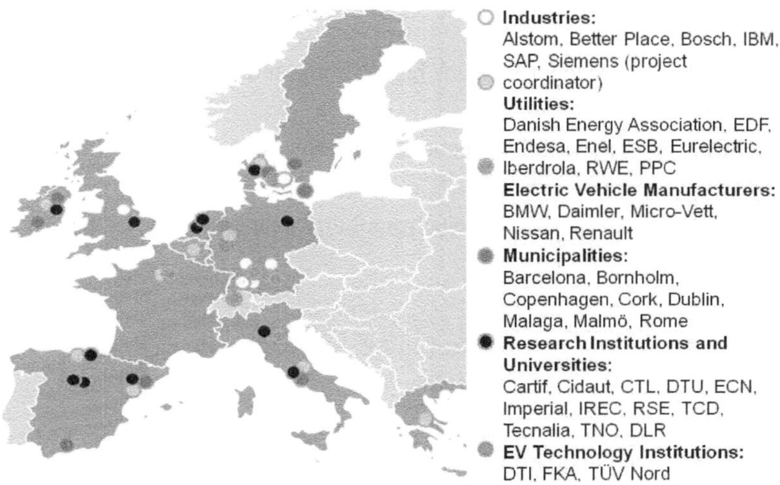

Figure 1.1. *The GeM partners. For a color version of the figure, see www.iste.co.uk/jacob/energy.zip*

GeM aims to establish a unique and user-friendly framework for green electromobility in the EU. Based on existing regional demonstration projects, feasibility and scalability of an interoperable and standardized mobility system including a clearing house[2] will be demonstrated to establish the best conditions to

2 Clearing house: this enables roaming for electricity customers across Europe (comparable with mobile phone services).

allow businesses and consumers to drive EVs and, thus, enable their mass-market introduction.

As a result, the user-friendly, interoperable, and scalable framework demonstrated will serve as a benchmark for further regions and their interconnection in Europe. In short, GeM will provide policymakers, urban planners, and electric utilities with the necessary tools to facilitate the Europe-wide rollout of EV.

1.3. GeM achievements so far

In the first two years of the project GeM partnership work covered four different areas:

– development of a marketplace as common innovative service platform: Europe-wide roaming and other services connecting the electromobility market players;

– improvement in standardization: highlighting and addressing the most urgent "new" needs;

– technical reports and policy recommendations: lessons learned in the practical implementation of electromobility;

– extensive amount of valuable electromobility data: use patterns, environmental impact (e.g. CO_2).

1.3.1. *Development of a marketplace as common innovative service platform*

GeM's first step was to create an ICT system concept for a European marketplace[3] for electromobility services and to demonstrate its operability in practice. For the first time, such a Business To Business (B2B) marketplace is bringing together all the actors involved in this segment: providers of charging services, charging station operators, power utilities, and suppliers of other electromobility-related services.

Such a marketplace permits associated service offerings – for instance, roaming between different charging point operators – to be provided on a pan-European

3 An electronic marketplace (for eMobility) is the virtual place where all participants in the ecosystem can connect and operate (process) their business (transactions). The marketplace in this proposal is not to be understood as the electricity or vehicle market. The "electromobility (mass-) market" refers to the whole set of products (vehicles, electricity and infrastructure) and services associated with electric vehicles and their operation.

scale, similar to current practices with cellphones. Furthermore, since the marketplace is open to other service providers, this would allow for innovative electromobility concepts to be made available more easily (see Figure 1.2). As the central technological component, the marketplace serves the following two functions:

– to reduce the number of interfaces between participants;

– to prove easy access to all offerings through standardized software services.

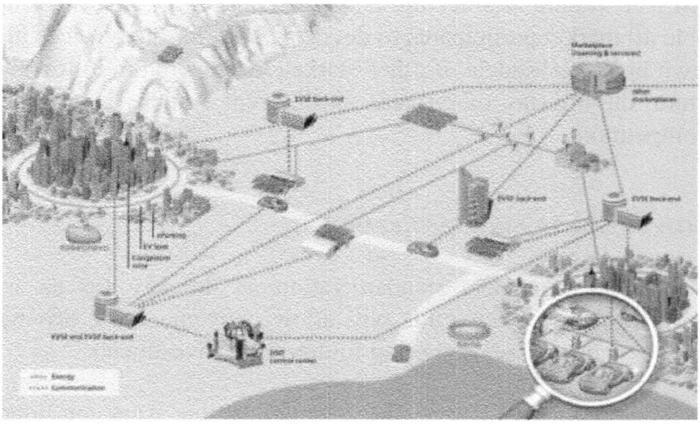

Figure 1.2. *The "Big Picture" of electromobility shows the complexity of a real interoperable system*

Right now, the first demo regions are connecting up. Subsequent phases will include energy-related services for optimizing grid utilization and new, added-value services for end customers, e.g. reservation of charging points.

Drivers of electric cars can make use of such services, which provide access to a variety of electromobility offerings. The result is a global system that allows for competition and individual solutions. Furthermore, the GeM marketplace for electromobility communicates with other marketplaces, such as pre-existing regional solutions, allowing broader integration beyond the GeM partners.

Currently two added value services are available through the marketplace:

– the roaming service allows the EV drivers who are customers of one GeM partner to access the charge points of the other partners easily. This is supported by a clearing house made available as a service on the marketplace. The clearing house service is in the background, transparently validating contracts of business partners involved in roaming;

– the search for EVSE service offers the charge spot data from most of the GeM demo regions. B2B partners connected to the marketplace can utilize this data and offer EV drivers to search for the charge spots through either a website, a mobile phone application or an in-car display.

The marketplace information hub is operational since August 2012 (see also Figure 1.3) and the integration of project partners using the enabled services is in progress. Successful tests between the first GeM demo regions have taken place.

The open ICT system architecture and the standardized interfaces and business objects enable all market participants to develop and offer the services in accordance with their own business models. If other marketplace operators follow the published IT architecture, the offering of services on different systems is enabled leading to enhanced competition.

1.3.2. *Improvement in standardization*

GeM partnership considers standardization as the key for an interoperable electromobility system, because this will allow convenient EV driving all across Europe, create new businesses and bring future-proof investments.

For this reason GeM focused the standardization work on identifiers and interfaces for the electromobility ICT systems, to enable roaming and allow a smooth system communication, independently from manufacturers, to allow an open competition.

Figure 1.3. *Commissioner Günther Hermann Oettinger (Energy) started operation of the first charging point at the European Commission on 19 June 2012*

To extend its results beyond Europe, GeM, with many other market players, initiated the electromobility ICT Interoperability Innovation Group (40 participants covering Asia, Europe and the Americas) to define industry standards for interfaces and user IDs for smooth communication between all the different electromobility ICT systems (for more information see www.eMI3group.com).

As the final standardization activity GeM proposed a New Work Item for the communication protocol between EVSE (EV Supply Equipment) and backend systems, leading to more competition for charging infrastructure and IT systems by avoiding proprietary protocols.

1.3.3. *Technical reports and policy recommendations*

Reports and recommendations were developed with the mass market for electromobility in mind. Obstacles encountered by the project partners in the electromobility policymaking process were described, as were the factors that made electromobility projects successful. This covers financial and non-financial incentives, infrastructure, EV issues and also the political environment.

Technical reports cover areas from infrastructure planning on different voltage levels to the performance of EVs in fleets.

1.3.4. *Extensive amount of valuable electromobility data*

GeM is also running an extensive data collection program in 10 demo regions, monitoring 600 electric vehicles and 1,800 charging points; 85% of the electric vehicles and 70% of charging points provide dynamic data sets today. These dynamic data are used by GeM teams to generate further results in the near future. Other external European mobility projects are being included in the data collection, and their data will be compared with GeM results.

Some results from the tests on site include:

– the mean distance run between EV charges is 37 km, but 75% of the users run a distance of up to 51 km;

– the maximum distance run between two charges has been 145 km;

– the mean battery level when starting a charging process is 65%, which means that most of the EV drivers charge their vehicle before the battery is drained;

– up to now, GeM EV drivers have run more than 65,000 km, saving more than 60 tons of CO_2.

1.3.5. *Developments in other areas*

In addition to building selected charging systems, infrastructure partners in GeM have issued preliminary findings on the effects of different charging strategies on the grid both in power quality and required reinforcement costs.

A deterministic network investment tool has been developed and successfully tested. It analyses the effects of different load charging profiles on real grids and calculates the total investment cost required in the network for hosting a specific percentage of EVs. The tool does not optimize the EV charging directly but is able to consider the value of control strategies by comparing the investment for different load profile shapes. For the first time, such a tool can be applied by the network operator itself. Furthermore, this tool will be publicly available.

GeM has also emphasized the design and development of EV-related hardware infrastructure.

This culminates in the completion of three important ongoing tasks:

– fast charging station, a novel charging station with second life buffer battery for integration of fast charging capability in weak grids;

– design and development of an inductive charger to provide wireless charging to a converted Nissan Leaf. This new design will allow easy conversion of EVs with CHAdeMO[4] interface;

– a 50 kW AC/DC/AC converter unit for grid re-enforcement application to support the grid in case of overloads due to a high number of charging EVs[5].

GeM is working on business models for electromobility. Key business model environments have been established, with all relevant interactions among actors defined. The goal is to evaluate the profitability of the different business models under various market considerations.

Concluding, GeM has done a broad set of measurements on various types of EVs. The goal is to evaluate the usability of EVs in everyday life. A new test setup for standard usage pattern (driving cycle) suitable for EVs was developed to show which range can be achieved realistically. Also, the influence of driving behavior on the driving range of EVs is under evaluation.

4 CHAdeMO is a standard for DC fast charging defined in Japan.

5 This project is running under the responsibility of one GeM partner. It is a grid buffering system realized with second life EV batteries to stabilize the grid and able to charge EVs.

1.4. Next steps

In the first two years of activities, GeM has already achieved important results and expects more of the same over the next two years, especially in:

– assessment of impact on EVs' diffusion of support/regulatory measures at local level;

– assessment of potential market penetration of EVs;

– demonstration of new forms of EV sharing and assessment of their transport and environmental impacts.

1.4.1. *Assessment of impact on EVs' diffusion of support/regulatory measures at local level*

The first activity is related to the understanding and assessment of the impacts on EVs' diffusion of support/regulatory measures that may be implemented at local level. In this regard, it is also important to understand and assess whether and how the impact of such measures varies according to local conditions such as population or gasoline prices. The assessment will be carried out by analyzing the support/regulatory measures already implemented on a sample of cities involved in GeM and the impacts they have had or are having on the diffusion of EVs. The methodology for the assessment has already been defined. Three main steps are foreseen for the assessment:

– survey of cities involved in GeM;

– analysis of data collected;

– review of the results and draft of a report.

The survey will be carried out by means of a questionnaire to the city representatives. The data collected will be of three types.

The first set of data is related to the city being investigated (population trends, area of the city and gasoline cost evolution).

The second set is related to the possible measures implemented to support the diffusion of EVs. The data will be related to how and when the measures have been implemented. Some of the measures considered in the survey are limited traffic zones in inner cities open/free for EVs, exceptions to road pricing for EVs in urban areas, the possibility for EVs to use bus lanes and reduction of local/national vehicle tax for EVs. In total, 11 possible measures will be investigated in the survey.

The third set of data is related to the evolution of EVs and internal combustion engine vehicles in recent years.

Once all the data are available, analysis will begin. Different analysis methods (e.g. time series analysis) will be used, and cities will be clustered according to selected data (e.g. population or area of the city). The results of the analysis were expected in summer 2014.

These results will be very important because the measures proposed can support the diffusion of electromobility at little or no cost to the local administrations. For this reason, these results will be used to support cities in selecting suitable support/regulatory measures to foster the diffusion of EVs according to local conditions and will be synthesized in a report available to the public.

1.4.2. *Assessment of potential market penetration of EVs*

To support full deployment of EVs in Europe, it is important to understand their potential market penetration, especially as regards recharging locations. To assess the driving factors, a study is underway in GeM Demo regions using a Stated Choice survey. The survey asks householders to indicate their most likely car purchase given multiple options, including lease of car or just batteries.

An efficient design is built and customized on the most likely next car purchase of each household. In particular, respondents are presented with different car classes and are asked to assume that they are in the purchase situation (previously defined in the survey) assuming that at the car dealership there are both the conventional vehicle and a comparable EV available with the characteristics shown for a given car class. Other than that, the respondent must assume that the two cars are identical.

The data collection is already complete in Denmark, and those results will soon be available. The survey is still in progress in the other GeM Demo regions.

1.4.3. *Demonstration of new forms of EV sharing*

Another interesting activity to be carried out within GeM is a small demo of micromobility in Rome. Micromobility is a new concept of EV sharing in cooperation with public transport. It aims to solve the last-mile problem for individual transport by integrating conventional collective PT with EV sharing. This represents a real opportunity for PT to compete with private transport. Micromobility stations spread over the area allow users to rent an EV (e-bike, e-moped or e-car) in A to reach B and leave the EV in the B station. This concept, which integrates sharing of EVs with public transport, can completely alter the approach to urban mobility, and users can use conventional collective PT for covering the long part of the trip and cover only the last "mile" by e-vehicle. Moreover, although the business model of the micromobility has not yet been defined, micromobility e-vehicles will not pay parking fees and will be able to enter limited traffic zones.

An example of the potential coverage of such a service, integrated with the underground lines of Rome, is shown in Figure 1.4. Of course, it needs to be tested. Accordingly, Rome will implement a small demo with three stations and a total of 24 EVs (eight vehicles per station) of different types (e-bicycles, e-mopeds, quads and city cars). The demo aims to shed light on such aspects as:

– costs and financial sustainability of the service;

– user acceptance;

– technical and administrative problems that can hinder the diffusion of such a system.

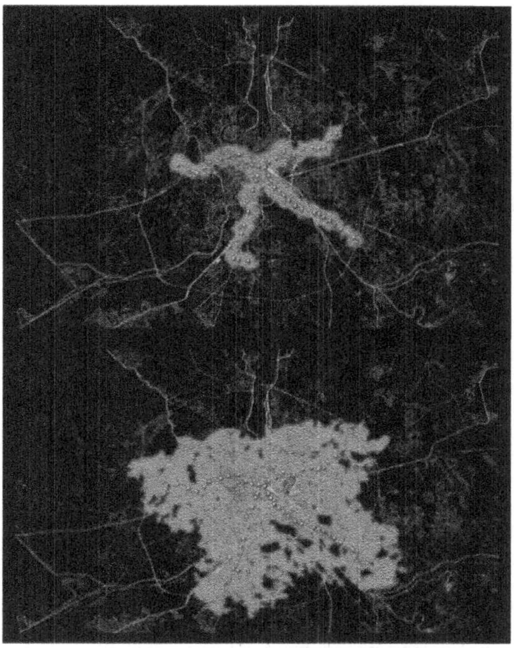

Figure 1.4. *Distance covered from underground station in Rome (above: distance covered by foot in 15 min; below: distance covered in 15 min with EVs, Vcom= kph)*

In parallel with the demo, which will provide important technical and financial information, a study will be carried out in Rome to:

– select the best area of Rome to deploy such a system and scale the system itself;

– estimate the number of private vehicle users that would shift to a combined use of PT and micromobility (considering also different scenarios of implementation);

– assess the potential reduction of pollutant emissions.

The results of this study are expected for autumn 2014 and will be used to evaluate a full deployment of micromobility in Rome.

1.5. Conclusions

In the first two years of work, GeM has already achieved important results. Its first step was to set up an ICT system concept for a European marketplace for electromobility services and to demonstrate its operability in practice. The marketplace is now open, and two added value services are available through the marketplace:

– the roaming service allows the EV drivers who are customers of one GeM partner to access the charge points of the other partners easily;

– the search for EVSE service offers the charge spot data from most of the GeM demo regions. B2B partners connected to the marketplace can utilize these data and allow EV drivers to search for charge spots through a website, a mobile phone application or an in-car display.

It is important to stress that the marketplace being developed is open. If more operators follow the published IT architecture, services can be offered on different systems, leading to greater competition.

To set up the marketplace and support the rollout of electromobility, GeM has carried out activities in:

– standardization of different protocols, procedures and tools;

– definition of recommendations to support policy makers and technicians in a successful rollout of electromobility;

– extensive data collection program in 10 demo regions;

– preliminary analysis on the effects of different charging strategies on the grid in power quality;

– design and development of EV-related hardware infrastructure;

– set-up of key business model environments with all relevant interactions between actors defined in order to evaluate the profitability of the different business models under various market considerations.

In the next two years, other important activities will be carried out:

– assessment of impact on EVs' diffusion of support/regulatory measures at local level;

– assessment of potential market penetration of EVs;

– demonstration of new forms of EV sharing and assessment of their transport and environmental impacts.

All these activities and the related results will allow the project to provide the EU, and elsewhere, with a wide and comprehensive set of tools, studies and measures to support the diffusion of electromobility, making Europe a leading area on the subject.

1.6. Bibliography

[ENE 09] ENERGY AND TRANSPORT DG, *EU Energy and Transport in Figures*, Statistical pocket book, 2009.

[EUR 09] EUROPEAN PARLIAMENT, *Regulation (EC) No 443/2009 of the European Parliament and of the Council of 23 April 2009*, Regulation, 2009.

[EUR 11] EUROPEAN COMMISSION, *Roadmap to a Single European Transport Area – Towards a competitive and resource efficient transport system,* White paper, 2011.

[GRE 12a] GREEN EMOTION CONSORTIUM, *Recommendations on grid-supporting opportunities of EVs*, Deliverable 4.2 of the Green eMotion project. Seventh Framework Programme, European Commission, 2012.

[GRE 12b] GREEN EMOTION CONSORTIUM, *ICT Standards & Protocols*, Deliverable 3.9 of the Green eMotion project. Seventh Framework Programme, European Commission, 2012.

[GRE 12c] GREEN EMOTION CONSORTIUM, *ICT Reference Architecture*, Deliverable 32 of the Green eMotion project. Seventh Framework Programme, European Commission, 2012.

[JEN 12] JENSEN A., CHERCHI E., ORTÚZAR J. de D., "A long panel survey to elicit preferences for electric vehicles and charging infrastructure", *XVII Panamerican Conference*, Santiago, Chile, 2012.

[JEN 13] JENSEN A., CHERCHI E., MABIT S., "On the stability of preferences and attitudes before and after experiencing an electric vehicle", *Transportation Research D*, vol. 25, pp. 24–32, 2013.

2

Advanced Services for Electromobility: the Integration of the SmartCEM Project Platform for the Reggio Emilia Pilot Site

This chapter presents the Italian test site currently in progress within the smartCEM project (www.smartcem-project.eu). The EC co-funded project smartCEM is aimed at implementing an ICT platform for the integration of five electromobility services (i.e. EV-sharing management, EV-efficient driving, EV-navigation, EV-trip management and EV-charging station management) to be piloted in four different pilot sites across Europe: Barcelona, San Sebastian and Gipuzkoa county, Newcastle and Reggio Emilia. Services in Reggio Emilia will be implemented on a fleet of 10 EVs and involving users among the municipality's employees. The smartCEM project provides added value by setting up a common framework for combining the five electromobility services into a more complex integrated system. The Reggio Emilia pilot site main objectives are focused on increasing stakeholder awareness and users' confidence in electric vehicles, by assessing the acceptance of efficient driving, navigation and monitoring management services designed for EVs.

2.1. Introduction

2.1.1. *Electromobility in Europe*

Electrification of transport (i.e. electromobility) is a priority in the Community Research Programme. It also figures prominently in the European Economic Recovery Plan presented in November 2008, within the framework of the Green Car Initiative.

Chapter written by Mauro DELL'AMICO, Guido DI PASQUALE, Leandro GUIDOTTI and Pietro MASCOLO.

Electricity as an energy vector for vehicle propulsion offers the chance to substitute oil with a wide diversity of primary energy sources. This could ensure security of energy supply and a broad use of renewable and carbon-free energy sources in the transport sector, which could help the European Union targets for CO_2 emission reduction.

Electric vehicle "tank-to-wheels" efficiency is about a factor of 3 higher than internal combustion engine vehicles. Electric vehicles emit no tailpipe CO_2 or other pollutants such as NO_x, NMHC and PM at the point of use. Electric vehicles provide quiet and smooth operation and consequently create less noise and vibration. Global CO_2 emissions have been steadily increasing over the last decades [BRI 13, DOD 09, FRA 13] making environmental sustainability a major area of interest. About 25% of CO_2 global emissions are generated by transport and 75% of these emissions are specifically due to road transport [INT 09, UNI 09, NAT 09]. The use of electric vehicles in place of vehicles running on fossil fuel may considerably reduce CO_2 emissions worldwide in the forthcoming years. By 2050, electric vehicles, if recharged using renewable resources, may be expected to have their emissions (CO_2 eq/km) 50–75% lower than regular vehicles (NRC 2009 and GEA 2012).

Intelligent transport systems (ITSs) – including both safety and telematic applications – have been the subject of significant research and development in Europe in recent years. Several models of passenger cars are now equipped with these ITSs as options, such as adaptive cruise control, lane departure warning and blind spot monitoring. In this context, most of the world's developed countries and large technological companies have focused their research, trialling and deployment actions to facilitate more energy-efficient and fewer petrol fuel-dependent vehicles and transport systems.

In this context, the smartCEM research project started with the main aim to demonstrate the effectiveness of and encourage the deployment of intelligent vehicle systems specifically designed for electric vehicles on European roads. Through the combination of technology and electromobility, smartCEM intends to increase the awareness of EVs and encourage the use of electric vehicles as part of everyday life. The smartCEM services will be tested by means of field operational tests (FOT) in three different European countries, namely Italy, Spain and the UK.

2.1.2. The *SmartCEM project*

Four European cities/regions (Barcelona, Gipuzkoa-San Sebastian, Newcastle and Reggio Emilia) have gathered together to demonstrate the role of ICT solutions for cities and citizens in addressing shortcomings of electromobility. Electric power solutions have proven to offer a potentially ground-breaking solution for more energy-efficient and less petrol-dependent vehicles and transport systems. The consortium aims to integrate regional initiatives which have been established all over Europe. However, it is known that EVs are under-used due to their limitations and lack of confidence in this type of transport mode. The smartCEM project aims to minimize the current EV limitations, by applying advanced mobility services (EV-navigation, EV-efficient driving, EV-trip management, EV-charging station management and EV-car sharing management) to existing multimodal electromobility transport modes. The combination of technology and electromobility aims to increase awareness and standardize the use of electrical modes. The objectives of the project are listed below:

– prove that user acceptance of electrical vehicles can be increased by at least 15% due to smartCEM services;

– evaluate how much the efficiency of transport can be optimized, taking into account environmental sustainability;

– develop tools for measuring, monitoring and assessing carbon emissions;

– identify and address all deployment elements such as business models, legal aspects and privacy;

– optimization of the energy use in the vehicle and infrastructure;

– support pan-European interoperability by standardization between operations performed by different services and facilitating the interoperability between different systems and vehicles;

– a complete integration of new services such as car sharing within the public transport system;

– expand the pilot to more cities, operators or service providers.

The smartCEM project kicked-off in early 2012 includes the piloting of services in different European cities. The implementation of the platform for ICT services is scheduled to be completed by 2013 and vehicles will be running for 9 months. Preliminary results will be available in 2014.

2.2. Reggio Emilia pilot site

The smartCEM services will be tested in Reggio Emilia on a local EV sharing fleet managed by TIL (Trasporti Integrati e Logistica), which is a Reggio Emilia-based long-term rental and sharing company. Vehicles are used by the employees of the Municipality of Reggio Emilia for day-to-day activities. From the fleet, composed of more than 30 electric vehicles, 10 EVs have been selected to be part of the smartCEM project. The EVs comprising the Reggio Emilia Municipality fleet are Piaggio Porter minivans, used for personnel and lightweight goods transportation in two specific project configurations (Figure 2.1). Users will drive the vehicles normally in the first configuration, i.e. with no access to smartCEM services, and will then be given access to smartCEM services.

Figure 2.1. *Piaggio Porter minivans used by the Municipality of Reggio Emilia*

In Table 2.1, some technical details of the vehicles are reported.

Piaggio Porter technical specifications	
Loading capacity (kg)	440–560, depending on the model
Autonomy, suburban (fast/economic mode)	90/112 km
Autonomy, urban (fast/economic mode	75/93 km
Max speed (fast/economic mode)	52/48 km/h
Recharging time	8 h, 2 h (fast)
Charging socket	10 A–200/230 V–50 Hz (4 poles)
Charger	2 A on board
Type of battery	Lead
Max slope	18.00%
Power (kW)	9.2

Table 2.1. *Piaggio Porter technical specifications*

2.3. Research hypotheses

The methodology to assess smartCEM services' impact has been defined in strict accordance with the FESTA project research hypotheses [FES 08]. Specific hypotheses have been defined in the Italian test site in order to be tested. Some of the research questions based on hypotheses are briefly listed hereafter: does EV-efficient driving contribute to reduced energy consumption? Does EV-navigation contribute to reduced energy consumption? What is the impact of EV-efficient driving on energy efficient charging behavior? Do the smartCEM services contribute to reducing range anxiety? What is the impact of smartCEM services on users' willingness to pay? What is the impact of smartCEM services on user acceptance?

Performance indicators, i.e. the quantifiable way to detect how these hypotheses can be assessed, have been defined. The main subjective performance indicators are perceived usefulness, acceptance, range anxiety, perceived ease of use and willingness to pay for the use of the services.

Participants' responses to questionnaires will be statistically analyzed. Repeated rounds of questionnaires will highlight changes in users' perceptions over time and, according to the design of the experiment, they will allow analysts to test the hypotheses. Finally, data collected will be checked for integrity and analyzed using inferential statistical methods.

2.4. Technical architecture and data exchange

Vehicles will be equipped with a data acquisition system (i.e. DAS) and an Android device on which smartCEM services will be installed. The DAS will gather data (e.g. state of charge, current and tension) from the vehicle electric network and will communicate with a remote server located at the University of Modena and Reggio Emilia (Italy) facilities. Data will be sent to this "on-site" server via 3G-GPRS. There, they will be postprocessed and then they will be available to be exchanged with partners in charge of smartCEM services execution. Data will also be sent to the central project DB located at the University of Newcastle (UK).

The device that will be mounted on the vehicles will be a 7" Android tablet which will be connected via 3G-GPRS with the on-site server and via bluetooth technology with the DAS. It will be mounted in landscape orientation, probably on the vehicle central dashboard (Figure 2.2).

Figure 2.2. *Piaggio Porter interiors and mount place for Android device*

After processing and if needed, data will be sent back to the Android device which will visualize the information requested by the user. The smartCEM apps will be accessible through the tablet mounted onboard, which will display information and data.

2.5. Experimental design, data and questionnaires

According to the experimental design, the Reggio Emilia pilot site (PS) has the objective of recruiting up to 20 drivers from the municipality employees who use the municipality's EVs on a regular basis for their day-to-day working activities across Reggio Emilia city centre and its outskirts. Usually, drivers get vehicles in the morning and use them all day long for moving wherever required. The vehicle is finally left back at municipality or rental service company parking areas at the end of the work shift.

Drivers will fill in questionnaires to assess their perception of EVs and their use. They will then be asked to drive the vehicles without the availability of smartCEM services for a maximum period of 3 months; usage data will be collected from the vehicles during this period. These months will serve as a baseline to verify any significant change in users' perception of EVs and their driving and recharging style.

Following the baseline period, the experimental phase will take place. Tests will be carried out over a maximum period of 6 months, during which users will be given access to smartCEM services and will be asked as much as possible, when they deem it useful or practical (Table 2.2).

Pilot overview	
Baseline start	January 2014 (expected)
Baseline duration	3 months
Test start	March 2014 (expected)
Test duration	6 months
Number of vehicles	10
Number of users	20 (maximum)
Users recruited from	Municipality of Reggio Emilia

Table 2.2. *Reggio Emilia PS overview*

Data will still be collected from the vehicles as well as from questionnaires administered to drivers, in order to identify any change in driving behavior or users' perception. The use of only subjective data provides a different level of reliability than objective data gathered from data loggers installed onboard in vehicles. However, given the aims of this project, the use of self-reported questionnaires is recommended because it permits collection of data about phenomena which are not directly observable, such as user acceptance, driver reactions, users' willingness to pay for the services, users' acceptance system and so on.

Three specific questionnaires have been designed for the users concerning user uptake, specifically:

– a general EV user uptake short questionnaire;

– a specific short questionnaire about smartCEM services user uptake;

– a longer and more detailed version of the specific questionnaire about smartCEM services user uptake.

All these questionnaires consist of the constructs of user acceptance, users' willingness to pay and users' range anxiety. The first questionnaire will be administered to assess users' knowledge, acceptance and attitude toward EVs in general as a mean of transportation, it still being not well-known and underestimated in Italy. The second questionnaire has been prepared for the assessment of the general perception of the users' attitude toward smartCEM services. It consists of very few questions, since in other PSs within smartCEM it is going to be administered after each use of an EV. The last questionnaire is a longer, more detailed version of the previous one. It is going to be administered in focus groups and users will have to go into more detail about their attitude toward smartCEM services and how users think their attitude toward EVs might be modified by them.

After piloting review of questionnaires and testing every operational procedure, the FOT will start with the recruiting ramping up phase (see Table 2.2 and Figure 2.5). Questionnaires will be administered as defined above. FOT experimental design includes a within subjects analysis, with a baseline period (i.e. drivers will not use the LDW system) and a treatment period (i.e. drivers will use the LDW system). Also, a control group (i.e. drivers of Lancia Delta cars without the LDW system) will be included in the test to ensure a between subjects analysis.

2.5.1. *Services to be tested*

Three smartCEM electromobility services will be tested in the Reggio Emilia pilot site. The services integrated through the smartCEM platform in the Reggio Emilia pilot site are:

– EV-efficient driving: on-board device and mobile application connected to the vehicle will be used to collect data about performance (i.e. SoC, speed and use of services);

– EV-navigation: it starts from eco-navigation and integrates charging stations. The mobile application shows real time availability of EV stations, and includes a battery management system that shows drivers which charging points are available on the basis of autonomy that is influenced by driving style and topography;

– EV-sharing management: this service aims to minimize constraints of EVs, such as the management of the battery, in order to ensure a successful sharing user experience. A vehicle-sharing platform allows municipality employees to book an EV, check in and check out through a smartphone app.

The smartCEM consortium consists of 31 partners from all over Europe. Partners involved in carrying out the Reggio Emilia pilot site (Figure 2.3) are listed below:

– University of Modena and Reggio Emilia (UNIMORE), responsible for integration and operations in the PS in collaboration with ICOOR.

– Interuniversity Consortium for Optimization and Operations Research (ICOOR) collaborating with UNIMORE in leading the PS.

– Pluservice srl, supporting activities related to EV-car sharing management service in Reggio Emilia.

– Centro Ricerche Fiat (CRF), provider of the EV-efficient driving service in Reggio Emilia.

– Teamnet, provider of the smartCEM common portal within the project and Charging Station Management service.

– PTV, provider of the EV-navigation service in the project.

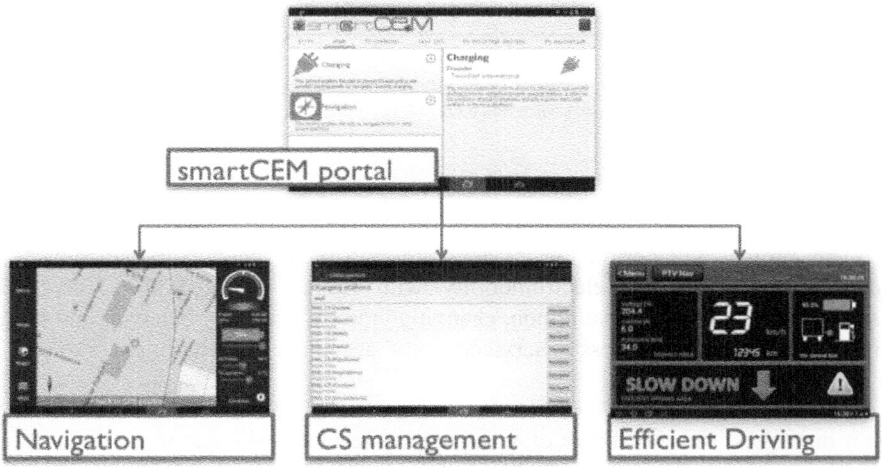

Figure 2.3. *SmartCEM services in the Reggio Emilia Pilot site*

2.5.2. *Facilities and key management*

The EVs are stored at the municipality premises, where they are also recharged overnight and maintained. The keys of the vehicles are stored in an electronic key manager which can grant access to keys upon user login (Figure 2.4). Users needing to use a vehicle need to swipe their magnetic badge into the key manager to obtain access to a subset of keys, based on that particular user's privileges.

The key manager has been programmed to give access to smartCEM-equipped vehicles only to authorized users participating in the project.

Figure 2.4. *Electronic key manager used by the municipality of Reggio Emilia*

2.6. Conclusions

This project will permit the investigation and understanding of the impact of smartCEM services for electromobility with respect to several aspects such as reduction of energy consumption, changing in users' charging behavior and range anxiety, perceived usefulness of services, users' acceptance and willingness to pay.

This assessment will offer stakeholders (e.g. public administrations or fleet managers) the possibility of extrapolating results applicable to a larger set of vehicles and users, and to evaluate the adoption of technology as standard equipment for EVs.

The Reggio Emilia pilot site will be focused on the study of a fleet of EVs operated by a municipal administration, giving the opportunity to evaluate the potential of light electric vehicles' integration in an administration car sharing fleet. This will give insight on possible future usage of EV fleets by other municipal administration all over Europe.

The complete results will be available at the end of the project in all deliverables regarding assessment and evaluation of services [SMA 13]. All project deliverables will be public and will be accessible at the official project website at http://www.smartcem-project.eu.

2.7. Bibliography

[BRI 13] BRITISH PETROLEUM, *Statistical Review of World Energy*, 2013.

[CAR 10] CARPENTER S., "Nissan Leaf's Promise: An Affordable Electric", available at http://articles.latimes.com/2010/mar/30/business/la-fi-nissan-leaf31-2010mar31, 2010.

[COM 08] COMMUNICATION FROM THE COMMISSION TO THE EUROPEAN COUNCIL, A European Economic Recovery Plan, 2008.

[DOD 09] DODMAN D., "Blaming cities for climate change? An analysis of urban greenhouse gas emissions inventories", *Environ. Urban*, vol. 21, pp. 185–201, 2009.

[FES 08] FESTA CONSORTIUM, *FESTA Handbook*, deliverable D6.4, 2008.

[FRA 13] FRAG KIAS M., LOBO J., STRUMSKY D. *et al.*, "Does size matter? Scaling of CO_2 emissions and US urban areas", *PLoS ONE*, p. 8, 2013.

[GLO 12] GLOBAL ENERGY ASSESSMENT, *Towards a Sustainable Future*, Cambridge University Press, 2012.

[INT 09] INTERNATIONAL ENERGY AGENCY, Transport, Energy and CO_2; Moving Toward Sustainability, 2009.

[NAT 09] NATIONAL RESEARCH COUNCIL, Driving and the built environment: the effects of compact development on motorized travel, energy use and CO_2 emissions, 2009.

[SMA 12a] SMARTCEM CONSORTIUM, Deliverable 4.1 – Evaluation Framework, 2012.

[SMA 12b] SMARTCEM CONSORTIUM, Deliverable 4.2 – Evaluation criteria and performance indicators, 2012.

[SMA 12c] SMARTCEM CONSORTIUM, Deliverable 4.3 – Experimental Design, 2012.

[SMA 13] SMARTCEM CONSORTIUM, Description of Work 2nd amendment, 2013.

[UNI 09] UNIONE INTERNATIONALE DES TRANSPORTES PUBLIQUES, Public Transports and CO_2 emissions, 2009.

[VAN 97] VAN DER LAAN J.D., HEINO A., DE WAARD D., "A simple procedure for the assessment of acceptance of advanced transport telematics", *Transportation Research Part C*, vol. 5, pp. 1–10, 1997.

[WES 92] WEST R., ELANDER J., FRENCH D., Decision making, personality and driving style as correlates of individual accident risk, TRL Contractor Report 309, Transport Research Laboratory, Crowthorne, United Kingdom, 1992.

3

Cognitive Mapping and Multi-criteria Assessment for the Design of an Electric Car Sharing Service

The chapter presents a model for the design of an electric car sharing service for the city of Milan. Several options of service configurations have been analyzed and evaluated according to indicators measuring the performance of such options in respect to relevant dimensions (i.e. economic and financial costs and revenues, mobility, social benefits and environmental effects). We set up a multi-criteria assessment, structured by means of cognitive maps. Causal networks to estimate the effects of the options have been identified and instantiated by means of simulation techniques and other qualitative and quantitative models. The focus of the chapter is on the development and use of the causal maps and their integration with a multi-criteria assessment, based on the ELECTRE TRI rating method. The use of cognitive maps allowed to us capture the multiple values of the problem and the value trees of stakeholders' objectives. The proposed method can be useful for design of mobility services, especially at a strategic level.

3.1. Introduction

The earliest origins of car sharing, understood as an organized form of shared use of the car, can be traced back to 1948 when an association started a service in Zurich [SHA 99]. In the following years, particularly in the 1970s and 1980s, several car sharing systems were set up, but still on a small scale, with poor results [SHA 98]. The original idea has been gradually replaced by an offer structurally

Chapter written by Alessandro LUÈ, Alberto COLORNI and Roberto NOCERINO.

organized according to strict business criteria, in order to achieve economies of scale, which has resulted in increased benefits to users in terms of low rates and diversification of the available fleet. The concept of vehicle sharing has been characterized in different ways over the years and depending on local conditions [BAR 02, SHA 07]. The most important factor for the success of car sharing is represented by the difficulty of reaching a critical mass of users to ensure an economic balance [BRO 04]), hence the frequent need for public support. The two main barriers for the use of the service consist of the distance from the origin and the stations, and the concern of not always having an available vehicle [KAT 03].

Within this framework, the Green Move project designed and tested an electric vehicle-sharing system for the city of Milan, in an attempt to overcome the obstacles that limit the diffusion of both electric vehicles and traditional car sharing initiatives [LUE 12]. The project aims to involve and integrate a multiplicity of key players of the Milan mobility system and to balance different dimensions of sustainability. Green Move[1] involves eight different departments and research centers of the Polytechnic University of Milan: DEIB (information communication technology), DIG (economic and stakeholder analysis), DAStU (travel demand analysis), DICA (geographical information systems), Fondazione Politecnico (administrative management), Design (service design and communication), DMAT (mathematical models), and Poliedra (evaluation and environmental analysis). The main idea behind Green Move is to create a flexible service of vehicle sharing, based on electric cars and open to a wide range of different users. The system will be made easily accessible due to an add-on device, the Green e-Box [ALL 12], a bridge between the user, the vehicle and the control center, that allows any vehicle to join the service network. In this way, single users, private companies and associations may share their personal car or fleet (peer-to-peer approach).

Figure 3.1 presents the overall articulation of the Green Move project. Within the first phase, the research team performed three main activities: best practice and literature review, context analysis and service idea identification [VIL 13]. The second phase of the project, the strategic design, focuses on the strategic design of different service configurations, chosen as the most effective in terms of user need fulfillment. Based on such activities, four specific service configurations (condominium-based service, network of services, the new business fleet and peer-to-peer) have been identified [ARE 12]. The third phase concerns the

1 The Green Move project has been co-financed by Regione Lombardia through the "Fondo per la promozione di accordi istituzionali – Bando accordi istituzionali".

technology necessary to manage the physical system [ALL 12, BIA 13, PAN 12], and the elaboration of the most effective and efficient solutions, taking into account different dimensions of sustainability (environment, finance, mobility and social) [ARE 15]. Regarding the latter phase, we set up a multi-criteria decision analysis, structured by means of cognitive maps. Causal networks to estimate the effects of the options have been identified and instantiated by means of simulation techniques and other qualitative and quantitative models. The last phases concern the implementation of a pilot in Milan to test the results and the elaboration of final recommendations concerning the implementation of a car sharing service.

An earlier version of this chapter has been presented in Luè *et al.* [LUE 14], which specifically focuses on the development and use of the causal maps and their integration with a multi-criteria method. Section 3.2 synthesizes a literature review that we made on decision aiding methods and causal maps. Section 3.3 describes how we applied such methods for the design of an electric car sharing service in Milan. Section 3.4 concludes the chapter.

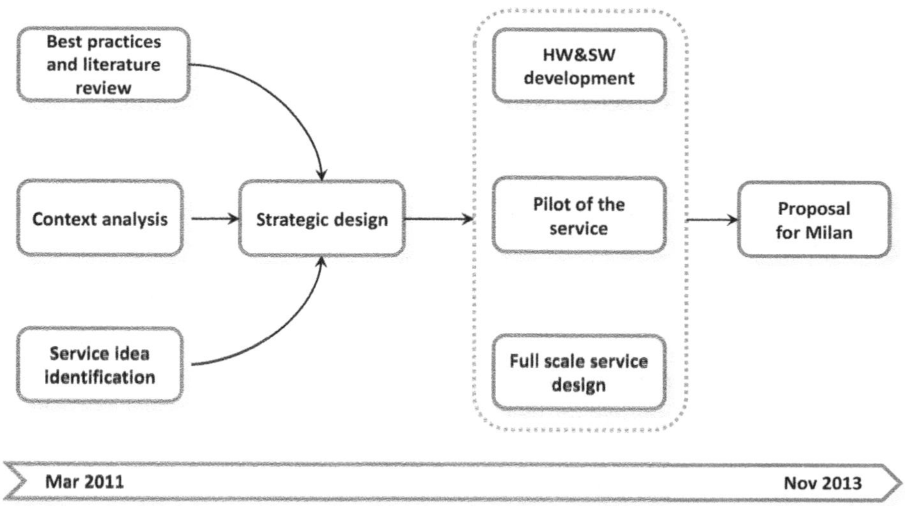

Figure 3.1. *Phases of the Green Move project*

3.2. Decision-aiding methods and causal maps

Decision aiding methods aim at supporting the decision maker in the process of making a choice [COL 13]. In an idealized decision problem, one person (the decision-maker) chooses among specific and well-described alternative options, evaluated according to a unique and defined objective (evaluation criterion). Moreover, the decision maker has all of the information, i.e. he knows the performance for each option, according to the evaluation criterion. Operational research classically deals with this kind of problem [HIL 05], formalizing a decision problem through a mathematical problem and finding the optimal solution (or an approximation, if the computation time is too long for finding the optimal solution).

The problem of designing a vehicle sharing service, however, is characterized by the presence of many objectives to be pursued that are in conflict with each other. For example, the expansion of territorial coverage (an objective to be maximized) will increase investment and operational costs (an objective to be minimized). It is not easy to compare and choose in a situation characterized by many evaluation criteria. Different methodologies have been designed, in order to classify and rank the options according to the set of criteria that the decision maker considers relevant [FIG 05]. In a first approximation, it is possible to identify two main categories of methods, characterized by a different approach:

– definition of a utility function (or of an overall satisfaction function) to aggregate the different criteria. Two examples are represented by the Multi Attribute Utility Theory [KEE 76], and the Analytic Hierarchy Process [SAA 90];

– pairwise comparison of the different alternatives, using concordance and discordance criteria. An example is represented by the outranking (ELECTRE) methods [ROY 96].

In addition, non-deterministic variables may change the outcome of the evaluation. These variables refer to:

– the information regarding the reference scenario (e.g. what will be the regional transport policies to improve air quality in Milan? Is the cost of an electric vehicle going to decrease in the next decade?);

– the estimation of the effects of alternatives (e.g. magnitude of the reduction of air pollutants due to the introduction of the service).

So, the classical operational research methodologies are not suitable for this process, because they need a completely-structured problem and a fully rational decision maker. In fact, in this particular case, even the decision maker (and the analysts that support the decision maker) do not have all the information necessary for a complete definition of the problem, which is a fundamental step for the decision process. Even the use of sophisticated models is not effective if the definition of the problem is imprecise or incomplete and can, paradoxically, help to solve a problem that is not the decision maker's one. Within such a framework, decision aid methods can support the *structuring process* of the service design in order to link the process to quantitative, or, at least clearly expressed, criteria.

The preliminary questions that can help in the design of the service in this preliminary phase are: who is the decision maker? Who is involved? Who will be affected or influenced by the decision? Moreover, the problem involves many stakeholders with different interests. In the first instance, we can identify:

– public administration, interested in minimizing the impacts on the environment and to improve the accessibility of an area;

– users, interested in a good quality of service, in a high capillarity of the service and in low fares;

– the service provider, primarily interested in profit.

In addition to these key figures, we can identify other stakeholders who may be more or less interested depending on the kind of configuration of the service (e.g. producers of energy from sustainable sources, in the case of an electric car sharing service).

The second step of the structuring process consists of the definition of reference scenarios, evaluation criteria and options to be evaluated. The option identification (in this case: the definition of the options for the service configuration) is one of the most sensitive aspects. It has been verified that mobility experts often define and choose the alternatives in coherence with their own experience of "what has already worked" [HUL 05]. A representative number of meaningful alternatives is the precondition for a consistent decision process [COL 99]. Depending on the specific decision problem, different approaches are possible, such as the ones proposed by Bana e Costa [BAN 01], Kelly *et al.* [KEL 08] and Jones *et al.* [JON 09].

Within such an approach, we used the *cognitive maps* tool, which so far has been used mainly in psychology and behavioral sciences, management [FIO 92, LAN 92], political sciences [EDE 04] and business [KWA 99]. A prime aim of cognitive maps is to graphically represent the ideas of an individual through a network of interrelated concepts. Moreover, it allows a shared vision of the decision problem to be built in a group of people and facilitate the identification of values and their conflicting elements that may have an impact on the consequence of a decision [EDE 88]. Cognitive mapping, widely used in problem-structuring [MIN 04], permits a rich representation of decision makers' views through modeling of complex chains (causes and effects) [EDE 04], and may help in identifying possible new better actions [MON 06, NEV 09]. In this use, cognitive maps have a limited capacity of inference. The following methods are examples to improve it:

– fuzzy cognitive maps, which extend in the direction of semi-qualitative outputs [GLY 12, VAN 10];

– qualitative probabilistic networks, where the links of the map denote a probabilistic dependence [WEL 94];

– Bayesian networks, probabilistic directed acyclic graphical models to draw inferences [NAD 04];

– decision maps, a framework to integrate scenario and multi-criteria analysis [COM 11].

Our objective was to develop a multi-criteria evaluation structured by means of cognitive maps (see Figure 3.2). In this sense, previous attempts for integrating cognitive maps have not reached the expected results [COM 11].

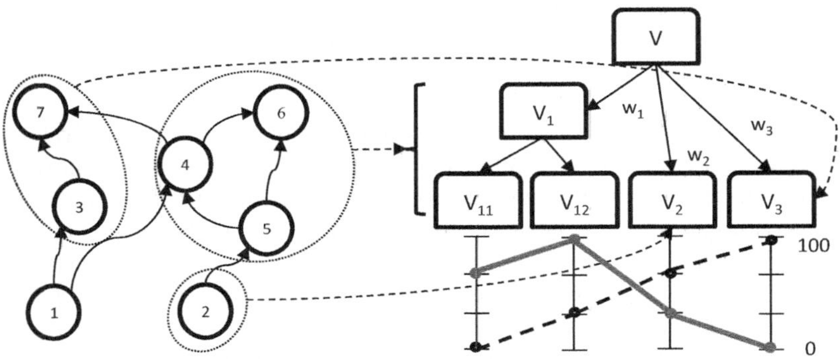

Figure 3.2. *An example of a value tree of a multi-criteria analysis, structured from a cognitive map. Elaborated from Montibeller and Belton [MON 06]*

3.3. Assessment of the configuration options

The design of a full-scale electric car sharing service for the city of Milan has been carried out by identifying different configuration options, taking into account parameters related to performance (e.g. capillarity, price and accessibility), additional services (e.g. baby car seat and navigation system), technology (e.g.

booking services, road toll device, lock and unlock system), etc. In order to support the identification and selection of the most promising options in terms of the overall level of sustainability (economic, social and environmental), we studied the possible effects of such parameters on the mobility system. In order to identify such effects, we used cognitive maps, elicited by means of interviews and workshops put together with researchers of different expertise from the Green Move team and territorial stakeholders. The result is shown in Figure 3.3, which represents cause–effect chains, and tracks the complex interactions among the parameters for each option and the effects.

The left part of the graph is represented by the configuration parameters, whose combination identifies different service options. The configuration parameters are connected to some intermediate nodes, which depend on the specific service configuration (e.g. number of vehicle stations and number of users), and, in turn, determine the value of performance indicators. Performance indicators measure the extent to which the objectives of the system are achieved; in this way, different service configurations can be compared based on their results in relation to performance indicators and the consequent level of fulfillment of the objectives.

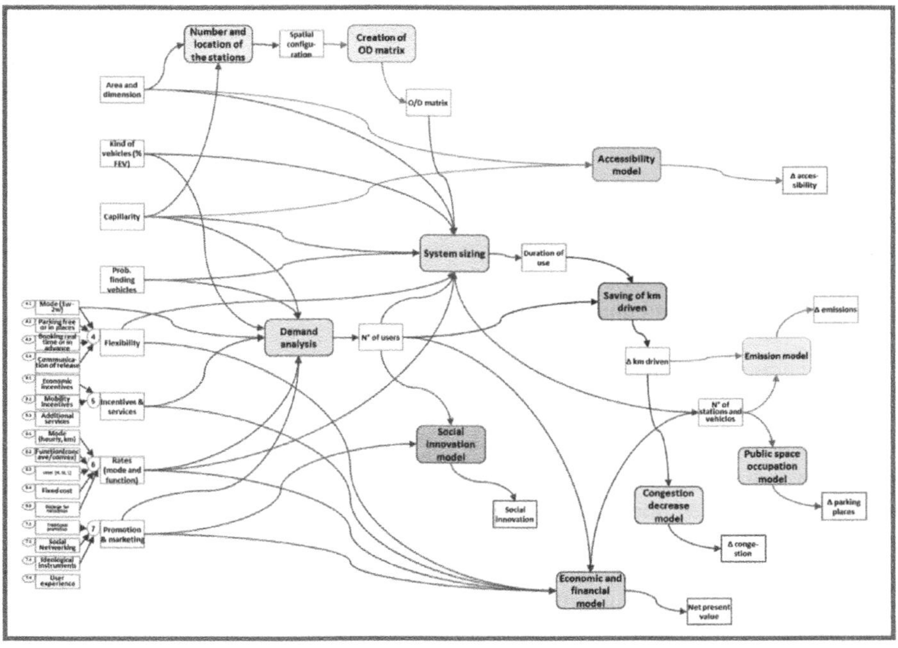

Figure 3.3. *Theoretical system modeling of service configuration. For a color version of the figure, see www.iste.co.uk/jacob/energy.zip*

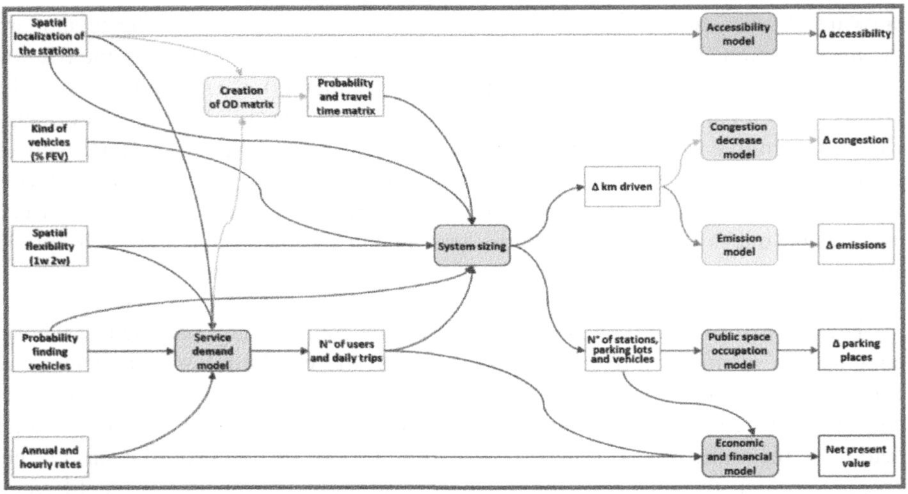

Figure 3.4. *System modeling of the implemented service configuration*

For the estimation of the indicators, the formalization of a set of models has been required. In the system model presented in Figure 3.3, for example, the nodes *capillarity*, *probability of finding a vehicle*, *flexibility*, *incentives and services*, *rates* and *promotions* influence the node *number of users*. However, not all the influences identified in the theoretical design have been converted into mathematical relations because of the complexity of the relations in reality. The result of this simplification is presented in Figure 3.4. Considering the example above, we were able to identify and quantify – through a stated preference model [BEN 85] – the influence of a limited number of parameters, i.e. spatial localization, spatial flexibility, probability of finding a vehicle, and annual and hourly rates.

The parameters used in the implementation of system modeling are briefly described below:

– *spatial localization of the stations*: represents the spatial configuration of the service in terms of localization of the stations in the urban area. The capillarity, i.e. the distance between one station and another, is a function of this parameter as well as the number of users potentially served by the service: in fact, the attractiveness of a station toward a potential user is a function of his distance to the station. The spatial localization directly influences the accessibility by the users of the urban territory;

– *kind of vehicles* (% FEV): characterizes the composition of the car sharing fleet in terms of engine technology (electric and internal combustion engine). Electric vehicles need not negligible time for recharging (and dedicated recharge stations) and the use of them imply a difference sizing of the vehicle fleet;

– *spatial flexibility* (1w 2w): with a one-way service, the user is allowed to leave the vehicle in a station other than where the vehicle was picked up; with a two-way, he is forced to make round-trip travel, bringing the vehicle back to the station of pick-up. The parameters influence the system sizing;

– *probability of finding vehicles*: an expression of the level of service given to the user. The higher the probability of finding a vehicles for users, higher, obviously, is the level of the satisfaction of users' requests;

– *annual and hourly rates*: price charged to the users for the vehicle's usage. It consists of two different components: an annual rate (i.e. a subscription cost) and an hourly rate (i.e. a usage fee), paid for each use.

For the definition of the spatial localization of the stations, three different areas have been outlined: City Centre, Ring Road Area and Municipality Borders. For each area, we defined three values of the average distance between stations: 500, 1,000 and 2,000 m. Not all the possible configurations have been evaluated because they have been preliminarily evaluated as of no interest (for example, the configuration with stations every 2,000 m in City Centre, no station in Ring Road Area and stations every 500 m in Municipality Borders). Table 3.1 lists the 18 considered combinations and Figure 3.5 depicts one of the possible spatial configurations.

For the annual and hourly rate, the combinations considered in the evaluation phase are listed in Table 3.2. As it is possible to see in the table, not all the possible combinations of rates have been considered, even if quite extreme alternatives have been evaluated, such as the number 1 and the number 8.

The performance indicators identified for measuring the achievement of the options' objectives are:

– accessibility – variation of the level of accessibility to the urban mobility system;

– congestion – variation of the congestion level on the road network;

– local and global emissions – quantity of pollutant and greenhouse gas emissions;

– parking places – variation of the public space occupied by private car;

– net present value – economic performance of the car sharing service.

N°	City Centre	Ring Road Area	Municipality Borders
1	500	–	–
2	500	500	–
3	500	500	500
4	500	500	1,000
5	500	500	2,000
6	500	1,000	1,000
7	500	1,000	2,000
8	1,000	–	–
9	1,000	1,000	–
10	1,000	1,000	500
11	1,000	1,000	1,000
12	1,000	1,000	2,000
13	1,000	2,000	2,000
14	2,000	–	–
15	2,000	1,000	500
16	2,000	1,000	1,000
17	2,000	2,000	–
18	2,000	2,000	2,000

Table 3.1. *The considered spatial configurations*

N°	Annual rate (€/year)	Hourly rate (€/hour)
1	0	7
2	0	15
3	50	5
4	50	7
5	50	15
6	100	3
7	100	5
8	100	7

Table 3.2. *The considered combinations of annual and hourly rates*

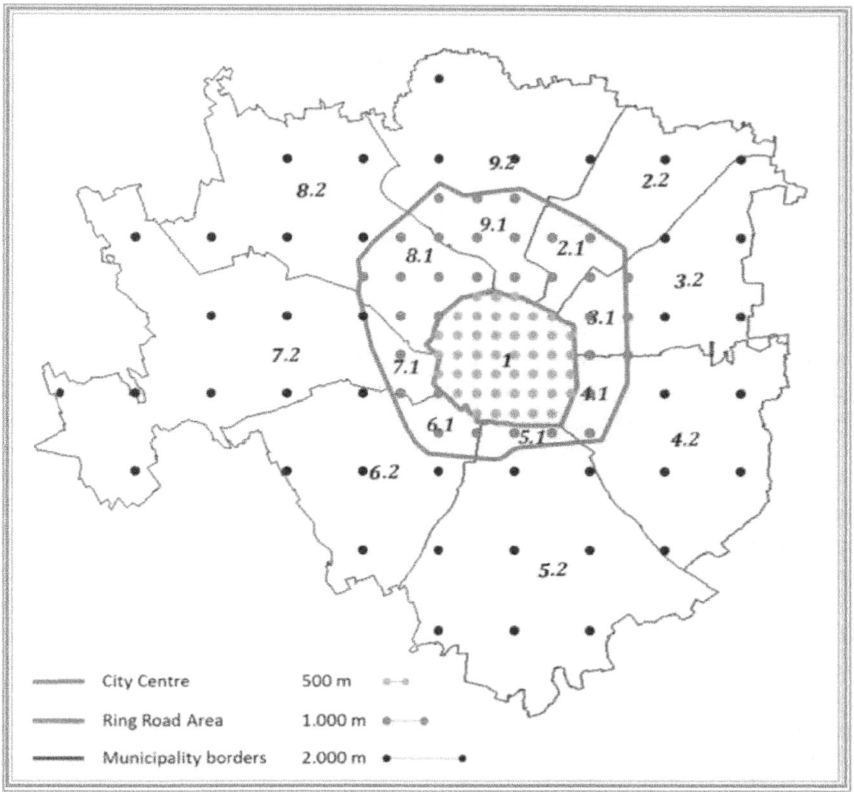

Figure 3.5. *Example of one spatial configuration. Density is higher in City Centre (the distance between stations is 500 m) than in Ring Road Area (1,000 m) and within Municipality borders (2,000 m). For a color version of the figure, see www.iste.co.uk/jacob/energy.zip*

The final step consists of the option evaluation, which is carried out by means of a multi-criteria analysis, in order to take into account possible conflicting objectives. Different service options can be compared based on their effects and the consequent level of fulfillment of the objectives. We implemented the ELECTRE TRI sorting method [DIA 02, MOU 00], in order to sort the options into categories and to elaborate a proposal. Moreover, evaluation of the options will be conducted taking into account the points of view of different stakeholders.

The implemented model is able to design and size the main dimensions for a car sharing service, considering both internal combustion engine vehicles and electric ones. The model, starting from a limited number of significant configuration parameters, estimates a set of indicators and evaluates the economic, environmental and social performances of the designed service. At the present stage of the project, the evaluation model still lacks the implementation of the engines of some models. So, we have not yet implemented the actual evaluation of the options for the city of Milan.

3.4. Conclusion

This chapter has presented a model for the design of an electric car sharing service for the city of Milan. This decision situation comprises a comprehensive set of stakeholders with possibly conflicting and unstructured views. In such an ill-defined decision context, it is crucial that the decision problem is structured in order to build consensus among stakeholders' objectives. In order to support the design of the configurations, we set up a multicriteria decision analysis, structured by means of cognitive maps. This structuring phase supported the mutual understanding between researchers with different expertise and the territorial stakeholders, and the identification of causal networks to estimate the effects of the options. The use of cognitive maps allowed us to capture the multiple values of the problem and reduce their conflicting aspects. Moreover, they were useful to define the estimation models and multi-criteria analysis by the mean of value trees of stakeholders' objectives. The considered multi-criteria assessment is able to take into account the points of view of the territorial stakeholders, expressed mainly by their quantification of the importance of different dimensions (criteria) and the possible support they may provide for the setup of the service. The proposed method can be useful in general for the design and planning of a mobility service, especially at a strategic level. Moreover, from the research point of view, we verified the use of decision-aiding methods within a service design problem [COL 12].

3.5. Bibliography

[ALL 12] ALLI G., BARESI L., BIANCHESSI A. *et al.*, Green Move: towards next generation sustainable smartphone-based vehicle sharing, *Proceedings of the Sustainable Internet and ICT for Sustainability (SustainIT)*, pp. 1–5, 2012.

[ARE 15] ARENA M., AZZONE G., COLORNI A. *et al.*, Service design in electric vehicle sharing: evidence from Italy, *IET Intelligent Transport Systems*, vol. 9, pp. 145–155, 2015.

[ARE 12] ARENA M., AZZONE G., COLORNI A. *et al.*, Innovative configurations for electric vehicle sharing services, *Proceedings of the 19th Intelligent Transport Systems World Congress*, Vienna, 2012.

[BAN 01] BANA E COSTA C.A., The use of multi-criteria decision analysis to support the search for less conflicting policy options in a multi-actor context: case study, *Journal of Multi-Criteria Decision Analysis*, vol. 10, pp. 111–125, 2001.

[BAR 02] BARTH M., SHAHEEN S., Shared-use vehicle systems: framework for classifying carsharing, station cars, and combined approaches, *Transportation Research Record: Journal of the Transportation Research Board*, vol. 1791, pp. 105–112, 2012.

[BEN 85] BEN-AKIVA M., LERMAN S.R., *Discrete Choice Analysis: Theory and Application to Predict Travel Demand*, The MIT press, 1985.

[BIA 13] BIANCHESSI A.G., ONGINI C., ROTONDI S. *et al.*, A flexible architecture for managing vehicle sharing systems, *Embedded Systems Letters, IEEE*, vol. 5, pp. 30–33, 2013.

[BRO 04] BROOK D., Carsharing: start up issues and new operational models, *Proceedings of the 83rd Transportation Research Board Annual Meeting*, Washington, DC, 2004.

[COL 99] COLORNI A., LANIADO E., MURATORI S., Decision support systems for environmental impact assessment of transport infrastructures, *Transportation Research Part D: Transport and Environment*, vol. 4, pp. 1–11, 1999.

[COL 12] COLORNI A., LUÈ A., Decision-aid tools for design problems, in RAMPINO L. (ed.), *Design Research: Between Scientific Method and Project Praxis. Notes on Doctoral Research in Design*, FrancoAngeli, pp. 153–167, 2012.

[COL 13] COLORNI A., TSOUKIÀS A., What Is a Decision Problem? Preliminary Statements, in PERNY P., PIRLOT M., TSOUKIÀS A., *Algorithmic Decision Theory*, pp. 139–153, Springer, 2013.

[COM 11] COMES T., HIETE M., WIJNGAARDS N. *et al.*, Decision maps: A framework for multi-criteria decision support under severe uncertainty, *Decision Support Systems*, vol. 52, pp. 108–118, 2011.

[DIA 02] DIAS L., MOUSSEAU V., FIGUEIRA J. *et al.*, An aggregation/disaggregation approach to obtain robust conclusions with ELECTRE TRI, *European Journal of Operational Research*, vol. 138, pp. 332–348, 2002.

[EDE 88] EDEN C., Cognitive mapping, *European Journal of Operational Research*, vol. 36, pp. 1–13, 1988.

[EDE 04] EDEN C., Analyzing cognitive maps to help structure issues or problems, *European Journal of Operational Research*, vol. 159, pp. 673–686, 2004.

[EDE 04] EDEN C., Ackermann F., Cognitive mapping expert views for policy analysis in the public sector, *European Journal of Operational Research*, vol. 152, pp. 615–630, 2004.

[FIG 05] FIGUEIRA J., GRECO S., EHRGOTT M., *Multiple criteria decision analysis: state of the art surveys*, Springer, 2005

[FIO 92] FIOL C.M., HUFF A.S., Maps for managers: where are we? Where do we go from here? *Journal of Management Studies*, vol. 29, pp. 267–285, 1992.

[GLY 12] GLYKAS M., Performance measurement scenarios with fuzzy cognitive strategic maps, *International Journal of Information Management*, vol. 32, pp. 182–195, 2012.

[HIL 05] HILLIER F., LIEBERMAN G., *Introduction to Operations Research* (8 ed.), McGraw-Hill, 2005.

[HUL 05] HULL A., TRICKER R., An assessment of the barriers to the delivery of sustainable local surface transport solutions, *Proceedings of the ETC 2005 – European Transport Conference*, Strasbourg, France, 2005.

[JON 09] JONES P., KELLY C., MAY T. *et al.*, Innovative approaches to option generation, *European Journal of Transport and Infrastructure research*, vol. 9, pp. 237–258, 2009.

[KAT 03] KATZEV R., Car sharing: a new approach to urban transportation problems, *Analyses of Social Issues and Public Policy*, vol. 3, pp. 65–86, 2003.

[KEE 76] KEENEY R.L., RAIFFA H., *Decision analysis with multiple conflicting objectives*: John Wiley & Sons, 1976.

[KEL 08] KELLY C., MAY A.D., JOPSON A., The development of an option generation tool to identify potential transport policy packages, *Transport Policy,* vol. *15*, pp. 361–371, 2008.

[KPO 12] KPOUMIÉ A., DAMART S., TSOUKIÀS A., Integrating cognitive mapping analysis into multi-criteria decision aiding, *Proceedings of the 76th MCDA EURO Working Group*, Portsmouth, UK, 2012.

[KWA 99] KWAHK K.-Y., KIM Y.-G., Supporting business process redesign using cognitive maps, *Decision Support Systems,* vol. 25, pp. 155–178, 1999.

[LAN 92] LANGFIELD-SMITH K., Exploring the need for a shared cognitive map. *Journal of Management Studies*, vol. 29, pp. 349–368, 1992.

[LUE 12] LUÈ A., COLORNI A., NOCERINO R. *et al.*, Green Move: an innovative electric vehicle-sharing system, *Procedia-Social and Behavioral Sciences*, vol. 48, pp. 2978–2987, 2012.

[LUÈ 14] LUÈ A., COLORNI A., NOCERINO R., Cognitive mapping and multi-criteria analysis for decision aiding: an application to the design of an electric vehicle sharing service, In *Proceedings of the Transport Research Arena*, Paris, France, 2014.

[MIN 04] MINGERS J., ROSENHEAD J., Problem structuring methods in action, *European Journal of Operational Research*, vol. 152, pp. 530–554, 2004.

[MON 06] MONTIBELLER G., BELTON V., Causal maps and the evaluation of decision options: a review, *The Journal of the Operational Research Society*, vol. 57, pp. 779–791, 2006.

[MON 07] MONTIBELLER G., BELTON V., ACKERMANN F. *et al.*, Reasoning maps for decision aid: an integrated approach for problem-structuring and multi-criteria evaluation, *J. Oper. Res. Soc.*, vol. 59, pp. 575–589, 2007.

[MOS 00] MOUSSEAU V., SLOWINSKI R. ZIELNIEWICZ P., A user-oriented implementation of the ELECTRE-TRI method integrating preference elicitation support, *Computers and Operations Research*, vol. 27, pp. 757–777, 2000.

[NAD 04] NADKARNI S., SHENOY P.P., A causal mapping approach to constructing Bayesian networks, *Decision Support Systems,* vol. 38, pp. 259–281, 2004.

[NEV 09] NEVES L.P., DIAS L.C., ANTUNES C.H. *et al.*, Structuring an MCDA model using SSM: a case study in energy efficiency, *European Journal of Operational Research,* vol. 199, pp. 834–845, 2009.

[PAN 12] PANIGATI E., RAUSEO A., SCHREIBER F.A. *et al.*, Aspects of pervasive information management: an account of the Green Move system, *Proceedings of the Computational Science and Engineering (CSE), 2012 IEEE 15th International Conference on*, pp. 648–655, 2012.

[ROY 96] ROY B., *Aiding Multicriteria Methodology for Decision-Aiding*, Springer, 1996.

[SAA 90] SAATY T.L., How to make a decision: the analytic hierarchy process, *European Journal of Operational Research,* vol. 48, pp. 9–26, 1990.

[SHA 07] SHAHEEN S., COHEN A., Growth in worldwide carsharing: an international comparison, *Transportation Research Record: Journal of the Transportation Research Board, 1992*, pp. 81–89, 2007.

[SHA 98] SHAHEEN S., SPERLING D., WAGNER C., Carsharing in Europe and North America: past, present, and future, *Transportation Quarterly,* vol. 52, pp. 35–52, 1998.

[SHA 99] SHAHEEN S., SPERLING D., WAGNER C., A short history of carsharing in the 90's, *The Journal of World Transport Policy & Practice,* vol. 5, 1999.

[VAN 10] VAN VLIET M., KOK K., VELDKAMP T., Linking stakeholders and modellers in scenario studies: the use of Fuzzy Cognitive Maps as a communication and learning tool, *Futures,* vol. 42, pp. 1–14, 2010.

[VIL 13] VILLARI B., LUÈ A., Design research on sustainable mobility: an electric vehicle sharing service for Milano, Italy, *Proceedings of the 10th European Academy of Design Conference – Crafting the Future*, Gothenburg, 2013.

[WEL 94] WELLMAN M.P., Inference in cognitive maps, *Mathematics and Computers in Simulation,* vol. 36, pp. 137–148, 1994.

4

Eco-driving for Clean Vehicles: Optimizing Energy Use for Trams and e-buses

The introduction of alternative clean fuels and new improved vehicle technologies (CNG, electricity, hybrid, etc.) will help to reduce energy consumption, greenhouse gas emissions and noise pollution. These new technologies require adaptation of how to use them in the best way. Professional driver training in terms of eco-driving and the correct vehicle handling has to consider and react to these new technologies. Therefore, training and education services for the safe eco-driving of clean vehicles in the electric public transport sector have been developed for trams, hybrid buses, trolleybuses or e-buses (equipped with super capacitors) in the framework of the Intelligent Energy Europe Project ACTUATE (Advanced Training and Education for Safe Eco-driving of Clean Vehicles). This chapter will outline the development of a training programme for safe eco-driving of clean vehicles and the impact evaluation approach within the framework of ACTUATE.

4.1. Background

The introduction of new clean vehicles, such as hybrid buses and trolleybuses, is accompanied by increasing demands into the qualification of professional drivers concerning driving behavior or new safety standards. However, so far, no specific requirements and vocational trainings for safe eco-driving of clean vehicles or new safety regulations of how to ensure safety in cases of accidents through appropriate handling of high-voltage technology exist. Currently, only the manufacturers of

Chapter written by Wolfgang BACKHAUS.

clean vehicles brief public transport operators about the specifics of recently purchased clean vehicles and those briefings differ significantly in terms of time and content.

The situation relating to further training for drivers employed at public transport companies in the 27 EU Member States has improved with the Directive 2003/59/EC on the initial qualification and periodic training of drivers of certain road vehicles for the carriage of goods or passengers (which came into force in September 2008). The directive requires compulsory periodic training to be provided to professional drivers (35 h per 5-year period). The directive lists eco-driving as a subject of minimum qualification and training requirements (Annex I of EU-Directive 2003/59/EC) and due to its strategic relevance for public transport operators in terms of economic operation of their diesel bus fleets, more than three-fourths of public transport operator companies train their drivers in eco-driving as per UITP's observatory of employment in public transport of 2012 [UIT 12].

Among experts, electric public transport is widely acknowledged to have the potential to promote carbon neutral and environmentally friendly transport for the 21st Century. Besides technical improvements, the correct vehicle handling in terms of safety and eco-driving of clean electric drive vehicles has a distinctive influence on both aspects of environmental protection and the economics due to energy savings and optimized operational costs of clean vehicles.

Against this background, training programmes and education services for safe eco-driving of clean vehicles within the electric public transport sector have been developed for trams, hybrid buses, trolleybuses or e-buses (equipped with super capacitors) in the framework of the Intelligent Energy Europe Project ACTUATE: Advanced Training and Education for Safe Eco-driving of Clean Vehicles.

This chapter will outline the development of the training programmes for safe eco-driving of clean vehicles and the evaluation framework of ACTUATE. The evaluation of the impact of the training programme is based on a model designed by Kirkpatrick [KIR 94]. The model delineates four levels of training outcomes: reaction, learning, behavior and results.

4.2. Advanced training and education for safe eco-driving of clean vehicles: ACTUATE

The European ACTUATE project is funded by the Intelligent Energy Europe programme. ACTUATE focuses on the importance of safe eco-driving of clean vehicles and provides public transport operators with the opportunity to introduce and test safe eco-driving training. The ACTUATE consortium is composed of five public transport companies (Salzburg, Leipzig, Parma, Brno and Eberswalde), one further education organization (Leipzig), one clean vehicle manufacturer (Van Hool), an international action group tasked with promoting e-bus systems with zero emission (trolley: motion) and the project coordinator organization (Rupprecht Consult) from four European countries: Germany, Austria, Italy and Czech Republic. Starting from the principle functions of clean vehicles, the ACTUATE partners developed training programmes, with the aim of raising awareness for the knowledge, skills and competencies necessary to perform safe eco-driving of clean vehicles such as trams, hybrid buses, trolleybuses or e-buses equipped with supercapacitors.

Eco-driving is described as driving practices that optimize energy use and contribute considerably to reducing emissions as promoted by the European Commission's White Paper "Roadmap to a Single European Transport Area – Towards a competitive and resource efficient transport system" [EUR 11]. Identifying and reducing inefficient driving habits (e.g. harsh braking and hard accelerating) and energy-wasting driving maneuvers (e.g. idling and hard turning) have the potential to reduce operating costs and emissions.

However, highly skilled drivers are necessary to introduce clean vehicle technologies in an optimized way and to realize the energy efficiency of these technologies by maximizing the impact of the investment into these new clean vehicles. But due to the lack of training options, there is still a considerable difference between the level of driving performance required to successfully exploit the clean vehicle technologies and the actual skill levels in the driver's workforce. The purpose of the development of the safe eco-driving training programmes within the ACTUATE project is to close the skills, knowledge and competencies gaps – which were analyzed during the initiation stage by the ACTUATE partners – between "what should be" and "what is" the driving behavior of the driver's workforce regarding the ability to optimize energy consumption (or diesel fuel consumption in case of hybrid buses) of clean vehicles. By this, ACTUATE matches clean vehicle technology market and skills needs.

The introduction of safe eco-driving training programmes to public transport companies is of strategic relevance. By involving the human resources development of public transport companies – drivers represent the majority of the staff in public transport companies – and also the management level and the personnel and labor policy of the company, the safe eco-driving training programmes for clean vehicles make the link tangible and demonstrate the direct relationship between technology direction (clean vehicle technologies) and the training needs of drivers of newly introduced clean vehicles.

The experience made in the ACTUATE project during the planning, design and implementation phases of the safe eco-driving training programmes for clean vehicles showed that it is of particular importance to involve different levels of determining factors from the beginning to ensure a sustainable positive impact of the training programmes. The introduction process is based on several jointly defined – among the ACTUATE partners – input- and output-related factors on the determining levels: (i) human resources development, (ii) management of change and (iii) personnel and labor policy. On the one hand, there are input-related factors such as available financial resources for the purpose, legal regulations (e.g. through the implementation of the Directive 2003/59/EC) or existing number of personnel in the field of human resource development, which influence the introduction process of safe eco-driving trainings for clean vehicles in a public transport company.

For example, eco-driving is one of the minimum qualification and training requirements (Annex I) of the EU-Directive 2003/59/EC on the initial qualification and periodic training of drivers. This could be seen as an entry point for introduction and standardization of safe eco-driving training for clean vehicle drivers (e.g. in Germany, eco-driving is a mandatory subject of the periodic training of professional drivers regulated by the national law "Berufskraftfahrer-qualifizierungs-gesetz" (BKrFQG)), as national authorities acknowledge the validity of ACTUATE's training programmes.

But the different national implementation approaches of EU-Directive 2003/59/EC do not necessarily lead to a common level of initial qualification or training quality and professional drivers' abilities [BAL 11]. Therefore, further input-related process factors and output-related factors are necessary and should be defined by the human resources development department or the involved persons like trainers, instructors in order to reach a sufficient level of training quality. Process-oriented input factors are related to didactical concepts and teaching methods, e.g. frontal or group learning settings or share of practical training parts during a training session. The ACTUATE partners agreed on Kolb's learning cycle as the main educational principle [KOL 84], which supports an experience-based learning approach. This means that learning is relating to or resulting from the users "own" experience. Following Kolb's approach, the ACTUATE partners integrated a practical session as a basic part of the trainings to let the participating drivers experience the newly-learned driving style. To experience the difference and the impact of the new safe and eco-friendly driving behavior, each driver has two short practical driving sessions to enable a comparison between the "old" driving style and the "new" eco-friendly driving style.

The output-related factors describe "learning outcomes", these are the skills, knowledge and competencies that the drivers will gain through the training. The learning outcomes are preferentially defined by the following three categories "skills", "knowledge" and "competencies" as suggested by the European Qualification Framework (EQF)[1].

Furthermore, the commitment from senior-level management must be a driving force; this initiative is not simply introducing a training programme, but a management of change process related to the behavioral change of the drivers and the learning culture of the company as well as regarding the alignment of corporate goals and management strategies. The eco-driving initiative should be integrated into existing management strategies, e.g. ISO certifications for environmental management (ISO 14000) or energy management systems efficiency (ISO 50001) in order to become part of the company's energy management plan.

Figure 4.1 shows a matrix of the levels of determining factors at public transport companies and exemplary relevant activities (on an operational level) for the introduction of safe eco-driving training programmes based on the experiences of the ACTUATE partners and from the results of an expert survey (DELPHI process) in the course of the project.

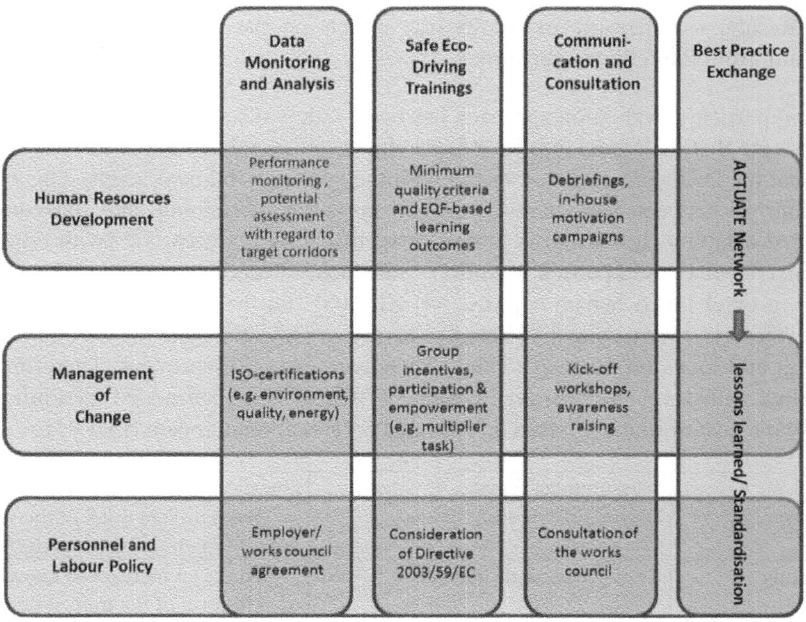

Figure 4.1. *Matrix structure for the introduction of safe eco-driving of clean vehicles training programmes*

1 More information about the European Qualifications Framework for Lifelong Learning (EQF) is available at: http://ec.europa.eu/eqf/documentation_en.htm.

4.3. Evaluation of training programmes for safe eco-driving of clean vehicles

Against the background of key questions in designing evaluation programmes, as suggested by Patton [PAT 97], the ACTUATE partners considered the purpose, target groups, methods and sources that could be used to obtain data, time and financial resources of the evaluation approach. As a result, the ACTUATE project built its training programme evaluation on the model developed by Kirkpatrick [KIR 94]. In the late 1950s, Kirkpatrick introduced a framework for evaluating training using four levels of measurement:

– Reaction – a measure of satisfaction (what the trainees/fellows thought and felt about the training);

– Learning – a measure of learning (the resulting increase in knowledge or capability as reflected in end of course assessment);

– Behavior – a measure of behavior change (extent of behavior and capability improvement as reflected in on the job performance);

– Results – a measure of results (the effects on the institutional environment resulting from the fellows' performance).

Kirkpatrick's evaluation approach has been widely discussed in the literature and critics note that the model implies a hierarchy of values related to the different levels and that the levels are each associated with the previous and next levels. This causal relationship between learner reactions and measures of training, and subsequently changed behavior, has not always been established by research and (with regard to the first level of Kirkpatrick's model) "satisfaction" does not necessarily lead to learning or changed behavior [ALL 89, ZIN 08]. Furthermore, within the results level, there is a criticism that benefits to the organization should be made more explicit and focus on monetary values such as return on investment. Therefore, the so-called fifth level "Return on Investment" (ROI) has been added by Phillips to justify the cost of training based on ROI and organizational impact [PHI 03].

Level	Measurement focus	Questions addressed
Reaction	Driver's perception	What did drivers think of the safe eco-driving training programmes?
Learning	Knowledge/skills gained	Was there an increase in knowledge or skill level of the trained drivers?
Behavior	Worksite (the driver's workplace, line operation) implementation	Is the new knowledge/skill being used on the job/in line operation by the trained drivers?
Results	Business impact on organization (Public Transport Company)	What effect did the safe eco-driving training have on energy consumption of clean vehicles?

Long-term impact	Return on investment and intangible outcomes	Were the benefits/impacts greater than the cost for the safe eco-driving training programmes? Was there a reduction of greenhouse gas emissions through safe eco-driving of clean vehicles? What effect did the safe eco-driving training have on the drivers' workforce satisfaction? Are there corporate image benefits through improved passenger satisfaction?

Table 4.1. *Kirkpatrick's four evaluation levels, measurement focuses and questions addressed in ACTUATE*

The following provides a summary of the five levels of the ACTUATE evaluation framework for safe eco-driving training programmes for clean vehicles.

Notwithstanding the critics of Kirkpatrick's evaluation model, the ACTUATE evaluation approach follows the model considering the four levels not as interdependent hierarchical levels, but as different levels evaluating the safe eco-driving training programmes for drivers of clean vehicles from different "angles". The combination of different quantitative and qualitative data collection methods used on the levels of Kirkpatrick's model provides a triangulation that increases the validity on the results level for ACTUATE. In addition, the ACTUATE evaluation approach includes a fifth level: "long-term-impact". Besides the suggested ROI view by Phillips (see above), this level also includes intangible outcomes such as corporate image benefits through improved passenger satisfaction and even effects beyond the organization like environmental protection through reduced emissions, as the ACTUATE project also has clear targets on reduction of greenhouse gas emissions defined in the common performance indicators of the Intelligent Energy Europe programme.

Transferred to the ACTUATE context, the essential questions the evaluation needs to address on the five levels of the ACTUATE evaluation framework are illustrated in Table 4.1.

Level 1 – Reaction

The involved public transport companies measure the drivers' satisfaction through a feedback questionnaire (five questions) to identify the trained drivers' reaction of what they thought and felt about the safe eco-driving training.

Figure 4.2 showcases the feedback of trolleybus drivers on "satisfaction" with the safe eco-driving trainings carried out at Salzburg AG.

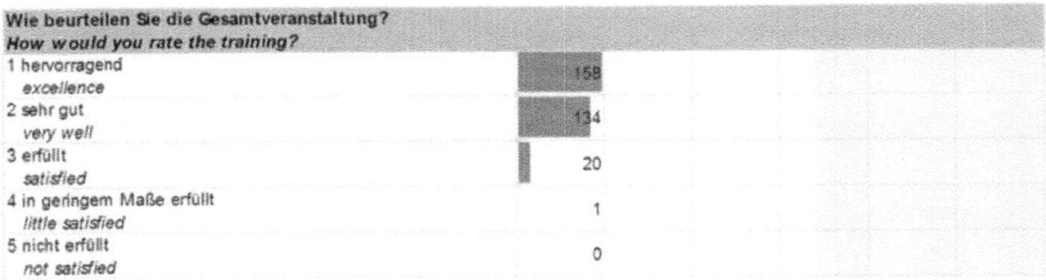

Figure 4.2. *Results of "reaction" feedback by trained trolleybus drivers in Salzburg (n = 313)*

Even though the positive feedback as demonstrated in Figure 4.2 does not guarantee that the drivers have learned new skills or gained new knowledge, it is certainly a filter for bad training as negative reactions can certainly hinder learning.

Level 2 – Learning

The learning outcomes of safe eco-driving training in the EQF-categories skills and knowledge, pre-defined by the ACTUATE partners, form the basis for the evaluation of how much the drivers have learned. As the safe eco-driving training does not include any tests, the main method to evaluate whether the driver has learned new skills in terms of safe eco-driving techniques is by the measurement of energy consumption during the practical part of the trainings, where the energy consumption of the clean vehicle is measured and recorded by customized software and subsequently displayed on a protocol. Thus, learning success through the application of a new skill and knowledge with regard to eco-driving can be measured during two practical driving sessions related to the comparison of driving behavior and its impact on energy-efficiency/consumption before and after theoretical input of safe eco-driving principles.

The immediate before/after comparison as part of the ACTUATE training sessions for trolleybus drivers in Salzburg showed that more than 90% of all drivers were able to cut down energy consumption (by 20% on average) by applying safe eco-driving techniques. Again, although it is useful to know that the drivers have gained new skills and knowledge through the safe eco-driving trainings, this positive outcome described above does not mean that the drivers will use the newly learned skills when they are back on the job, i.e. in line operation. In addition, there is a certain risk that the transfer of new skills and knowledge is only effective on the driver's short-term memory. To limit this "risk", the ACTUATE partners developed in-house motivation campaigns in order to motivate drivers to apply and retain their newly acquired eco-driving skills. All ACTUATE partners design motivational in-house campaigns, which range from bonus programmes to internal team competitions, e.g. "driver of the month", and incentives related to short knowledge tests (partly based on e-learning modules).

Level 3 – Behavior

This level measures the extent to which the drivers apply the new knowledge and skills on the job. According to Kirkpatrick [KIR 94], measuring at this level is difficult as it is often impossible to predict when the change in behavior will occur and thus requires a well-thought-out evaluation approach, e.g. when to evaluate, how often to evaluate or how to evaluate.

Therefore, all public transport companies involved in the ACTUATE project have measured the energy consumption of their clean vehicle fleets over a period of several weeks to months, either on a single vehicle or on line basis (measuring the energy consumption between two selected network substations), before the implementation of the safe eco-driving training programmes. By this, the energy consumption (where possible shown separately for the energy use of the drive train) of the clean vehicle fleets measured before the training programmes for safe eco-driving started could be used as a baseline for comparison to post-training effects. For this, public transport companies measure energy consumption of trained drivers on chosen vehicles/lines in operation again after the training programmes. Even though these long-term before and after comparisons of energy consumption are influenced by several dependencies at this evaluation level, the results will be a sound indicator for the usage of safe eco-driving by the trained drivers on the job.

Level 4 – Results

At this level, the ACTUATE partners evaluate the success of the safe eco-driving programmes for clean vehicles in terms of reduced energy consumption and consequently cost savings for the operation of the clean vehicles. Besides the cost savings for public transport companies, the indicator of reduced accidents through an anticipatory and forward-looking driving style, which is an elementary factor of safe eco-driving, could be measured as a "result" of safe eco-driving training programmes for clean vehicles. The "results" were evaluated by the ACTUATE partners after the completion of long-term measurements of energy consumption in line operation, which were expected for the end of 2013.

Level 5 – Long-term impact

For the long-term impact, the ACTUATE partners started to assess the monetary benefits of the training programmes for safe eco-driving of clean vehicles (see level 4 "results"), including further tangible benefits such as reduction of abrasion effects on clean vehicles, which are of importance due to high cost per unit for clean vehicles, and on the needed infrastructure in case of trolleybus or tram systems. These monetary benefits will be compared to the cost for the development and implementation of the safe eco-driving trainings.

The long-term impact evaluation on reduced greenhouse gas emission will also be based on the energy consumption measurements before the implementation of the training programmes in the course of the ACTUATE project. As a baseline, 2020 scenarios have been developed for each involved transport company: an "optimistic" (6% reduction of energy consumption) and a "realistic" (3% reduction of energy consumption) scenario. For example, assuming an overall reduction of energy consumption of 6% ("optimistic" scenario) as a result of the eco-driving training programmes, primary energy savings of 9.360 tons of oil equivalent and 50.135 tons of CO_2 reduction could be reached by 2020 for the clean vehicle fleets of the five public transport operators involved in ACTUATE.

4.4. Conclusions

The described framework for evaluating the impact of safe eco-driving training programmes for clean vehicles is based on the theoretical model of Kirkpatrick's [KIR 94] four levels of training evaluation. The model has been extended to a fifth level of "long-term impact" in the ACTUATE evaluation framework and together with the pretraining phases planning, design and implementation, which have been outlined in Chapter 2 (see also Figure 4.1), the conceptual approach of the ACTUATE project follows a logical pathway and benefit chain toward the impact evaluation of safe eco-driving training programmes (see Figure 4.3). Underpinning this evaluation approach is the "triangulation" of several evaluation methods focusing on both quantitative and qualitative evidence concerning the impact of eco-driving training programmes on the different levels of the evaluation framework.

The described evaluation approach involves evaluation activities not undertaken to date in the course of the ACTUATE project. Nevertheless, the data collected so far as "evidence" for positive "reaction" and "learning" of professional drivers in the implemented training programmes at the involved public transport operators is a promising start. However, level 1 and 2 evaluations can lead to a false sense of (short-sighted) success. It will need the described evaluations on the remaining levels 3, 4 and 5 to achieve an ideal full-scale impact evaluation of ACTUATE's safe eco-driving training programmes.

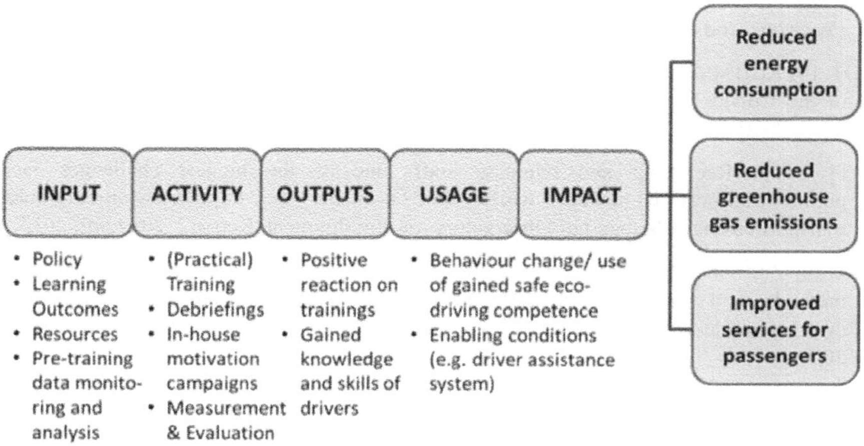

Figure 4.3. *Logic pathway and benefit chain of safe eco-driving training programmes for clean vehicles (according to [ROT 10])*

4.5. Bibliography

[ALL 89] ALLIGER G., JANAK E., "Kirkpatrick's levels of training criteria: Thirty years later", *Personnel Psychology*, vol. 42, pp. 331–342, 1989.

[BALL 11] BALL C., KONINGS H., VAN RIJN J., VET schemes for professional drivers in Europe, available at: http://www.project-profdrv.eu/fileadmin/Dateien/Downloads_front/ProfDRV_WP2_del06_12_01_CB.pdf, 2011.

[EUR 11] EUROPEAN COMMISSION, Roadmap to a Single European Transport Area – Towards a competitive and resource efficient transport system, White Paper, available at: http://eur-lex.europa.eu/LexUriServ/LexUriServ.do?uri=COM:2011:0144:FIN:EN:HTML, 2011.

[EUR 03] EUROPEAN PARLIAMENT AND THE COUNCIL OF THE EUROPEAN UNION, Directive 2003/59/EC on the initial qualification and periodic training of drivers of certain road vehicles for the carriage of goods or passengers, amending Council Regulation (EEC); No 3820/85 and Council Directive 91/439/EEC and repealing Council Directive 76/914/EEC), available at: http://eurlex.europa.eu/LexUriServ/LexUriServ.do?uri=CELEX:32003L0059:EN:HTML, 2003.

[KIR 94] KIRKPATRICK D., *Evaluating Training Programs: the Four Levels*, Berrett-Koehler San Francisco, 1994.

[KOL 84] KOLB D.A., *Experiential Learning: Experience as the Source of Learning and Development*, Prentice-Hall, Englewood Cliffs, N.J., 1984.

[PHI 03] PHILLIPS J.J., *Return on Investment in Training and Performance Improvement Programs*, 2nd ed., Butterworth-Heinemann, Burlington, 2003.

[ROT 10] ROTEM A., ZINOVIEFF M., GOUBAREV A., "A framework for evaluating the impact of the United Nations fellowship programmes", *Human Resources for Health*, vol. 8, no. 7, 2010.

[UIT] UITP, Recruiting and retaining staff: one of the biggest challenges facing public transport companies worldwide!, available at: http://www.uitp.org/public-transport/business_human/pdf/Observatory_of_Employment_Report2_2012.pdf, 2012.

[ZIN 08] ZINOVIEFF M.A., ROTEM A., "Review and analysis of training, impact evaluation methods, and proposed measures to support a United Nations system fellowships evaluation framework", *WHO's Department of Human Resources for Health*, Geneva, 2008.

5

The Role and Activities of SMEs in EU R&D Transport Programmes: the Case of Electric Vehicles

This chapter discusses some major results from the EU-funded INTRASME (Innovative Transport SME Support Action) project. Incremental improvements to currently dominant vehicle propulsion technologies are unlikely to reduce carbon emissions sufficiently to meet EU targets over the medium term. The development of the market for electric vehicles (EVs) and associated technologies opens up opportunities for new actors to enter the market. SMEs currently employ 55% of the EU workforce in transport, and their important role in the value chain is expected to expand. However, SMEs entering the EV space face a number of challenges, notably the dominance of the vehicle supply chain and R&D by a small number of existing large players, intense competition and the sunk costs of major investments in current production and transport infrastructure.

5.1. Introduction

The EU's 2011 Transport White Paper [EU 11] identifies the aim of growing transport and mobility while reaching a 60% emissions reduction target by 2050. In order to meet this target, innovation in transport sector technology is essential. However, incremental improvements to currently dominant vehicle propulsion technologies are unlikely to reduce carbon emissions sufficiently to meet EU targets over the medium term [FON 10, ITR 11]. The development of the market for electric vehicles (EVs) and associated (supporting) technologies opens up opportunities for new actors to enter the market. SMEs currently employ 55% of the

Chapter written by David MORRIS.

EU workforce in transport, and their important role in the value chain is expected to expand. However, SMEs entering the EV space face a number of challenges, notably the dominance of the conventional ICE (internal combustion engine) vehicle supply chain and R&D by a small number of existing large players, the intense competition characteristic of mature industries and the sunk costs of major investments in current production and transport infrastructure.

Recent studies (for example [WIE 11]) note that transport market segments vary in the structure of their preexisting relationships, the capital intensity required to meet development, production, standards, certification and regulatory requirements and both vertical and horizontal competitive pressures. The ability of SMEs to acquire new technologies and develop them new into marketable products is contingent on, among other factors, the position of the firm in the market and the nature of interaction with other actors. In broad terms, these factors can be categorized into market, product and network (the patterns of relationships between actors in getting products to market) types. This chapter concentrates on a particular field of innovation (land EVs) and views the innovation landscape through the specific lens of SMEs. The focus is on the effectiveness of EU R&D programmes as a means of promoting innovative SME participation in the emerging EV industry. Electric vehicle technology is highly disruptive in that it displaces existing vehicle fuels and components and may lead to structural change in the design and purpose of the vehicle. This challenge to the existing model creates gaps and opportunities at different stages of the product development process where manufacturers may seek collaboration with different market actors to overcome commercial and technical pitfalls.

5.2. Innovation in the transport market

Innovation is recognized as one of the central requirements toward the achievement of the EU's White Paper goals:

> *Innovation is essential for this strategy. EU research needs to address the full cycle of research, innovation and deployment in an integrated way through focusing on the most promising technologies and bringing together all actors involved.*

The acquisition of technology, whether from internal industry development or from external sources, is the first stage in a two-part process of innovation. Innovation encompasses the whole cycle of invention, product development, commercialization and the opening up of new markets [OEC 05]. Acquiring technology focuses the early process of invention and product development. This process is characterized by the matching of technical know-how with a knowledge of market needs, and an element of risk and experimentation in marketing a product, where demand may be unproven for the technology or product in its new context [PAV 05]. The process of acquiring technology is, therefore, shaped by the technology type, its limitations and application, the capacity of the organization(s) involved and the position of the agent in the overall market and supply chain.

This requires a focus on three factors:

1) *Acquisition and application of technology*: the type and maturity of technology, the production processes involved and likely demand for the product application have implications for the length of time required to place new products on the market. Transport components, infrastructure and assembly of vehicles are capital intensive activities and research indicates it is typically hard for new entrants to enter the market and manage cash flow during the development and early production phase [JRC 11]. This development period is often termed the "valley of death" where new ideas face financial barriers and product revenue is low or non-existent. In a successful innovation model, this gap in financing is offset by discounting the expected surplus revenues generated in the later mass production and commercialization stage. However, the route to market for electric vehicle technology is less clear than for mass-market products such as conventional automotives; for EVs, there are few existing models in the market and demand at their current performance-to-price ratio remains uncertain [JRC 11].

It is reasonable to suggest, therefore, that the "valley of death" period for electric vehicle technology will be longer and that companies will seek to develop alternative route-to-market strategies, in addition to focusing on the current supply chain. Furthermore, the degree to which a company will seek to diversify its product applications will be associated with the extent to which an EV supply chain is established within a sector.

2) *Company competencies*: in assessing the competencies of companies seeking to develop new technology, it is useful to employ the well-known and commonly accepted classification system of "Technology Developers" and "Technology Networkers" [SME 09] to assess SME positions in the innovation process. This model separates SMEs that have the development of a particular technology as their primary objective, from those where it functions as a secondary objective to primary interests such as diversifying, increasing competencies or reducing costs. Businesses are then classified by their level of R&D intensity. Technology developers are often subdivided into strategic innovators (R&D expenditure

accounting for more than 30% of turnover), exploitation seekers (those who seek to exploit results of R&D and spending less than 10% of turnover on R&D activity) and technology translators (who translate technologies to the market, typically spending between 10 and 30% of turnover on R&D). High research-intensive firms focused on the development of a new technology are likely to be at a precommercial stage where the development of functioning product application is the principal priority and customer revenues are low. Low research intensive firms looking to diversify their existing product line (technology translators) are more likely to be larger, more established companies seeking to acquire technology from elsewhere. This distinction is helpful for examining relationships within the transport sector, where new entrants interested in promoting innovative EV technologies must establish ways to acquire or access mass-production capabilities and a route to market.

Thus, it is likely that company priorities, size and the extent to which resources are allocated to either R&D or production will influence the position of the firm within the supply chain and the relationships it will seek to create. In addition, emerging technology developers will seek collaboration with organizations that have a role in exploiting technology. These may be large commercial organizations higher up the value chain, retailers and distributors, universities or regional authorities where they have a role in promoting new technologies.

3) *The innovation system structure*: the existing transport supply structure [JRC 11] can be divided between consumer vehicles, commercial fleet vehicles and public mass transit; the structures present in the market are well established, contract-based and dominated by a relatively small number of large firms. For new entrants, there are four issues: (1) the relationship of the firm to its primary customer(s); (2) the contexts of the industrial structure, for example, are there open standards, or a closed system of non-disclosure contracts?; (3) the relationships between SMEs and other SMEs/suppliers during the exchange of know-how, market information and services; (4) the role of the industrial support services in brokering relationships and providing subsidies or investments where there is evidence of market failure (for example, where the development of new emergent technology presents short-term financial risks that form a barrier to commercial partnerships).

The development of clusters in a technology type is likely to be dependent on demand from the primary customer(s), the willingness of the customer(s) to promote an open standard of exchange, the extent to which SMEs adopt collaborative relations with other SMEs/suppliers in the area and the alignment of industrial support structures with the needs of the primary customer. Where a primary customer does not exist, or demand from the end-user is not proven, SMEs may seek to establish a supply chain that is independent of current structures.

5.3. Methodology

This chapter reports some major results from the EU-funded INTRASME (Innovative Transport SME Support Action) project. The overall aim of INTRASME is to help improve the capacity of European SMEs to develop and implement products in the low carbon transportation sectors more rapidly. The project conducted 40 interviews with EV industry SMEs in four EU regions: Ruse (Bulgaria), Piedmont (Italy), Warsaw (Poland) and the West Midlands in the UK. Seventeen of the companies were directly involved in electric land vehicles and 23 in enabling technologies. This latter group comprises technologies with applications across a range of sectors and reflects the transversal properties of EV technologies. The group has representation from a mixture of areas such as battery development, electric motors, charging systems and so on. The study adopts an agent-based network analysis approach where SMEs were questioned about their strategies for partnering with other organizations with the aim of getting their products to market. Key aspects of this approach are to examine technology types, definitions of success, sources of ideas and innovation, product development processes, the nature of innovation network relationships and the benefits of the chosen approaches. On the basis of this information, we identify a number of different networking strategies and emerging models of innovation. A key question is how new technology may change relationships in the wider transport innovation system. In turn, this has implications for the types of policy intervention that will best help SMEs take a full and active role in the developing land EV marketplace.

Data on SME decisions, actions and experiences were captured using a semi-structured interview questionnaire. This approach provides a number of beneficial features:

– Agent based: the information is given from the perspective of the SME and the strategies they followed.

– Subjective: the data reflects the decisions, experiences and perceptions of the interviewee, as well as the actions taken and consequences.

– Quality of data: there are limited sources of statistical data on SMEs in the EV sector, particularly for very small or very new companies; an interview enabled the study to acquire both significantly more detailed information and also provided scope for follow-up questions in real-time.

– Time span: although taken from a single standpoint in time, interviewees were able to reflect on progress-to-date and future plans.

– Narrative: the semi-structured approach enables interviewers to follow and interrogate narratives constructed by the organization, enabling interviewers to form an understanding of how the interviewee views their motivations.

The semi-structured design of the questionnaire gave interviewers flexibility in following relevant themes, and also provided a sufficient consistency of data to allow for comparison. The structure is also intended to prevent the gathering of data becoming an entirely individual and subjective experience. Although questionnaires are intended to interrogate the viewpoint of the company, the approach inevitably reflects that of an individual occupying a particular position within a company and at a certain time-point. By providing a structure that provides a consistency of results over a larger sample, it is possible to construct generalized observations about common experiences.

The sample of 40 SMEs was established through reference to INTRASME project consortium partners' own contacts and those of Business Innovation Centers in the target regions. Consortium partners were originally selected on the basis of their links to developments in the EV sector and predominantly occupy "hub" positions as universities, SME associations, leading market research and advice consultancies and globally renowned engineering and testing consultancies. These positions enabled partners to select SMEs that represented a cross-section of technologies. Interviews were conducted on a face-to-face basis or, where this was not possible, via phone. This gave interviewers the flexibility to explore topics of interest and, jointly with interviewees, develop a narrative of the SME's activities that might not be possible through the completion of a questionnaire alone.

5.4. SME involvement in EU R&D programmes

SMEs reported that involvement in EU R&D projects had both advantages and disadvantages. On the one hand, participation may have the following direct benefits:

– increase the investment funds available to develop new tools, products and services;

– allow companies to increase R&D funding in areas of strategic importance;

– establish supply chain partnerships and exploitation routes with end-users;

– help to consolidate market position and enter strategically important new markets.

As well as the direct benefits of increased funding for R&D, other benefits cited included:

– a higher profile and increased credibility from the moment a project is awarded and publicized;

– acquisition of new knowledge and skills;

– providing a broader market view;

– networking with potential customers and collaborators;

– strengthening collaboration with large industrial companies and research institutes;

– the possibility of developing future products/services and working with project partners who can help them;

– providing the funding to risk employing people to develop new solutions.

A key benefit often cited was that interacting with the researchers/engineers of OEMs/large companies at the working level enables SMEs with new concepts to learn what OEMs look for in terms of production-ready technology, for example the need to carry out formalized failure analysis, production tolerances, etc. The access to the network of partners is also cited as a major benefit to SMEs. Participation in R&D projects can lead to direct bi-lateral contracts for work outside the EU project.

However EU R&D projects:

– only make sense where sharing information with collaborators does not prejudice longer term commercial interests in terms of IPR or current customer/industry relationships;

– are not appropriate where speed is essential to seize a good opportunity;

– take a long time to get funded – up to 12 months from call to contract.

INTRASME considered representative R&D projects from a range of EC programmes which have different characteristics relevant to SME participation. These included: FP7 Cooperation (Transport, ICT, NMP, Green Car PPP), FP7 People (Marie Curie Fellowships and Industry-Academic Partnerships), FP7 Capacities (Research for benefit of SMEs), Joint Technology Initiatives (ARTEMIS, ENIAC, FCH, Clean Sky), Competitiveness & Innovation Programme (CIP), EUREKA (EUROSTARS) and ERANET+. R&D projects within the FP7 Cooperation Programme also include two main instruments relevant to SMEs, that is STREPS (small targeted research projects) and IPs (large-scale integrated projects).

Programme	Key Points
FP7 Cooperation – STREPs or IPs in ICT, Transport, NMP including Green Car Initiative	– Collaborative R&D with typically 5–8 partners (STREPs) or 10+ (IPs) – Further product development is needed after completion
Joint Technology Initiatives – Fuel Cells and Hydrogen (FCH), ARTEMIS (Embedded Computing), ENIAC (Nanoelectronics)	– Large projects with supply chain sub-projects – SME participation encouraged – Nearer market projects
EUROSTARS (EUREKA)	– For research-performing SMEs – At least one other participant from another EUROSTARS country (the main EU countries) – Participation rules and funding differ for each Member State – Product ready for market within two years of completion
FP7 Capacities – Research for the Benefit of SMEs and Demonstration Projects	– To produce next generation products/services for SMEs that are not research performing – R&D is outsourced to two R&D providers from two EU countries and SMEs own the IP – At least 3 SMEs from 3 EU countries – Product ready for market within two years of completion
Competitiveness Innovation Framework Programme (CIP): Eco-Innovation	– CIP does not fund RTD projects. It aims to bridge the gap between RTD and commercialization – Priority given to SMEs
Competitiveness Innovation Framework Programme: ICT Policy Support	– Pilot actions for take-up of ICT and use of ICT R&D by citizens, governments and businesses, e.g. car sharing schemes
Competitiveness Innovation Framework Programme: Intelligent Energy Europe	– Encourages uptake of intelligent energy solutions

Table 5.1. *Key features of EU R&D programmes*

The general principles of IP ownership and access rights for EU co-funded projects are:

– ownership of background is unaffected by participation in a collaborative project;

– foreground resulting from the project is owned by the participant generating it;

– background and foreground are made available royalty-free for other partners (solely for use in the project) if they cannot perform their part of the project without it;

– background and foreground are made available on fair and reasonable terms if needed for other partners to exploit their foreground outside the project;

– specific background can be excluded from the obligation to grant automatic access rights. SMEs made the following points related to exploitation problems with their EC projects:

- there is a need to strengthen the commitment to exploitation by large companies and make exploitation plans truly collaborative. In many projects, the exploitation plan is fully developed in the last 6 months of the project when it soon becomes contractually irrelevant. Exploitation plans should be clearly outlined and agreed by all partners at the start of projects, then reviewed and modified as appropriate throughout to focus the project outputs. If an OEM writes the exploitation plan, it may focus only on the OEM's needs and may not be collaborative and may not support pull-through of any innovations developed by SMEs. A more formal and inclusive review of exploitation plans is needed to support this.

- SMEs want to understand from OEMs if they expect the SME technology to be incorporated into their production plan, how likely this is, and when it might occur. SMEs seek an OEM obligation to move to the next level of cooperation if technology development is successful. Typically, when an OEM goes into production it goes out to tender, so even if an SME has built a relationship, goodwill and some IP there is no guarantee of subsequent production work. The exploitation path in EC projects needs to be more formal, offering a path that SMEs could be presented to potential investors.

INTRASME also collected the views of EU project coordinators. These provide a useful counterpoint to those of the SMEs themselves. Their main conclusions were:

– SMEs usually lack a clear understanding of the end-user's needs and their position in the value chain: hence being involved in large consortia helps them to do this by receiving mentoring from large companies which see the wider picture;

– SMEs need to understand their niche and be realistic about getting into the supply chain, for example, it may be better to focus on their core activity, the bulk of which they can control, rather than trying to develop complete systems where they can control only a small part;

– there are many examples of SMEs trying to develop products to enter the electromobility market going bankrupt (including SME partners in EC projects), usually because unanticipated challenges arise that delay market entry and are too expensive for the SME to resolve;

– it may be a good strategy for SMEs to focus on the technologies where they can control 80–90% and are hence more likely to be involved in a supply chain – the next part of the value chain will then take it up;

– there is sometimes a disincentive to involve SMEs in projects, as there are several cases of small companies going bankrupt or being taken over by larger companies whose strategy does not then fit continuing in the project. This causes administrative problems and the need to find a replacement partner or manage the SME's tasks within the consortium.

The majority of coordinators interviewed stated that SMEs are a good source of innovative ideas:

– SMEs can try things that large companies cannot do, but they need to be partnered with a more mature company to balance the risk taking inherent in innovation with an established product development path;

– SMEs are less hampered by more structured planned "innovation" from Tier 1/OEMs and can develop ideas that are more interesting;

– large companies often have business units that look for something different. Problems arise when an OEM does not want to damage its 'conventional' business by adopting an SME innovation;

– both coordinators and SMEs commented that there were few cases where EC programmes followed up successful EC projects to help take them forward toward the market, other than the research for SMEs demonstration instrument where successful projects can bid for follow-on funding. There are also relatively few cases where past project leaders in a particular area are proactively brought together by the EC to present outcomes of completed projects and identify ways that project results can be exploited further;

– IP is very important to both large companies and SMEs and so both sides have to be careful about collaboration and sharing IP in an appropriate way.

5.5. Strategies employed by SMEs to overcome barriers

It is useful to distinguish between firms that have commercialized their products and those which are at the precommercialization stage. Within our sample of 40 SMEs, 13 of the 17 electric land vehicle companies and 13 of the 23 companies specializing in enabling technologies had commercialized their products. In all cases, companies identified product performance (which often included innovative features) and price as essential elements in commercialization. However, a number of other key factors were identified:

– the importance of establishing an early lead in a niche market. Examples include: material handling vehicles, mobility vehicles, EV scooters, EV conversion kits, etc.;

– a strong emphasis on flexibility, customization and responding to customer needs. Successful firms were often able to offer a wide range of products and services either through a diversity of internal competencies or leveraging the benefits of local supply clusters. The importance of responding quickly to customer needs was seen as essential in winning contracts;

– exploiting comparative advantages of low labour costs and the competitive advantages of being a first mover in the field;

– an open approach with large customers and an emphasis on protecting IP through superior know-how and delivery.

For firms that are at a precommercialization or low volume production stage, the key challenges are:

– financing two distinct "valley of death" phases: the first relates to funding R&D and prototyping stages, unit cost capital products to potential suppliers, where the costs of development need to be offset before investments can be made in mass production;

– managing through uncertainty and promoting a convincing value proposition where all stages of the supply chain are unsure about the degree of market demand for EV products and likelihood that demand will offset the development costs of improving technology;

– expanding beyond a local region where supply and demand factors are favorable.

These observations have two significant implications for understanding the product development process in the electric vehicle market:

1) The product development process most frequently described by SMEs is strongly orientated around the technical development of the product. In firms where a product has been successfully commercialized, a focus on technical

performance is combined with the ability to quickly customize these competencies to the needs of the market. This is reflected by the need expressed by companies for more marketing/brokerage support.

2) An understanding of a firm's position within the surrounding industrial structure is essential. Almost all of the companies that had commercialized, a product supplied B2B (business-to-business) and half of the original product ideas came from commercial partners. Clusters of similar companies can enable SMEs to share know-how and market knowledge as well as draw on a diverse range of competencies and quickly customize an offer. The presence of sympathetic regional support can help offset development and promote confidence in the region about the technology's future. The presence of a developed and localized supply chain enables SMEs to diversify their product range and reduce risk (examples include supplying to the commercial and higher education sectors or to the EV and ICE sectors) where SMEs are able to establish continuing relationships.

5.6. Summary

INTRASME interviews with electric vehicle SMEs have highlighted a number of issues. First, companies developing new technologies in the EV field possess a set of business objectives and technical competencies that can be applied in different product contexts. If a particular sector is not considered viable, they may leave it. Second, at an early stage EV, SMEs form horizontal networks to exchange technical and market information, but often lack access to the vertical supply chain structures of production and distribution. Third, production of vehicles and components is a capital-intensive activity that requires SMEs to possess very high levels of independent resources or to seek support which may derive from equity or financial support at the R&D and at the production phase, an alliance with a larger manufacturer in the EV or conventional supply chain or support for the development of a lead market by a public authority customer.

Participation in EU R&D projects is considered valuable by SMEs in the EV sector and can help to overcome the barriers many, if not all, of them face in commercializing their products. The findings of the INTRASME project should be helpful in improving the design and implementation of Horizon 2020 programmes in the transport arena.

5.7. Acknowledgments

The research leading to this chapter has received funding from the European Union Seventh Framework Programme (FP7/2007-2013) under Grant Agreement number 314587.

5.8. Bibliography

[EU 11] EU COMMISSION, Roadmap to a Single European Transport Area – Towards a competitive and resource efficient transport system, Brussels COM (2011) 144 Final, 2013

[FON 10] FONTARAS G., SAMARAS Z., On the way to 130 g CO_2/km—Estimating the future characteristics of the average European passenger car, Laboratory of Applied Thermodynamics, Aristotle University Thessaloniki, Thessaloniki, Greece, 2010.

[ITR 11] ITREN-2030, Experiences and results for integrated technology, energy and transport policy assessment, Final Report, available at: www.tmleuven.com/project/itren2030/D6_finalreport.pdf, 2011.

[PAV 05] PAVITT K., "Innovation processes", in FAGERBERG J., MOWERY D., NELSON R. (eds), *The Oxford Handbook of Innovation*, Oxford University Press, Oxford, 2005.

[OEC 05] OECD, *Oslo Manual, Guidelines for Collecting and Interpreting Innovation Data*, 3rd ed., OECD, Paris, 2005.

[SME 10] SMEPACT, Impact assessment of the participation of SMEs in the Thematic Programmes of the Fifth and Sixth Framework Programmes for RTD, 2010.

[WIE 11] WIESENTHAL T., LEDUC G., CAZZOLA P. *et al.*, Mapping innovation in the European transport sector, JRC Scientific and Technical Reports, EUR 24771 EN, 2011.

PART 2

Vehicle and Engine Technologies Development

Introduction to Part 2

One of the major objectives of current research on engine and vehicle development is to achieve further substantial reductions in fuel consumption, while optimizing power production and usage. This will be discussed in the chapters of Part 2. One chapter addresses the issue of high efficiency ship engine with ultra-low emissions. The aim is for marine engines to cost-effectively produce the required power throughout their lifecycle, with responsible use of the natural resources and while respecting the environment. Another chapter deals with the hybridization of the conventional vehicle drivetrain as an option for simultaneous reduction of fuel consumption and exhaust emissions and shows that the hybridization of heavy duty commercial vehicles (buses, trucks and vans) requires significant innovation that must be supported by well-structured testing activities on key components such as energy storage systems. A third chapter addresses the development of hydrogen infrastructure and fuel cell passenger vehicles in European cities. The activities which took place in Frankfurt, Germany and Mantova, Italy, as well as the evaluation of results and lessons learnt are presented and discussed. A fourth chapter discusses the results of a new high pressure fast opening rate shaping injector. The injector has been specifically designed for state-of-the-art injection pressures of up to 2,500 bar combined with completely flexible continuous rate shaping and shortest needle opening and closing times, thus avoiding needle seat throttling as far as possible. The last chapter is related to intelligent approaches, utilizing information and communication technologies that allow predictive energy management in modern vehicles.

Introduction written by Zissis SAMARAS.

6

HERCULES-1: The Long-Term (2004–2014) R&D Programme on Large Engine Technologies for Ships

In the year 2004, the Integrated Project HERCULES-A (High Efficiency Engine R&D on Combustion with Ultra Low Emissions for Ships) was initiated by the major engine makers MAN & WARTSILA, which together hold 90% of the world market. This was Phase I of the HERCULES R&D programme on large engine technologies. HERCULES-A involved 42 industrial and university partners, with a budget of 33 M€, partly funded by the European Union. HERCULES-B was Phase II of the Programme, from 2008 to 2011, with 32 participating organizations and a 26 M€ budget. The general targets for emissions and fuel consumption were retained in HERCULES-B. The HERCULES-C project (2012-2015), with 22 participating organizations and a 17 M€ budget, is Phase III of the HERCULES programme and adopts an extensive integration of the multitude of new technologies identified in Phase I and Phase II, for engine processes optimization, system control, as well as engine reliability and lifetime. This chapter provides an overview of the main achievements of the major R&D programme HERCULES in the 10 years of its existence and the future prospects of the related work.

6.1. Introduction

The three stages of the HERCULES R&D program on large engine technologies are shown in Figure 6.1, illustrating the progression from each phase to the next and the links of the R&D thematic areas.

Chapter written by Nikolaos KYRTATOS.

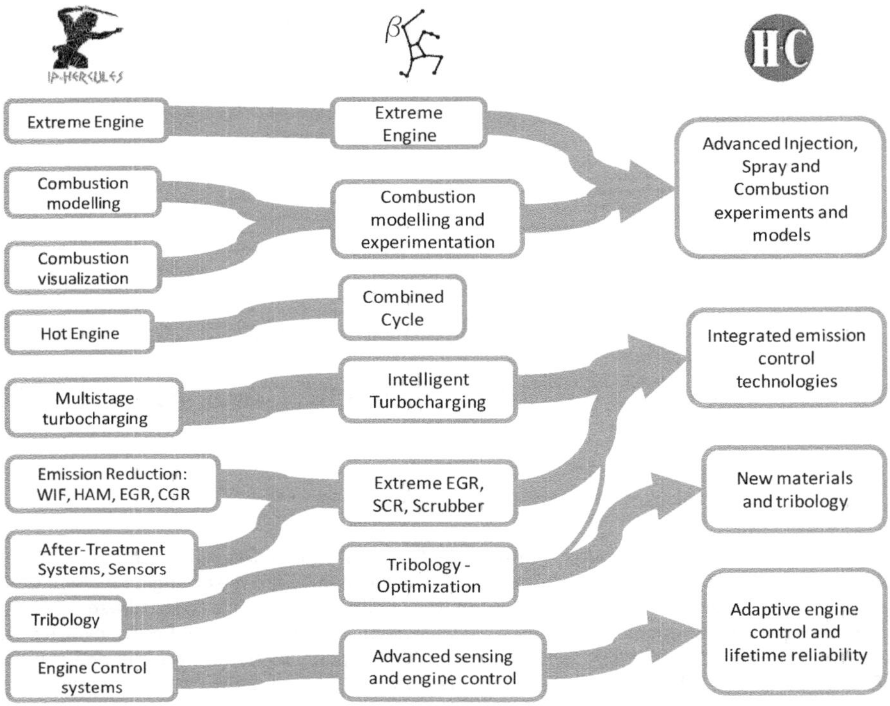

Figure 6.1. *Links from H-A to H-B to H-C*

The HERCULES-C (H-C) Project van over a 3-year period from 2012 to 2015 and had a budget of approximately 17 million euros, bringing the total combined budget of the HERCULES Projects (2004-2015) to approximately EUR 80 million.

The overall timeline of the three projects is presented in Figure 6.2. While there was a gap between Phase I and Phase II, Phase III started immediately after Phase II.

6.2. Achievements of Hercules-A and Hercules-B

The achievements of the previous projects have been widely published and are summarized in previous works, such as [KYR 12] and [KYR 13].

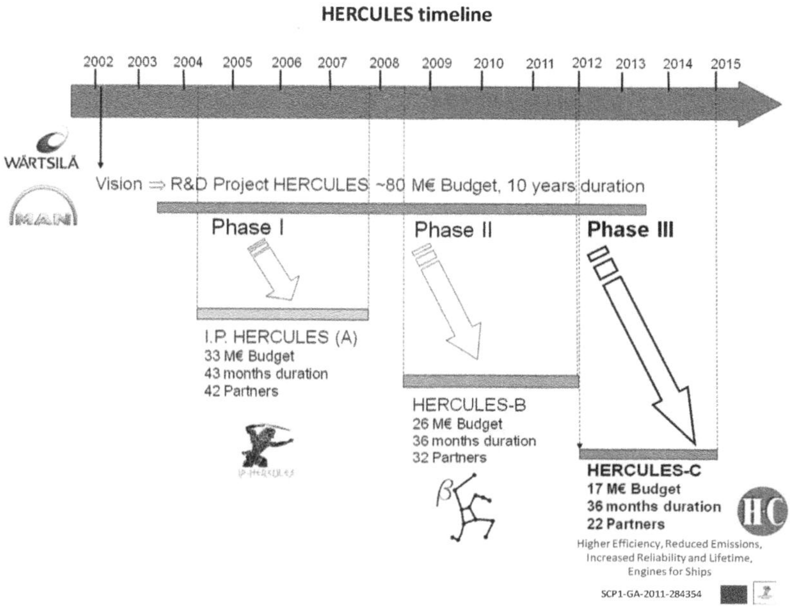

Figure 6.2. *Timeline of HERCULES programme*

The prototypes shown in Figure 6.3 are related to an experimental rig for large engine combustion visualization in Figure 6.4.

Figure 6.3. *Cylinder head for combustion process visualisation. For a color version of the figure, see www.iste.co.uk/jacob/energy.zip*

Figure 6.4 shows the advanced turbocharger components such as PTI/PTO and variable geometry compressor and turbine.

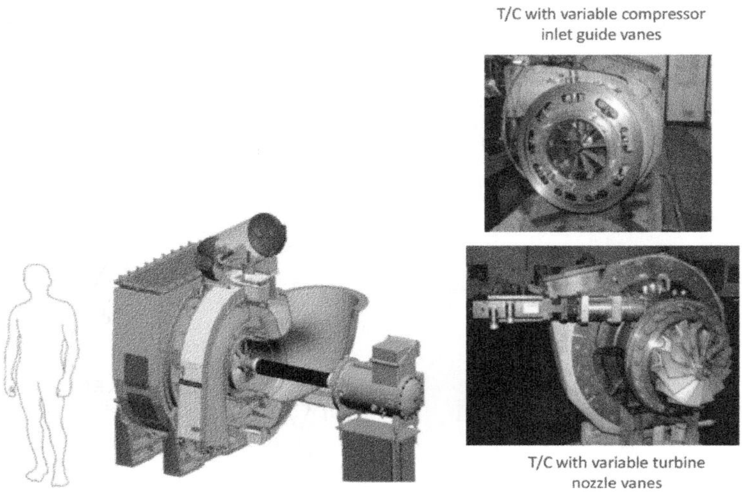

Figure 6.4. *Advanced intelligent turbocharger. For a color version of the figure, see www.iste.co.uk/jacob/energy.zip*

Figure 6.5 shows the prototype emission reduction units installed onboard for long-term testing under realistic operating conditions.

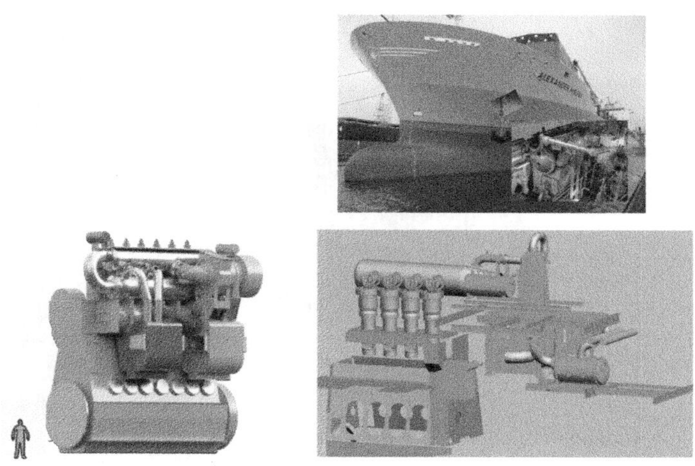

Figure 6.5. *Emission reduction- exhaust gas recirculation and after-treatment*

6.3. Objectives of Hercules-C

The first objective of HERCULES-C is to achieve further substantial reductions in fuel consumption, while optimizing power production and usage. This will be achieved through advanced engine developments in combustion and fuel injection, as well as through the optimization of ship energy management, and engine technologies supporting transport mission management. A second specific objective of HERCULES-C is to achieve near-zero emissions by integrating the various technologies developed from the previous collaborative research efforts. Furthermore, green product lifecycle technologies will be introduced and the third objective is to maintain the technical performance of engines throughout their operational lifetime. This includes advanced materials and tribology developments to improve safety and reliability, as well as sensors, monitoring and measurement technologies to improve the controllability and availability of marine power plants.

The HERCULES-C Project comprises five Work Package Groups (WPGs) in respective R&D Areas. Each WPG includes a pair of related work packages (WPs). Each work package contains several subprojects. Each subproject includes several activities. Shown in Figure 6.6 is an overview of the R&D thematic areas and respective WPGs.

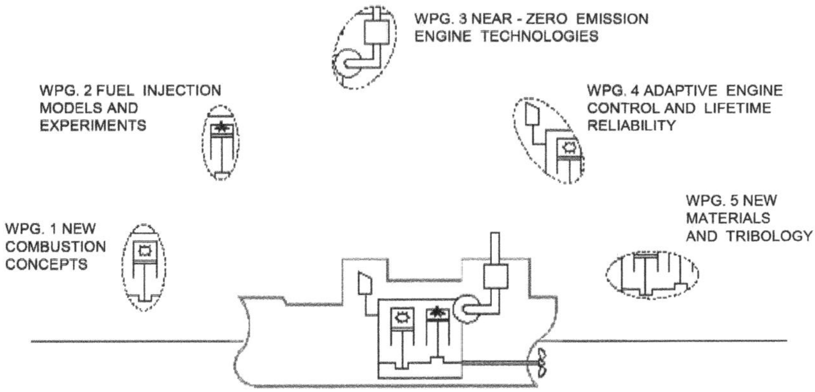

Figure 6.6. *Schematic overview of HERCULES-C Work Package Groups*

6.4. Project HERCULES-C results

At time of writing, this project has reached the middle of its duration. In the following sections, for each WPG, the objectives together with selected results are presented on per WP basis.

6.4.1. *WPG1: new combustion concepts*

6.4.1.1. *WP1: advanced combustion*

The WP1 objectives are:

– evaluation of direct injection gas combustion system concept allowing improved emissions and efficiency compared to current state-of-art;

– improvement of multi-fuel engines efficiency and fuel switching technologies beyond current state-of-art (ne>47%);

– investigation of partially premixed combustion with advanced fuel injection and air management systems included for control of combustion to reach concept for IMO Tier 3 emissions at part loads;

– optimization of cool combustion system for extreme Miller cycles in combination with exhaust gas re-circulation (EGR) to support targets of Work Package 5;

– during the first 18 months of the Project, important CFD simulation results were produced in relation with the diesel combustion system optimization, as well as the low-pressure gas direct injection for gas engines. Shown in Figure 6.7 are designs for an optimized gas engine.

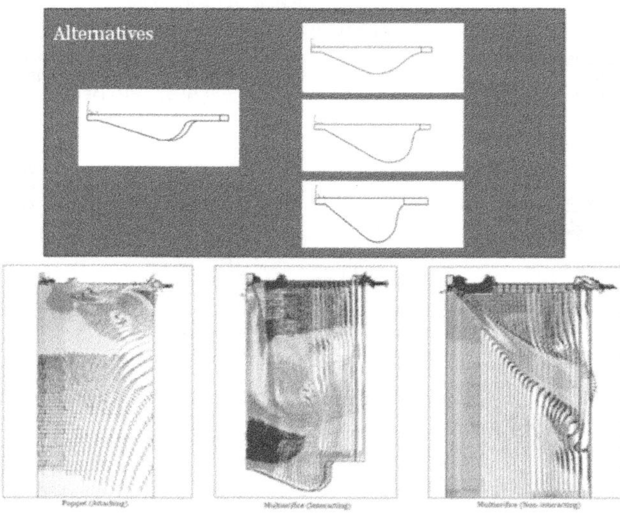

Figure 6.7. *Used piston top shapes in partly pre-mixed comb. simulations and velocity vector plots of characteristically different gas jets. For a color version of the figure, see www.iste.co.uk/jacob/energy.zip*

6.4.1.2. *WP2: computer aided combustion optimization*

The general objective of Work Package 2 is to develop innovative engine designs and optimized combustion strategies to fulfill future emission standards. More specific objectives are:

– improvement of engine design regarding combustion and engine process for both diesel and multi-fuel engines using advanced optimization techniques. The goal is to optimize fuel oil consumption within the IMO Tier III emission standard;

– conceptual optimization of various configurations of waste heat recovery systems leading to an overall adaptation of the ships thermal processes toward improved total energy efficiency;

– future combustion strategies for exhaust gas after treatment;

– validation of numerical models and tools and optimized engine components with measurements;

– future diesel combustion development based on computer aided engine design results on research engines and with support of optical investigations;

6.4.2. *WPG2: fuel injection models and experiments*

6.4.2.1. *WP3: injection, spray formation and combustion*

The specific objectives of the Work Package 3 are:

– to simulate and investigate by experiments the flow inside the injection system of large diesel engines and establish connections between design related factors, e.g. geometry effects and cavitation and the conditions at the nozzle exit, which in turn are decisive to the subsequent spray formation;

– to carry out detailed experimental studies on sprays and fuel–air mixing in order to establish the influence of different injection system design and operation parameters;

– to establish and validate methods and approaches for modeling such effects;

– to characterize the combustion process in large diesel engines and its dependence on the injection, spray and fuel air mixture formation processes;

– to apply the available knowledge to further develop, validate and apply CFD-based tools, techniques and methods for modeling combustion in large marine engines.

In this context, during the first half of the Project, important CFD results regarding the internal flow of the fuel injection system were produced. Spray experiments with Mie-scattering method were performed at the Spray Combustion Chamber both for symmetric and eccentric sprays (Figure 6.8).

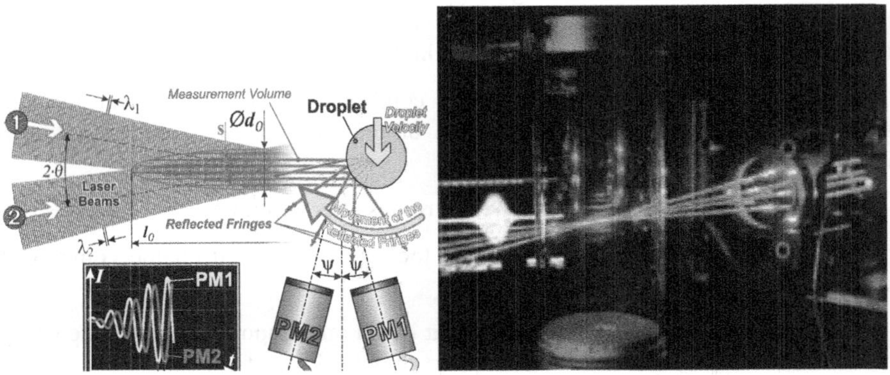

Figure 6.8. *Laser optic combustion investigations*

6.4.3. *WPG4: experimental and modeling studies of fuel injection systems*

The Work Package 4 objectives are:

– to develop and apply both experimental and computational techniques to the study of fuel injection systems for large engines;

– to develop advanced laser optical methods for characterizing in-nozzle flow together with the near-nozzle spray flow field;

– to develop CFD tools for describing the fuel injection system.

During the first 18 months of the Project, the design of three optical accessible rigs applicable to investigate different aspects of liquid fuel injection was completed. In Figure 6.9, the third rig, i.e. a pressurized vessel which facilitates simultaneous measurements of the near-nozzle spray and the in-nozzle cavitation pattern together with measurements of nozzle hole flow coefficients is presented.

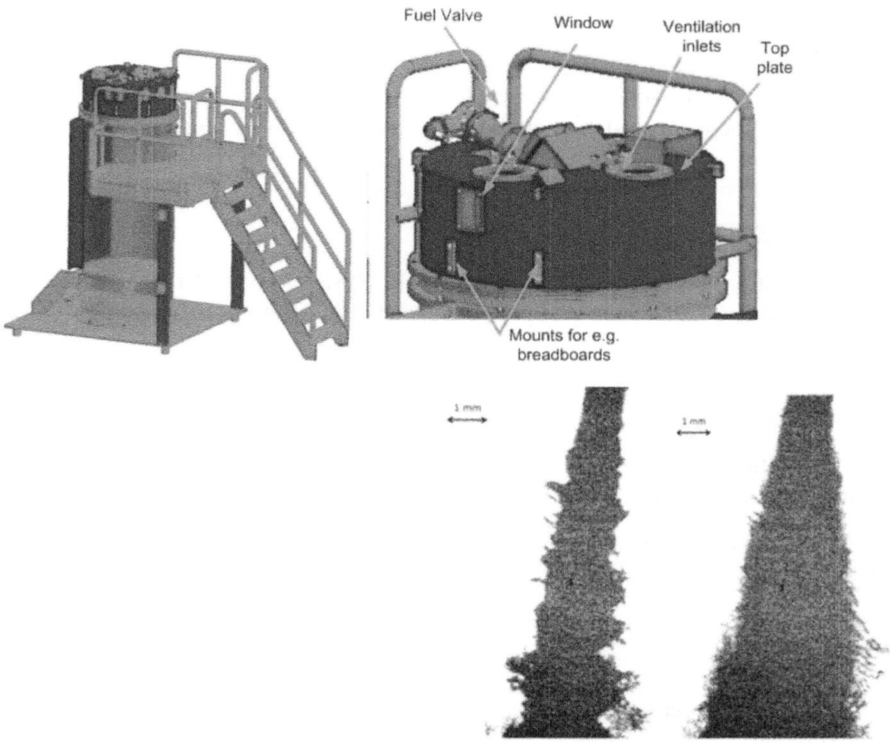

Figure 6.9. *Experimental studies of fuel injection*

6.4.3.1. *WPG3: near-zero emission engine technologies*

6.4.3.1.1. WP5: integrated emission control technologies

The Work Package 5 objectives are:

– to investigate options for further optimization, integration and flexibility of key components of advanced turbocharging and emission control technologies;

– to develop and test prototype(s) of integrated advanced turbocharging and emission reduction systems for IMO Tier III NOx emissions and best efficiency;

– development of sequential aftertreatment concept which enables SCR operation on high sulphur fuel and makes SCR integration with scrubber feasible allowing near zero emission level (>90% reduction in NOx and SOx emissions);

– development of scrubber concept capable of particulate reduction, up to 70% with specific focus on pre-treatment based on novel electrostatic precipitator.

6.4.3.1.2. WP6: near-zero emission combustion and DPF technologies

The Work Package 6 objectives are:

– provision of advanced turbocharging systems including exhaust gas recirculation (EGR) for low- and medium-speed diesel engines to achieve IMO Tier III NOx limits;

– PM reduction on medium speed diesel engines by sequential DPF after-treatment for US-EPA Tier4 equivalent 0.04 g/kWh limits;

– adaptation and integration of advanced DPF technologies for marine Diesel engines on ships including DPF regeneration strategy for higher fuel efficiency;

– toward zero NOx emission by combining EGR and water emulsified fuel (WIF);

– control system integration of combined EGR and WIF optimised performance on ships.

New turbocharger specifications involving Variable Turbine Area (VTA) at the High Pressure T/C are already produced, arriving at an optimized VTA design improving reliability at high gas exhaust temperatures (Figure 6.10). A fast and precise CO_2 EGR sensor is also procured and tested at 4T50ME-X test engine (Figure 6.11).

Figure 6.10. *Exhaust gas blower and VTA unit turbocharger*

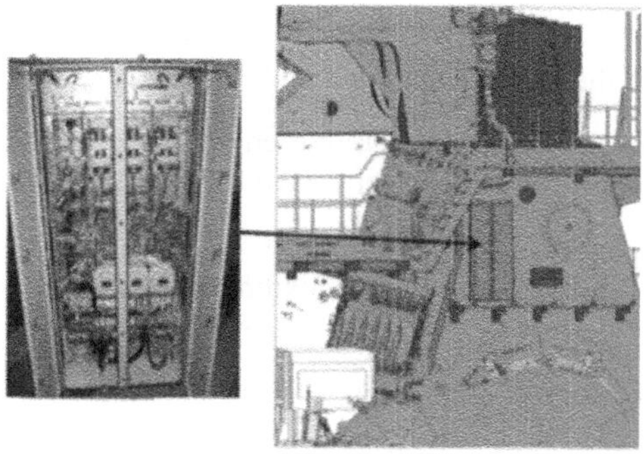

Figure 6.11. *EGR sensor on engine*

6.4.3.2. *WPG4: adaptive engine control and lifetime reliability*

6.4.3.2.1. WP7: advanced system and plant control

The Work Package 7 objectives are:

– development of integrated computer-based optimization methods and systems to improve the total control quality of ship propulsions systems;

– development of model-based diagnostic and health monitoring methods to increase system availability. Development of adaptive and advanced control applications for extreme operating conditions to eliminate the need for manual tuning;

– new concepts for ship propulsion supporting vessel asset management system and data management infra-structure.

A hybrid laboratory test engine is now set-up (Figure 6.12).

Figure 6.12. *Hybrid test bed*

6.4.3.2.2. WP8: intelligent engine

The objectives of the Work Package 8 are:

– development of an extended adaptive engine control with cylinder individual control algorithms applicable to diesel as well as multi-fuelled engines at extreme operating conditions;

– expected fuel savings due to an extended adaptive engine control up to 3g/kWh;

– development of signal evaluation criteria and diagnosis algorithms for new sensor types, e.g. in-cylinder pressure sensors;

– sensor prototype for health monitoring of core components;

– development and validation of algorithms to predict performance, faults and operating behavior.

6.4.4. *WPG5: new materials and tribology*

6.4.4.1. *WP9: cylinder lubrication concept for optimized emissions*

The general objective of the Work Package 9 is to develop an emission optimized lubrication concept. More specific objectives are:

– determine the amount of lube oil losses and its contribution to exhaust emissions, specifically carbon dioxide (CO_2), hydrocarbons (HC) and particulate matter (PM);

– establish a correlation between lubrication system parameters and exhaust emissions;

– develop a new cylinder lubrication system for 2-stroke engines based on an “inner lubrication system” and lube oil recirculation in order to reduce the lube oil consumption by 30%;

– to adjust lubrication system parameters on the basis of operational requirements as well as the health monitoring of re-circulated lube oil properties.

A cylinder lube oil recirculation concept developed to optimize lube oil transportation mechanisms, leading to reduced lube oil consumption is shown in Figure 6.13.

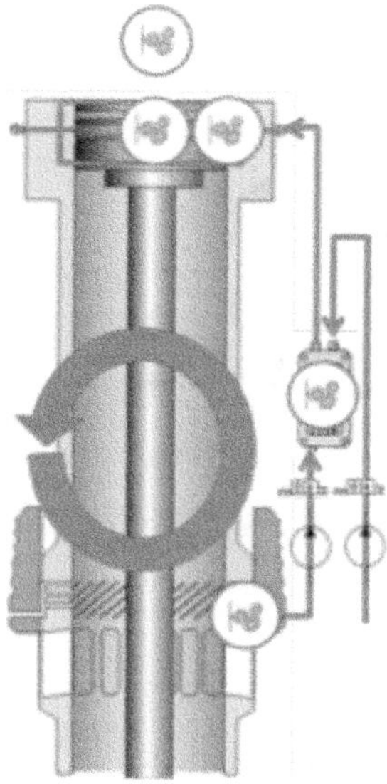

Figure 6.13. *Cylinder lubrication recirculation concept*

6.4.4.2. *WP10: advanced bearing and combustion chamber technology*

The Work Package 10 objectives are:

– optimize main bearing design for increased reliability by the use of advanced optimization, fluid-structure, thermal modeling tools and in “engine-like” test rig validation;

– lower the frictional loss in the main bearings of two-stroke engines by 10–15%;

– identify new piston ring – cylinder liner materials and surface technologies for large marine engines;

– development of low friction piston skirt with 20% lower frictional loss;

– reduce the piston ring wear by 10% and reduce piston ring-cylinder liner friction by 10% (2-stroke) and 20% (4-stroke);

– investigate the possibilities to reduce heat loss from piston crown cooling.

The Reciprocating Test Rig (RTR) used in testing of new material coatings is shown in Figure 6.14.

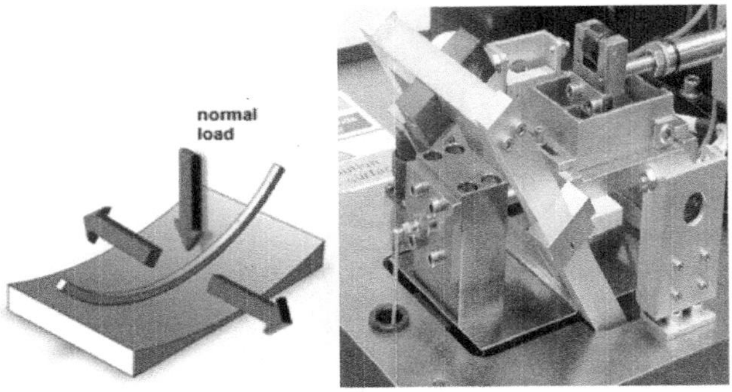

Figure 6.14. *Principle and appearance of reciprocating test rig (RTR)*

6.5. Conclusions

The HERCULES-C project was completed in 2015 with excellent progress and cooperation between the partners. The overall vision of the HERCULES programme is for sustainable and safe energy production from marine power plants. The technological themes of the HERCULES initiative, since its inception in 2002, are "higher efficiency", "reduced emissions" and "increased reliability" for marine engines.

In HERCULES-A (2004-2007), large-scale research platforms were established, with the main objective on screening the potential of a broad range of emission reduction technologies, in which area great improvements were achieved. In HERCULES-B (2008-2011), the quest for reduction in emissions was retained, focusing on several specific novel technologies, but at the same time more importance was placed on improved efficiency, hence reduced fuel consumption and CO_2 emissions. HERCULES-C (2012-2015) adopted an extensive integration of technologies developed in Phases I and II, with objectives to achieve further substantial reductions in fuel consumption, near-zero emissions and high performance of engines throughout their operational lifetime.

The aim is for marine engines to be able to cost-effectively produce the required power throughout their lifecycle, with responsible use of the natural resources and respect for the environment. The main added value of the HERCULES programme is that in no previous occasion a grouping of this magnitude has ever been assembled, involving world-leading organizations in marine engine development, pooling scientific /technological capacities and facilities, in a research programme of long-term scope and with a common vision in developing ultra-low emissions, efficient, reliable marine engines.

The group is already considering a follow-up stage within EU Horizon 2020, targeting at a futuristic fuel-flexible large marine engine, optimally adaptive to its surrounding operating environment.

6.6. Acknowledgments

The HERCULES Programme involved a large number of organizations and co-workers. The author acknowledges the fine work of the Task Leaders, the Members of the Project Steering Committee, the Task Controllers, the Project Management Team, the Partner Working Team Leaders and the more than 200 engineers and scientists involved in the various Phases and subprojects.

The HERCULES partners gratefully acknowledge the financial contribution of the European Commission, DG Research and the Swiss Federal Government (BBW & BFE) through EC- FP6 Contract TIP3-CT-2003-506676 with EC Project Officer Mr. Michael Kyriakopoulos, through the EC-FP7 Grant Agreement SCP7-GA-2008-217878, with EC Project Officer Mr. Joost De Bock and through the EC-FP7 Grant Agreement SCP1-GA-2011-284354 with EC Project Officer Mr. Gabriel Mialocq.

6.7. Bibliography

[KYR 07] KYRTATOS N., KLEIMOLA M., MARQUARD R., "The HERCULES Project: A major R&D effort for marine engines of high efficiency and low emissions", 25th CIMAC Congress, Vienna, 2007.

[KYR 10] KYRTATOS N., HELLBERG L., POENSGEN C., "HERCULES-B: The continuation of a major R&D effort towards the next – generation marine diesel engines", 26th CIMAC Congress, Bergen, 2010.

[KYR 12] KYRTATOS N., "HERCULES A-B-C, A 10-Year Major R&D Effort Towards the Next Generation Large Marine Diesel Engines", Transport Research Arena, Athens, 2012.

[KYR 13] KYRTATOS N., HELLBERG L., POENSGEN C., "Ten Years After: Results from the Major Programme HERCULES A-B-C on Marine Engine R&D", 27th CIMAC Congress, Shanghai, 2013.

http://www.ip-hercules.com

http://www.hercules-b.com

http://www.hercules-c.com

7

Energy Storage System Studies for Heavy Duty Hybrid Electric Vehicles in the EC HCV Project

The hybrid electric vehicle (HEV) is an excellent option for simultaneous reduction of fuel consumption and exhaust emissions in urban areas. However, substantial research efforts and industrial development are necessary to reduce cost and further improve high-efficient hybrid systems including key components such as the energy storage system, the electric machine, power electronics and electric auxiliaries. A large European project, the HCV (Hybrid Commercial Vehicles) project, started in January 2010 with the participation of 18 European organizations and with the scope to develop and demonstrate the current Heavy Duty (HD) HEVs in preparation for the next generation of hybrid commercial vehicles. A dedicated "Energy Storage Systems" Sub-project was planned to verify technical performances and reliability of the storage systems (lithium-ion and supercapacitors) in relation to specific HEV architectures and drivetrains. This chapter summarizes major achievements and experimental results, by describing the tailored testing procedures and the validated mathematical models.

7.1. Introduction

Public transport buses and commercial vehicles are being substantially improved to meet more demanding emission legislation and sharply increasing fuel prices. The proposed innovations for such vehicles are in some cases extremely challenging to achieve, aiming at technological and economical breakthroughs in research and development and subsequent industrialization. The conventional approach for introducing improvement measures usually brings along deterioration of engine

---

Chapter written by Mario CONTE, Francesco VELLUCCI, Massimo CERAOLO, Peter DAEMON, Giorgio MANTOVANI, Christian NIKLAS, Sophie TINTIGNAC and Giancarlo TROTTI.

efficiency and does not sufficiently pay off to satisfy all the mandatory targets in terms of emission reduction, increased energy efficiency and cost-effectiveness.

The hybridization of the conventional vehicle drivetrain is considered to be excellent option for simultaneous reduction of fuel consumption and exhaust emissions. Further research efforts are now performed to develop high-efficient hybrid systems including novel components such as energy storage system, electric machine, power electronics and electric auxiliaries. Cost is today a major obstacle for the market introduction of hybrid technologies in commercial vehicles.

The development of advanced and cost-effective heavy duty (HD) hybrid vehicles (buses, trucks and commercial vans) with significant improvements in drivetrain technologies and construction optimization is the key objective of a European Commission (EC) project aimed at using alternative or complementary storage systems to perform key functions (traction assistance to conventional internal combustion engine, regenerative braking and, eventually, pure electric traction mode for a limited range). This 4-year European project, named Hybrid Commercial Vehicles (HCV), started in January 2010 with the participation of 18 European organizations (leading HD vehicle manufacturers, components integrators and suppliers, research organizations) and with the scope to develop and demonstrate the current HD HEVs and to prepare for the next generation of hybrid commercial vehicles by using various types of storage systems: the final practical objectives were to reduce powertrain cost of about 40% and fuel consumption of 30% in a city bus cycle, compared with current hybrid bus technologies.

This project has a dedicated "Energy Storage Systems" sub-project (SP) to better assist suppliers/assemblers and vehicles manufacturers in designing and installing the various storage systems and optimize their use in different hybrid vehicle applications. This SP is dedicated to the experimental evaluation of the performance and reliability of the selected storage components (cells and modules) in relation to the specific HEV architectures and drivetrains, developed in the HCV project, before their integration in the final energy storage systems (ESS) to be installed in the HEVs. These activities aimed at analyzing electrical and safety performances and developing mathematical models for optimizing management and control of the EES made of two alternative energy storage technologies: lithium-ion (Li) and supercapacitor (SC) cells. Specifically, the main objectives of the activities on energy storage systems have been:

– to improve the reliability/safety and reduce the costs of the electric energy storage;

– to carry out basic characterization for evaluation and bench test of technologies/suppliers (including, for the power buffer type, supercapacitors) allowable for short-medium term industrial applicability;

– to carry out ageing, safety and life testing, and modeling implementation for control optimization (estimation of State of Health and State of Life).

These activities are jointly carried out by seven organizations (AIT-Austrian Institute of Technology, DimacRed, ENEA, IVECO-Altra, Magna Steyr, University of Pisa and Volvo) and are summarized in this chapter, by describing the testing procedures, adapted to the specific HEV configurations and targeted performances, and the chief experimental results achieved in various test activities and in different laboratories, including the developed mathematical models for both storage technologies.

7.2. The reference vehicles and the selected energy storage systems (ESS)

There are two second-generation HEVs under development in the HCV project that are used as reference HEVs (both parallel hybrid vehicles) with well-defined basic technical specifications applied in the design and test of the ESS. Altra and IVECO have developed a *daily* 5-ton delivery van, equipped with an ESS able to supply a maximum power (in charge and discharge) of 45 kW, a maximum voltage and current of about 300 V and 200 A and a delivered energy up to the end-of-life (EOL) of 800 Wh: these technical specifications apply to the two alternative ESS, one based on Li-ion technology and the other one on SC. Volvo has built up a 12 m city bus (a second-generation parallel hybrid) using only an Li-ion ESS with the following specifications: a peak power of 90 kW over the life (120 kW at Beginning-Of-Life, BOL), a rated voltage of 633 V, a peak current of 150 A and an available energy (for a complete discharge) also at end-of-life (EOL) of 1200 Wh.

These technical characteristics have been scaled down and adapted to the sample sizes (cells and modules) by using the concept of *battery size factor* (BSF), intended as an integer number, which is the minimum number of cells expected to be required to meet all the performance and life targets. For example, in the case of the Altra HEV Li-ion ESS, the BSF has been roughly rounded to 192, but if the ESS is based on SC, the BSF is 144.

These BSFs remain constant scaling factors for all subsequent performance and cycle life tests. Any test profile is then scaled by dividing the nominal profile power levels by the BSF. For example, by applying this BSF to the Altra ESS design, the 45 kW peak discharge power for HCV to be used in lifecycling test would then be reduced at a pulse power level of $45000/192 \approx 230$ W for Li-ion cells, while for SC cells a pulse power level 312.5 W must be used. In case of module testing, the achieved pulse power, scaled down with BSF at cell level, must be multiplied by the number of cells in the module.

7.3. Dedicated testing procedures

The approach used in specializing already available test procedures (and even existing or under development standards) to the experimental energy storage test needs in the HCV project is very simple and follows a four-step procedure:

– EUCAR and FreedomCar testing procedures and also various specific standards are carefully revised and adapted;

– the test sequences mostly containing operating conditions and general duty cycles in existing procedures are changed by introducing the operating conditions and performance characteristics identified for HCV commercial HEVs. These modifications have been made possible by defining the technical specifications of the HEVs and the storage systems by the two vehicle manufacturers at the beginning of the project;

– application-oriented duty cycles as identified by the vehicle manufacturers are used for the execution of life and ageing tests that are still underway;

– definition and use of additional tests for the development of mathematical models of HCV storage technologies and for the identification of performance characteristics or module design and control needs.

The HCV test procedures contain a dedicated set of electrical and abuse tests for cells and modules of Li-ion and SC technologies defined for the specific heavy duty applications.

For example, the specialization of electrical testing activities was achieved with the identification of the planned duty cycles for each HEV. The basic idea was to optimize the design of the ESS, by experimentally evaluating the behaviors of these devices by using realistic operating conditions, proposed by the HEVs manufacturers. For example, Altra–IVECO has proposed a power micro-cycle for the mild hybrid version of its *daily* van.

The overall duration of this micro-cycle is of 167 s, which corresponds to a travelled distance of 1 km and an ESS variation of SOC (state-of-charge) of 5%.

In addition, for life testing analysis, this micro-cycle had to be repeated 180 times in a day (with an idle time of 10 min every 18 micro-cycles) and in total 45,000 times in a year. To also take into account temperature effects and variations, three different temperatures (related to three periods of the year) were selected (20, 30 and 40°C) for carrying out yearly cycles and analyzing cooling/heating needs of the ESS and its BMS (battery management system). Similarly, Volvo has supplied a more detailed and experimental duty profiles (power, voltage and current) for a typical city urban route.

7.3.1. *Abuse testing*

Abuse testing is aimed at characterizing energy storage systems Li and SC cells and modules during uncontrolled or severe operating conditions/environments. The test program of abuse testing has the main objective of identifying, through controlled simulation testing, all the possible risk conditions. These conditions are then analyzed to clearly define mitigation measures to be used in design, control and usage of such storage systems. Abuse testing results are also useful to investigate failure pathways (without post-mortem analysis), so as to eventually propose new control parameters.

After the verification of compatibility of testing condition in existing safety test procedures and standards with HCV ESS specifications, the abuse testing and safety procedure was concentrated on:

– mechanical:

- controlled crush;
- penetration;
- immersion.

– thermal:

- thermal stability;
- elevated temperature storage;
- rapid charge/rapid discharge.

– electrical:

- overcharge/overvoltage;
- short circuit;
- overdischarge/voltage reversal.

– mixed (for example, mechanical and electrical together).

A safety testing matrix, based on key influencing parameters, was used to statistically define the required number and type of sample per test. Initially, three samples were proposed for each test for acceptable statistics, but this choice was hardly applied to module testing because of the cost of the sample and the total available number.

7.3.2. *Electrical testing*

In summary, the electrical test procedures of Li and SC cells and modules have been structured into five different test sequences aimed at different testing scopes:

– general preparation tests;

– basic characterizations for the designed operating conditions of the HCV demonstrators;

– ageing – accelerated life testing to estimate cell life under with degradation accelerating factors to give quick feedback to system design and road demonstrations phases;

– management-oriented tests on a small set of series-connected cells (or module with balancing electronic board) to study management needs in terms of voltage dispersion and thermal control;

– input measurements for modeling to collect operating data for the definition and validation of mathematical models.

Figure 7.1 presents the flowchart of the electrical test procedures for SC cells and modules, while for the Li-ion samples a similar test sequence was defined, with the extension of two different sizes and operating conditions, referred to the two reference HCV vehicles. Test matrices were also agreed to optimize the use of available number of samples (initially limited by budget restrictions) among the various tests (including safety) and laboratories.

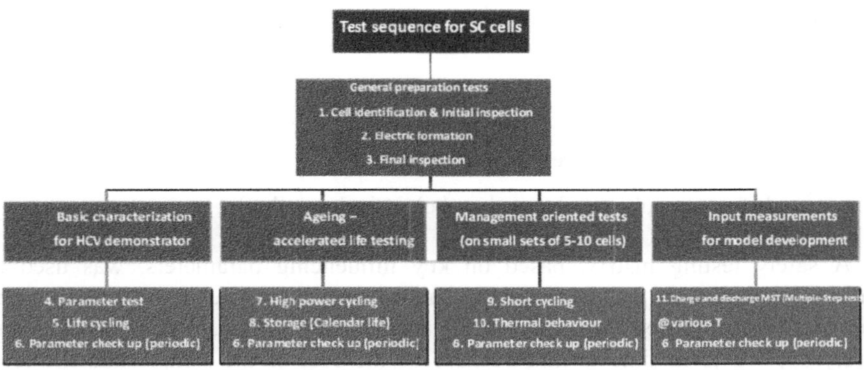

Figure 7.1. *Flowchart of electrical test procedure for SC samples (cells and modules)*

For modeling purposes, the definition and validation of mathematical models requires a specific set of electrical tests for measuring or calculating the model parameters.

The proposed test shown in Figure 7.2 was selected to give a lot of information on the cells behavior (and electrical model circuit parameters, such as internal resistances and capacitances), much more than the OCV-correlation curve. For its importance, it is given in this document a name of its own: it will be called *multiple-step test* (MST). This test can be performed exactly as reported in Figure 7.2, starting from a fully discharged battery or a fully charged battery and by using charging current steps or discharging current steps. Therefore, the MST can be charge-based and discharge-based, and consequently applied during both phases.

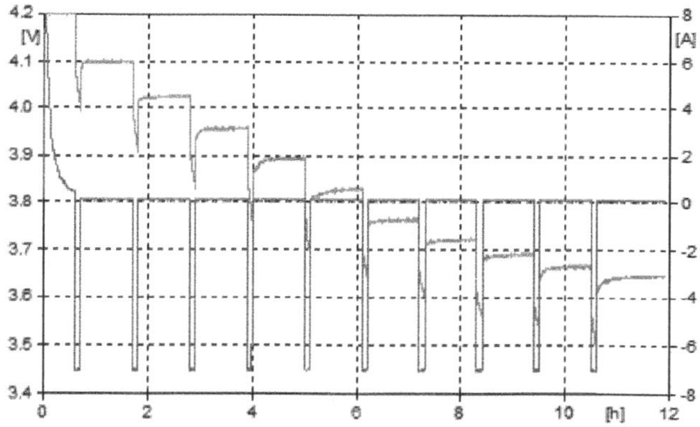

Figure 7.2. *The discharge-based multiple-step test. Green: current; red: voltage. For a color version of the figure, see www.iste.co.uk/jacob/energy.zip*

Each MST was performed, both in charge and discharge, at different temperatures. Moreover, it was basically performed on single cells to evaluate actual cell performance, independently of the effects of the battery management system (BMS).

7.3.3. *Sample under test*

Samples of cells and modules of both technologies (Li-ion and SC) underwent safety and electrical testing. Table 7.2 shows the basic characteristics of Li and SC cells.

Properties	Li Cell	SC cell
Capacity/Capacitance	4.4 Ah	3000 F
Nominal Voltage	3.3 V	2.7 V
Mass (average)	205 g	510 g
Diameter	32 mm	60.7 mm
Length	113 mm	138 mm
Energy	14.5 Wh	3.04 Wh
Specific Energy/Energy content	75 Wh/kg	5.96 Wh/kg
Power (25C, 50%SOC, 10sec)	578 W	3020 W
Specific Power (25C, 50%SOC, 10sec)	2979 W/kg	12300 W/kg
DCR (25C, 50%SOC, 10sec)	4.5 mΩ	0.29 mΩ

Table 7.1. *Technical properties of HCV Li-ion and SC cells*

During the progress of the project, an industrial assembling process was applied to the final generation of the Li-ion modules, while for the SC modules an initial generation was improved and a second generation was assembled in 2013. Figure 7.3 presents the second generation of SC module.

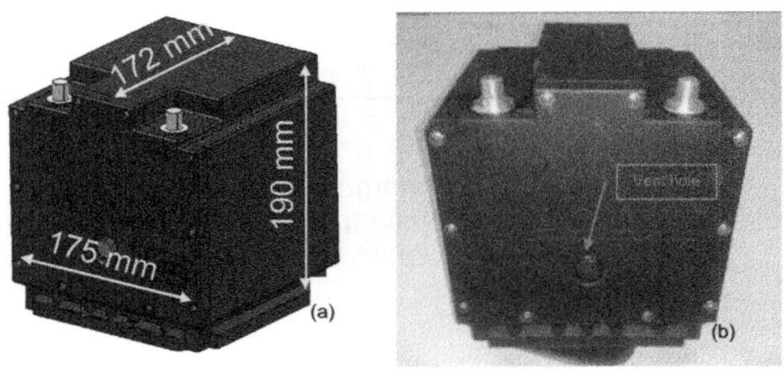

Figure 7.3. *Second generation SC module: a) drawing and b) real sample*

7.4. Testing results

7.4.1. *Abuse testing results*

The abuse testing results only refer to cell testing because this type of test on modules was delayed after the completion of electrical testing for economical reasons. Figure 7.4 shows the short circuit set up for a Li-ion cell.

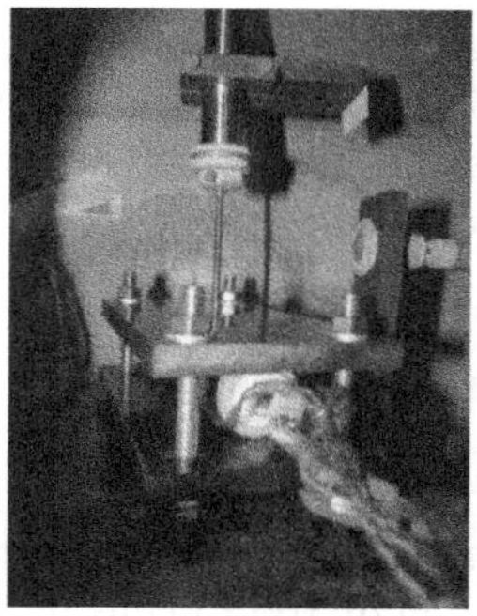

Figure 7.4. *Setup for internal short circuit abuse test. Cell with rod and pneumatic cylinder in pressure tank, voltage sensing and three thermal sensors (two surfaces and one valve). For a color version of the figure, see www.iste.co.uk/jacob/energy.zip*

During the short circuit test, an adequate and accurate test environment for measuring the gas emission was established. The measuring equipment developed by AIT was able to properly determine the gas released by a Li-ion cell, presented in Figure 7.5.

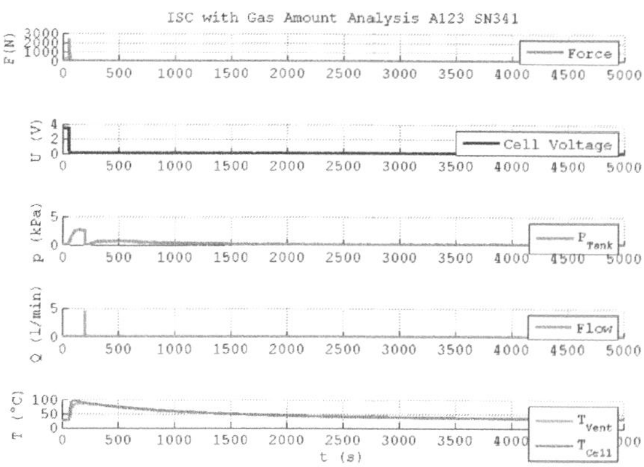

Figure 7.5. *Measured values of Li-ion cell during internal short circuit test with gas amount detection at SOC = 100%. For a color version of the figure, see www.iste.co.uk/jacob/energy.zip*

In the same way, SC cells were abuse tested, substantially confirming abuse characteristics declared by the manufacturer.

In general, it can be stated that all the cells of both Li-ion and SC technology show uncritical behavior. In none of the tests fire, rupture or explosion could be observed. Only in tests considered to be destructive, electrolytes were lost and in some cases smoke was visible. Additional gas emission measurements confirmed that, during abusive conditions, only very small amounts of gases were released and also only very low heat was produced.

7.4.2. *Electrical testing results*

Basic characterizations on cells and modules confirmed design performances by allowing for the accurate selection of the samples for the subsequent tests. Figure 7.6 exemplifies the voltage dispersion of 100 Li cells during storage: self-discharge and control needs were investigated. The voltage variation is small and within the recommended limits. In Figure 7.7, the discharge curves of an Li module are presented as a function of the current rate: the achieved values are still below the nominal capacity of 8.5 Ah at BOL, because normally it is necessary to carry out more complete charge/discharge cycles before reaching nominal capacity.

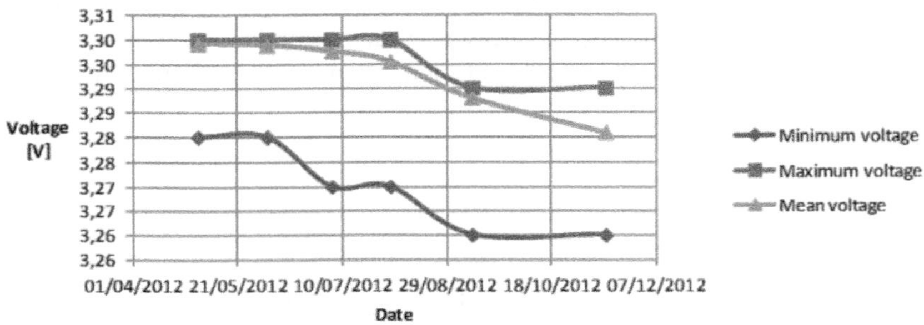

Figure 7.6. *Voltage variation of 100 Li cells during storage*

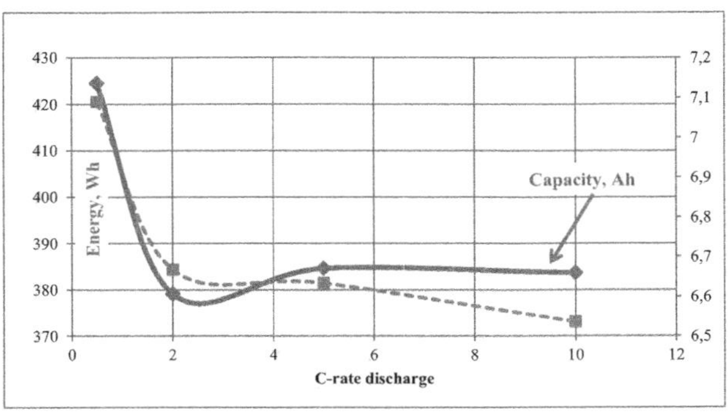

Figure 7.7. *Discharge capacity and energy of an Li module*

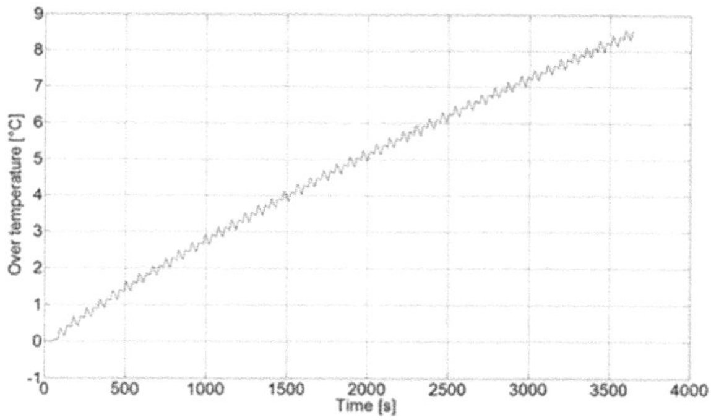

Figure 7.8. *Temperature increase during power cycle testing of an SC module*

Lifecycle testing with selected profiles is still underway, but the available results provide some preliminary feedback for manufacturers and assemblers. Figure 7.8 presents the temperature rise in an SC module, which is, at the end of 1 h of testing, increased to 8.55°C, well below the maximum acceptable value of 20°C.

In lifecycle testing, carried out with the proposed micro-cycle and planned yearly life, the same SC module performed 50,000 complete cycles, shown in Figure 7.9, by decreasing the module capacitance of about 11%.

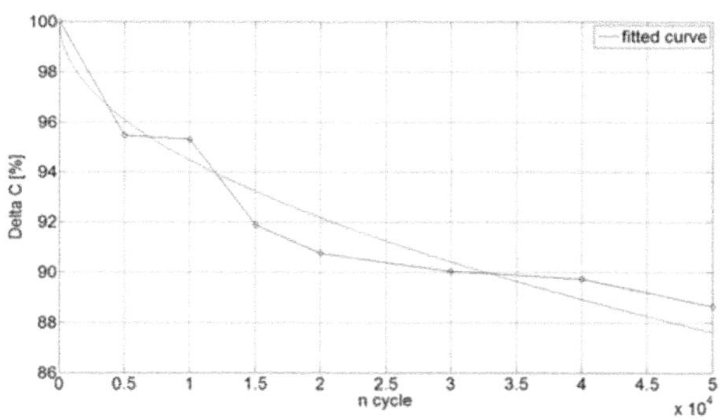

Figure 7.9. *Variation of module capacitance C during 50,000 lifecycles. For a color version of the figure, see www.iste.co.uk/jacob/energy.zip*

According to the experimental (accelerated) lifecycle testing, the extrapolated end of life due to capacitance reduction is 130,000 complete cycles. With respect to the projected lifecycle of 360,000 cycles, corresponding to an operational life of 8 years, there is still a factor of 2.8 to be improved.

After the performance determination tests of modules, some tests were devoted to verify module design and other aspects (polarization curves and cycle life at various energy windows).

The thermal behavior of modules was equally investigated by taking into account the cooling needs proposed by the HCV manufacturers. Infrared camera thermographs give detailed indications about the thermal behavior during cycling of modules, giving useful feedback for module assembly and thermal management needs.

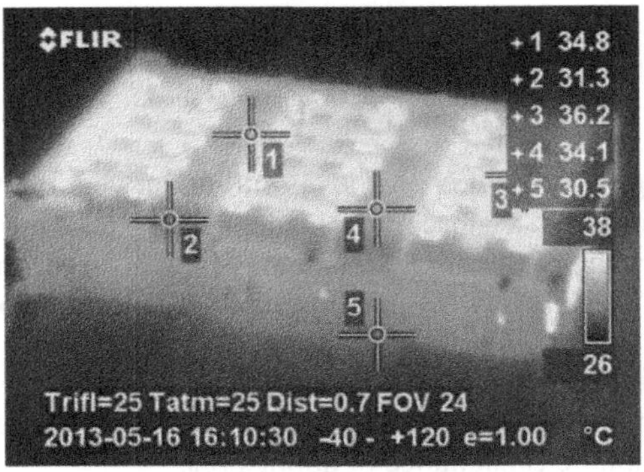

Figure 7.10. *Module thermograph almost at the end of test. For a color version of the figure, see www.iste.co.uk/jacob/energy.zip*

The module thermographs (see an example in Figure 7.10) clearly show a gradual increase in temperature during testing (without a cooling system or thermal management), but the temperatures reached are still considered in an acceptable window for the Li technology and for the working conditions recommended by the battery and vehicle manufacturers.

Finally, mathematical models for module behavior estimations were defined and validated with experimental activities. In particular, selected model parameters and an extended Kalman filter (EKF) algorithm (for estimating the ESS SOC in a simplified manner) were validated for Li-ion and SC technology by using various duty cycles. A long and complete experimental work was required.

Figure 7.11 summarizes all the work on the SOC estimator, by showing the practical errors developed by the EKF algorithm (for Li cell) for estimating the ESS SOC.

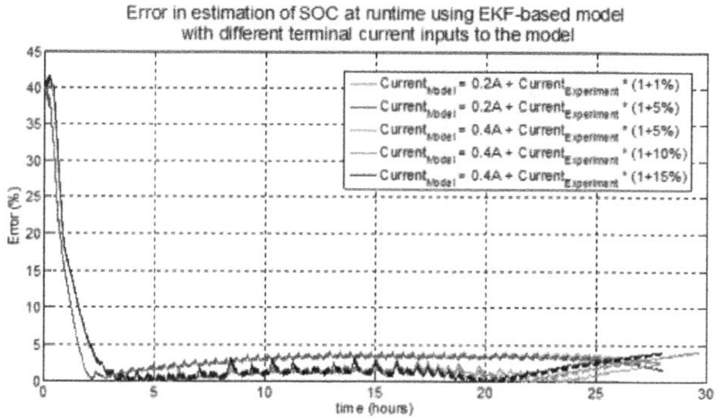

Figure 7.11. *Percentual error in SOC estimation by EKF-based model with different current inputs. For a color version of the figure, see www.iste.co.uk/jacob/energy.zip*

7.5. Conclusions

The hybridization of HD commercial vehicles (buses, trucks and vans) requires significant innovation that must be supported by well-structured testing activities on key components such as the energy storage systems. The experience in the HCV project of the European Commission demonstrates the good and useful feedback that the vehicle and component manufacturers may receive from the experimental characterization of ESS systems during the development phase. The adaptation of existing standards and procedures can be helpful in improving design and use of energy storage systems, by reducing time and cost through adequate sharing of expertise and equipment. The combination of abuse and electrical tests, as shown in HCV work and analyzed in this chapter, may improve their efficacy if adequately adapted to the specific design and duty cycles of the developed HD HEVs, which may reach their target cost with ESS optimization. Furthermore, the possibility to use additional test results for developing and validating mathematical models for implementing simplified and reliable ESS parameters estimators has been also positively presented and is also beneficial to improve ESS control and management and, consequently, increase cycle life and reduce operating costs.

7.6. Acknowledgments

The research leading to these results has received funding from the European Union Seventh Framework Programme (FP7/2007-2013) under grant agreement no. 234019 for the Hybrid Commercial Vehicle (HCV) project.

7.7. Bibliography

[AIT 10] AIT, Hybrid Commercial Vehicle (HCV) FP7-Project, Abuse Test Plan for Li Batteries and SC, Deliverable D3100.4, 2010.

[AIT 13a] AIT, Hybrid Commercial Vehicle (HCV) FP7-Project, 1st Intermediate Technical Report on Abuse Tests on Li cells, Deliverable D3300.1, 2013.

[AIT 13b] AIT, Hybrid Commercial Vehicle (HCV) FP7-Project, 1st Intermediate Technical Report on Abuse Tests on SC, Deliverable D3100.3, 2013.

[ALT 11] ALTRA, Hybrid Commercial Vehicle (HCV) FP7-Project, Technical specifications for HEV, Deliverable D3100.1, 2011.

[BER 12] BERTONI S., "State-of-charge evaluation of lithium cells showing voltage hysteresis", *Energy Efficiency Trends Annual Report*, Brussels, 2012.

[CER 12] CERAOLO M., "Challenges in state-of-charge estimation of high-power LFP cells", *Energy Efficiency Trends Annual Report*, Brussels, 2012.

[CER 13a] CERAOLO M., Hybrid Commercial Vehicle (HCV) FP7-Project, Algorithm for Li cell, Deliverable D3200.7, University of Pisa 2013.

[CER 13b] CERAOLO M., Hybrid Commercial Vehicle (HCV) FP7-Project, Algorithm for SC cell, Deliverable D3200.8, University of Pisa 2013.

[ENE 13a] ENEA, Hybrid Commercial Vehicle (HCV) FP7-Project, Electrical test plan for Li batteries, Deliverable D3100.2, 2013.

[ENE 13b] ENEA, Hybrid Commercial Vehicle (HCV) FP7-Project, Electrical test plan for SC, Deliverable D3100.3, 2013.

[EUC 98] EUCAR, Traction Battery Working Group: Specification of Test Procedures for Supercapacitors in Electric Vehicle Application, 1998.

[EUC 00] EUCAR, Traction Battery Working Group: Specification Of Test Procedures For High Voltage Hybrid Electric Vehicle Traction Batteries, 2000.

[EUC 06] EUCAR, EC LIBERAL Project: Operational Profiles & Accelerated Battery Life Test Procedures-Final Issue, 2006.

[FRE 03] FREEDOMCAR, Battery Test Manual For Power-Assist Hybrid Electric Vehicles, INEEL, DOE/ID-11069, 2003.

[FRE 04] FREEDOMCAR, Ultracapacitor Test Manual, INEEL, DOE/NE-ID-11173, Revision 0, 2004.

[IEC 09] IEC, Electric Double-Layer Capacitors For Use In Hybrid Electric Vehicles – Test Methods For Electrical Characteristics, 62576, 2009.

[IEC 10a] IEC, Secondary batteries for the propulsion of electric road vehicles – Performance testing for lithium-ion cells and batteries, 61982-4, 2010.

[IEC 10b] IEC, Secondary lithium-ion cells for the propulsion of electric road vehicles – Reliability and abuse testing, 61982-5, 2010.

[US 08] U.S. DEPARTMENT OF ENERGY, Vehicle Technologies Program, Battery Test Manual For Plug-In Hybrid Electric Vehicles, 2008.

8

Achievements and Lessons Learnt in the EU Zero Regio Project

The Zero Regio project within the 6th Framework Program of the EC is briefly described. Zero Regio aimed to develop and demonstrate hydrogen infrastructure and fuel cell passenger vehicles in European cities. Demonstration activities took place in Frankfurt, Germany and Mantova, Italy. Experience with driving fuel cell vehicles (FCVs) in real-life driving cycles and hydrogen refueling infrastructure is presented at both sites. Five Mercedes-Benz F-CELL vehicles including one with 700 bar storage have been tested in Germany and three fuel-cell Pandas from Fiat have been tested in Italy. Evaluation of data collected and analyzed during the demonstration is presented, providing important information on the performance, availability, maintenance requirements and consumption of the FCVs tested in comparison with conventional vehicles. Lessons learnt in the project such as in safety aspects, approval formalities and project management are described which should be useful for present and forthcoming larger demonstration activities in the area of sustainable mobility.

8.1. Introduction

Zero Regio was one of the first demonstration projects on low duty fuel-cell vehicles and the corresponding refueling infrastructure performed within the 6th framework program of the European Commission. The project had the overall objective of developing and testing low-emission urban transport systems based on hydrogen as an alternative motor fuel. The project addressed the complete well-to-wheel chain of hydrogen at two demonstration sites, Frankfurt in Germany and Mantova in Italy. In Frankfurt, hydrogen produced as a by-product in a chemical plant is used whereas in Italy, hydrogen is produced from natural gas in an on-site reformer. Infrastructure systems for transport, compression, storage and distribution of hydrogen have been developed and demonstrated at both sites. Pioneering fuel-

Chapter written by Heinrich LIENKAMP and Ashok RASTOGI.

cell vehicle fleets (the Mercedes-Benz A-Class in Germany, and the Fiat Panda in Italy) were operated over a period of 3 years under real-life conditions. An associated research program to evaluate the data and experience gathered from the demonstration of the infrastructure and the vehicle fleets was carried out. The Zero Regio project consortium (see www.zeroregio.com) consisted of 16 partners from four European countries. The project was coordinated and managed by the authors. An amount of 19.68 million Euros was spent on the project, out of which 7.46 million Euros was co-financed by the European Commission.

The present chapter describes the work performed and the results obtained within the Zero Regio project. Experience from the project is put forward as lessons learnt which should be valuable to similar demonstration activities in Europe and abroad. The hydrogen infrastructure developed is described in the next section followed by the description of the fuel-cell vehicle employed. Demonstration activities and the evaluation of results are described in the following section. Lessons learnt based on a variety of events and observations made during the project are put together in a separate section. Activities undertaken at the two sites after the project ended in May 2010 are briefly described and the chapter ends with some concluding remarks.

8.2. Hydrogen infrastructure

The hydrogen infrastructure described here consists mainly of three stages: production, transport to the station and refueling dispensers at the station.

8.2.1. *Hydrogen production*

Three different pathways have been employed in Zero Regio to obtain hydrogen for refueling the passenger vehicles: by-product hydrogen, industrial hydrogen and hydrogen produced on-site.

By-product hydrogen is produced in a chlorine-soda electrolysis plant at Infraserv Höchst in Frankfurt. The by-product hydrogen stream is cleaned and collected in a 1,000 m^3 storage vessel at 70 mbarg pressure. The total amount produced is over 50 million Nm^3/year. This hydrogen is compressed up to 900 bar pressure employing an ionic liquid compressor and transported to the station via a pipeline.

Hydrogen produced in an industrial steam reforming plant of Sapio has been used during the initial phase of the demonstration. Under normal circumstances, hydrogen produced on-site was used. Industrial hydrogen presented a back-up source for emergency situations or when the on-site plant was either in construction or not in operation.

An on-site production plant was developed and installed during the project in Mantova. On-site hydrogen production plant is based on Short Contact Time – Catalytic Partial Oxidation technology (SCT-CPO) of Eni [BAS 10] for converting natural gas together with air and steam, into Syngas. Syngas is processed inside a high temperature water gas shift reactor (HTS). Hydrogen is then purified from Syngas stream inside a pressure swing adsorption unit (PSA). The plant scheme is presented in Figure 8.1. The plant was mounted on skids with wheels located inside a bunker construction. The plant was designed to produce 20 Nm^3/h of hydrogen flow with a possibility of 60% turndown ratio.

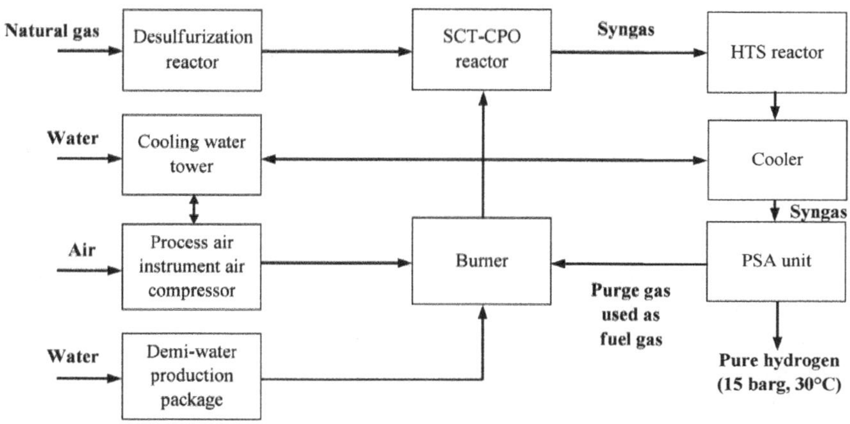

(Source: Eni S.p.A)

Figure 8.1. *On-site reformer process scheme*

Hydrogen quality [RAS 06] obtained from all the above sources met the values required for safe operation of fuel-cell vehicles specified in SAE [SAE 05].

8.2.2. *Hydrogen transport to station*

Different transport schemes such as transport via pipeline or via truck depending upon the hydrogen source have been developed.

A high pressure pipeline (design pressure of 1,000 bar) has been constructed (Figure 8.2) in Germany for transporting the by-product hydrogen from its production site to the refueling station located at 1.7 km distance. This pipeline feeds two refueling dispensers, one for 350 bar and the other for 700 bar vehicle storage, at the service station. It must be mentioned here that to date this is the only operating hydrogen transport pipeline at such high pressures worldwide.

The industrial steam reformer of Sapio is only a few kilometers away from the Mantova service station. Hydrogen was transported by truck in 200 bar cylinder packs to the refueling station. A storage room has been built at the service station in Mantova to store this hydrogen. Hydrogen produced on-site the Mantova service station (see section 8.2.1) could be directed to the compressor and stored in the high pressure storage system feeding the refueling dispenser.

8.2.3. *Refueling dispensers*

Both in Frankfurt and Mantova, public service stations have been built selling the conventional fossil transport fuels as well as hydrogen. Hydrogen refueling systems have been integrated fully in the stations. Main features of the hydrogen dispensing systems are described in the following.

8.2.3.1. *Hydrogen refueling in Frankfurt*

At the Frankfurt service station run by Agip, two refueling dispensers, one for 350 bar and the other for 700 bar vehicle storage, have been installed. The complete hydrogen pathway is shown in Figure 8.3. The two dispensers are supplied through the pipeline mentioned above. The feed line to 700 bar dispenser goes through an ultra low cold fill to precool hydrogen to –40°C in order to obtain fast filling rate in accordance with SAE [SAE 07] standards. An infra-red communication system between the vehicle storage and the dispenser has been installed at the 700 bar dispenser which allows a filling time of 3 min for 4 kg of hydrogen. The dispenser can be operated also without this communication, in which case the filling time is increased to 5 min. No precooling is employed for the 350 bar dispenser.

At the Agip service station in Höchst, Frankfurt, a distribution system for subcooled liquid hydrogen was also installed, which was delivered by Linde by truck. This dispenser was used for refueling BMW series 7 hydrogen cars for 2 years. Figure 8.4 shows the three hydrogen dispensers.

8.2.3.2. *Hydrogen refueling in Mantova*

In Mantova, a hydrogen gas refueling system with one dispenser was built for 350 bar vehicles. Hydrogen from cylinder pack storage or the on-site production facility is compressed with a two-stage membrane compressor and stored in a high pressure storage bank. The dispenser (see Figure 8.5) is equipped with two nozzles, one for cars and the other for buses. Each nozzle has a 3 m long hose and a breakaway fitting.

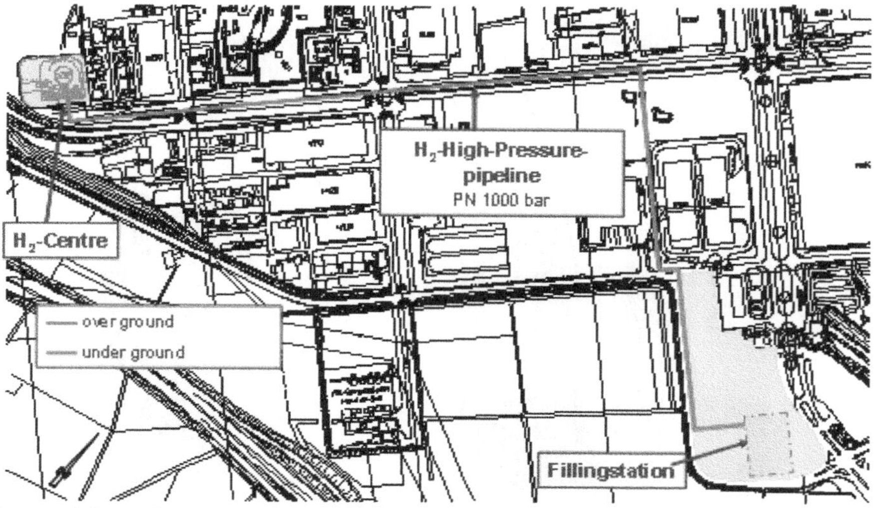

(Source: Infraserv)

Figure 8.2. *Pipeline layout in Frankfurt transporting hydrogen to the filling station. For a color version of the figure, see www.iste.co.uk/jacob/energy.zip*

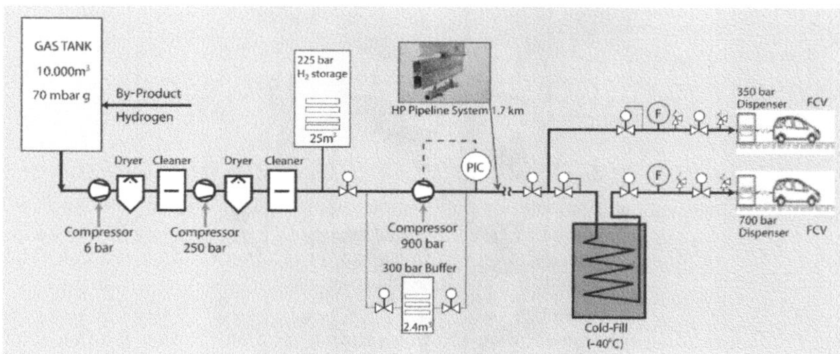

(Source: Infraserv)

Figure 8.3. *Complete hydrogen refueling system in Frankfurt, Germany*

Prior to the dispensing system at the service station in Mantova, refueling had been performed with a mobile refueling system supplied by Air-Products and operated by Sapio.

(Source: Infraserv)

Figure 8.4. *Hydrogen dispensers in Frankfurt*

(Source: Sapio S.r.l)

Figure 8.5. *Hydrogen dispenser in Mantova*

8.3. Fuel-cell vehicles

Fuel cell passenger vehicles were demonstrated at both sites over a period of 3 years under realistic urban conditions and different applications. The refueling infrastructure described in section 8.2 has been used to refuel these cars. The two fleets are briefly described in this section.

8.3.1. *FCV fleet in Germany*

Five A-Class F-Cell cars from Daimler have been demonstrated in Germany. One of them had a 700 bar storage tank. These vehicles used a hybrid (fuel-cell and battery) configuration. The vehicles were equipped with a data acquisition system. The homologation formalities were performed by the supplier prior to delivery for the project. These vehicles were leased to the project according to the demonstration time line. Some technical specifications of the vehicles are stated in Figure 8.6(a).

8.3.2. *FCV fleet in Italy*

In Italy, 3 FC-Panda from Fiat all with 350 bar storage have been demonstrated (see Figure 8.6(b)). These vehicles have been powered entirely by fuel-cells. A data acquisition system for these vehicles was developed within the project. These vehicles were acquired by Regione Lombardia according to the Italian regulations for material acquisition (tendering, etc.) and were used in Mantova for demonstration activities. Homologation formalities were performed within the project.

Minor repairs and maintenance for both fleets could be done on-site and for major repairs or transmission related problems the vehicles were taken to the original equipment manufacturers (OEMs). Data acquired on vehicles were transferred to the central server of the respective OEM for further evaluation. Both fleets were equipped with most advanced safety devices including hydrogen sensors.

8.4. Results of demonstration

The demonstration phase of the project lasted for 3 years (November 2006–October 2009). Major results obtained for the vehicles and the refueling infrastructure are briefly described here. Further details can be seen in the final report of the Zero Regio project [ZER 10].

(Source: Daimler AG) (Source: Fiat)

a) Range: 160 km (@350 bar
Maximum speed: 140 km/h
Ballard PEM FC 72 kW

b) Range: 250 km (@350 bar
Maximum speed: 135 km/h
Nuvera PEM FC 70 kW

Figure 8.6. *Fuel cell vehicles used in Frankfurt a) and Mantova b)*

8.4.1. *Vehicle performance*

Following the vehicle delivery plan, the four 350 bar vehicles in Germany were operated by Fraport, Infraserv and TÜV-Hessen at different times and for different periods of time. Two vehicles were employed by Fraport for the complete demonstration period. A number of drivers at these project partners were trained. The 700 bar vehicle was used by Daimler. Different driving cycles were employed by different partners. All in all, the vehicles showed an availability of 95% with no road calls. Toward the end of the demonstration period, one of the 350 bar vehicles was taken out of service as a result of an integrated power train (IPT) failure. Otherwise, not many fuel-cell and hydrogen-related failures were experienced showing a certain maturity of the technology during the project. Total hydrogen consumption during the demonstration in Germany has been 850 kg.

The fleet in Italy functioned very well (availability 86%) except for some data transmission problems with one car. The fleet has been used by Mantova town hall for different office work in city and the neighborhood. Due to problems in homologation (see Chapter 5), all the three cars could not be operated in accordance with the original demonstration plan. Cars were also permitted to be driven in Italy with only 200 bar pressure in the storage tank although the storage had been tested stringently by the supplier for 350 bars. Total hydrogen consumption in Italy was 350 kg. Figures 8.7 and 8.8 show the operating hours (total 2,671 h) and the mileage (total 94,601 km) covered by both the fleets in the project.

8.4.2. *Experience with hydrogen infrastructure*

Experience with all the dispensers was satisfactory. Apart from some minor incidents, such as faults in control electronics and valves which were able to be repaired within a day or sooner, the reliability and availability (95% in Germany and 78% in Italy) of the dispensers have been high. Dispensers can be used by car drivers in the usual manner after initial instructions.

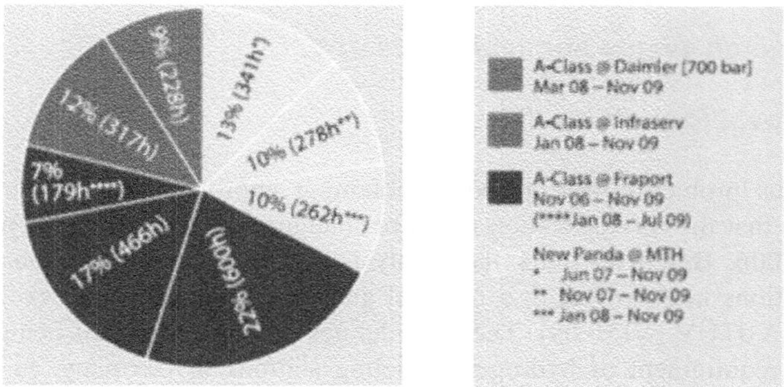

Figure 8.7. *Operating hours of the demonstration fleets. For a color version of the figure, see www.iste.co.uk/jacob/energy.zip*

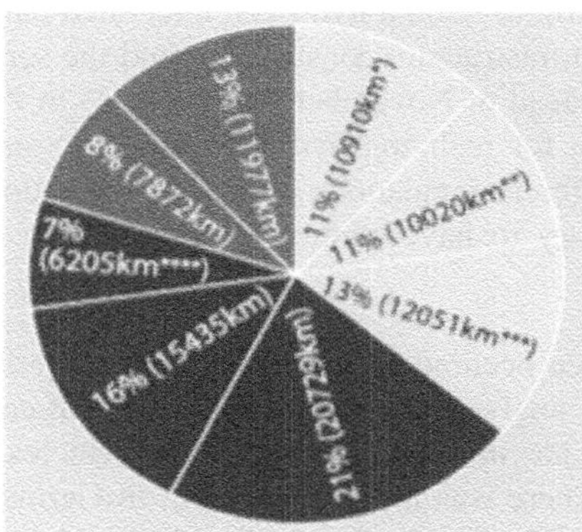

(Source: Zero Regio (2010) Final Report)

Figure 8.8. *Mileage acquired by the fleets. For a color version of the figure, see www.iste.co.uk/jacob/energy.zip*

Pipeline operation in Germany was without incident except in the first start-up phase when the pipeline was not sufficiently purged with hydrogen initially (see section 8.5). The pipeline provided an availability of 96%. The ionic compressor employed in Germany had an availability of 86%. Considering the fact that this was the first prototype operating at 900 bar pressure, the availability can be considered satisfactory.

The on-site hydrogen production plant installed in Mantova was operated in batch mode and provided an availability of 98%.

8.5. Unexpected events and lessons learnt

8.5.1. *Approval formalities in Italy*

No serious problems were faced in obtaining approvals for infrastructure in Germany, so that the service station could be inaugurated very timely as planned in November 2006. Based on the safety analysis and experience with the hydrogen refueling stations in Zero Regio, a national guideline 514 has been developed and published in VdTÜV [VDT 09]. These guidelines have the status of being national standard on requirement of hydrogen refueling stations in Germany. In Italy, the procedure was unclear to start with. An application was filed for a conventional service station which was approved. Hydrogen facilities were then added to this application in accordance with so-called "single authorization act" in Italy. In spite of all the safety related detailed documentation, the approval formalities took over 10 months for permission to build a hydrogen integrated public service station. This resulted in a delay in infrastructure construction and the service station was opened in September 2007, instead of November 2006. Efforts in Zero Regio have resulted in an Italian regulation for realizing hydrogen refueling stations in Italy. This has shown clearly how different regulations or lack of regulations exist in different EU countries. It is necessary to have common regulations in Europe to improve this situation.

A similar experience occured regarding the homologation of FCVs in Italy. In spite of detailed safety analyses and tests under different conditions, the authorities were not able to provide driving permission for vehicles with storage pressures higher than 200 bar. There was no such problem in Germany. FCVs with 700 bar storage are common today in Germany. For FC-vehicles, common harmonized registration practice Europe-wide is also required.

8.5.2. *Refueling infrastructure*

At the very beginning of demonstration in Germany, the FCVs indicated low cells and it turned out that due to insufficient purging at the beginning of the pipeline with hydrogen, there was more nitrogen present in the vehicle tanks than accepted by the fuel-cells. The cells were, however, not damaged. This shows that all of the equipment should be sufficiently purged with hydrogen and examined regarding hydrogen composition before starting.

The breakaway coupling at the 700 bar dispenser in Frankfurt was activated on two occasions, presumably by hooligans. The dispensers are camera monitored, however, data are overwritten after a certain time. A better monitoring procedure should be used for critical components such as the breakaway coupling.

During the first year of operation in Germany, there was an accident at the 700 bar dispenser which was hit by a station customer (driving a conventional car) while reversing his vehicle. Detailed examination revealed that it was entirely due to the negligence of the customer. The dispenser was damaged, but no hydrogen leakage was experienced. The dispenser was replaced by a new one. Such incidents, though undesirable, indicate the robustness of the dispenser equipment.

Experience with the ionic compressor in Germany has been less satisfactory. The first year of operation did not show any serious malfunction, although some incidents resulted in loss of ionic liquid and were unexpected. The compressor was out of operation for 45 days during the first year. During the second year, problems with the filter, phase separator, leakage of ionic liquid and hydrogen were also noticed on different occasions. In April 2008, it was decided to overhaul the compressor hydraulic system, replacing all the seals and renewing the hard chromium coatings on cylinder surfaces. The availability of the compressor during the second year of operation fell to about 70%. After the overhauling, the compressor operated without major outages. At this point, it must be stated that the ionic compressor employed in Zero Regio was the first prototype operating at 900 bar pressure. The ionic compressor technology has improved considerably in the meantime.

8.5.3. *Assessment framework*

An assessment framework was developed in the project to process the demonstration data collected for infrastructure and vehicles. JRC at Ispra performed the data evaluation [ZER 10]. Delivery of data to JRC, in particular regarding infrastructure, was not always timely which delayed the evaluation process. Management methods should be implemented in demonstration projects in order to avoid such delays, such as making the payment conditional to delivery of reports or evaluation results. Some data, such as vehicle costs, were treated as confidential which hindered certain economic evaluations. Incident reporting was also a part of the assessment framework and was performed more successfully than the data evaluation.

8.5.4. *Public acceptance*

Regarding acceptance, results have been derived from interviews with vehicle drivers and surveys on high-school students participating in demonstration events. Acceptance was generally found to be high. Users, although only a limited class were interviewed, were found to be very satisfied with the vehicles, except for their limited driving range. Reservations were observed with regard to societal costs and their funding, the energy sources used to produce hydrogen fuel, the user experience at the refueling stations and road safety problems due to low vehicle noise. Safety was not otherwise found to cause acceptance problems. This shows that not only technical performance but also social issues such as public acceptance do not present a problem for bringing the technology to market.

8.6. Post-project activities and future plans

After having invested a reasonable amount of funds and effort on a demonstration project such as Zero Regio, it is important to see how and to what extent the results and the hardware developed in the project can be used after the project. This is done in this section. Demonstration activities of the project ended in November 2009 and the project itself was concluded in May 2010 holding the project final conference [LIE 10] at the World Hydrogen Energy Conference, WHEC 2010 in Essen, Germany. The refueling infrastructure for compressed hydrogen in Frankfurt belongs to Infraserv Höchst and has been maintained by us after the project. Three OEMs (Daimler, Opel-GM and Honda) have been operating a limited number of fuel-cell vehicles in the Frankfurt area and have used the two dispensers regularly. Recently, the State of Hessen has become an associated member in the CEP (Clean Energy Partnership) light-house project in Germany. Within this membership, the OEMs plan to operate more FCVs with private and public customers in Hessen. Infraserv has leased two B-Class F-Cell vehicles and more customers will follow. In all, about 20 FCVs will be operated shortly in the Frankfurt area. The refueling dispenser system in Frankfurt has been upgraded in accordance with CEP requirements. Within this partnership, it is envisaged to install a second refueling station for hydrogen in the area.

The post-project situation in Mantova is unfortunately different. The FCVs and the hydrogen refueling infrastructure have not been used after November 2009. The on-site production plant has been relocated by Eni for further development and applications in other areas. The vehicles are maintained by Regione Lomabardia for possible application in the future. Attempts have been made to develop the refueling station in Mantova for Hythane (Methane + up to 10% Hydrogen) refueling but has not been materialized so far. The service station in Mantova can be employed for refueling industrial hydrogen as was done in the project. The authors hope and expect that this will happen in the future.

At the European level after Zero Regio, other similar but more ambitious demonstration projects have been launched and have provided useful results on use of fuel-cell vehicles and hydrogen refueling infrastructure. Two projects which should be mentioned here are “H2moves Scandinavia” and “HyTEC” which have employed a larger number of vehicles and filling stations for more realistic demonstrations.

8.7. Concluding remarks

Achievements made in the Zero Regio project have been presented along with the lessons learnt. The objectives set at the beginning of the in 2004 have been realized. Infrastructure for production, transport, storage and distribution has been developed following different pathways in two European cities. The infrastructure developed has been operated and demonstrated successfully for fuel-cell vehicle fleets. Two dedicated small FCV fleets have been operated and tested under real-life conditions over a period of 3 years. Experience with the fleet performance has been good and it is expected that the next generation of the FCVs (e.g. B-Class F-Cell) will perform even better and will come a step forward to real-life application. Some important lessons regarding approval formalities have been learnt in Italy which will help in planning future activities in this area in Italy and also other European countries. This experience must be taken as a strong incentive to streamline and harmonize the regulatory framework for hydrogen infrastructure and vehicles in Europe. Some lessons have also been learnt regarding the management of data evaluation process. Future plans look more favorable in Germany than in Italy at the present time.

8.8. Bibliography

[BAS 10] BASINI L., Operational experience and assessment of the on-site hydrogen production facility in Mantova, Project deliverable report D3.6, available at http://www.zeroregio.com/front_content.php?idcat=443, 2010.

[LIE 10] LIENKAMP H., RASTOGI A., “Zero Regio: recent experience with hydrogen vehicles and refueling infrastructure”, in STOLTEN D. (ed.), *Hydrogen Energy*, Wiley VCH Verlag, pp. 609–626, available at http://www.whec2010.com, 2010.

[RAS 06] RASTOGI A., BOENING A., LIENKAMP H., “Chemical by-product hydrogen as an alternative motor fuel”, *Proceedings Fuel Cell Seminar*, Hawaii, 2006.

[SAE 05] SAE, Development of a Hydrogen Quality Guideline for Fuel-Cell Vehicles, SAE J 2719, SAE International, Warrendale, PA, 2005.

[SAE 07] SAE, Refueling Protocols for Gaseous Hydrogen Surface Vehicles, Draft SAE J 2601, SAE International, Warrendale, PA, 2007.

[VDT 09] VDTÜV, Requirements of hydrogen filling stations in Germany, VdTÜV Merkblatt 514, available at http://www.vdtuev.de/publikationen/merkblaetter, 2009.

[ZER 10] ZERO REGIO, Demonstration of Fuel Cell Vehicles & Hydrogen Infrastructure, Summary of Results, Final Report Zero Regio, available at http://www.zeroregio.com/front_content.php?idcatart=186&lang=3&client=5, 2010.

9

FEV HiFORS Injector with Continuous Rate Shaping: Influence on Mixture Formation and Combustion Process

This chapter covers the investigations carried out with the HiFORS high pressure fast opening rate shaping injector on optical and single cylinder research engine test rigs. The optical investigations were conducted in an optically accessible high pressure, high temperature chamber using shadowgraphy and OH chemoluminescence detection, thereby capturing both liquid and gaseous penetration as well as combustion areas. The single cylinder investigations were carried out at a 0.4l research engine in different operating points from low part load to full load. Compared to a serial production piezo injector, base measurements using conventional square injection profiles already show reduced combustion noise at improved NO_x-soot trade-off as well as lowered HC and CO emissions. Detailed investigations at part load operating points explore the potential of different continuous rate shaping patterns, while the injector's high injection pressure capability of up to 2,500 bar and its influence on engine performance is tested at full load operation points.

9.1. Introduction

Increasing energy cost as well as fiscal incentives for low CO_2 emission vehicles is a strong driver toward efficient drivetrain concepts, resulting in a constantly high request for diesel engine driven passenger cars especially in Europe. Despite its outstanding thermodynamic efficiency, the intrinsic pollutant emissions of the diesel engine, mainly nitrogen oxides and particulate matter, pose a significant challenge against the background of strict European emission regulation. Exhaust gas after treatment systems, although technically mature and highly efficient, contribute considerably to drivetrain cost both for manufacturers and, in the form of increased

---

Chapter written by Paul GRZESCHIK, Hermann-Josef LAUMEN and Udo SCHLEMMER KELLING.

fuel or additional AdBlue consumption, and for customers. In addition, increasingly efficient combustion processes resulting in low exhaust gas temperatures are a limiting factor for after treatment effectiveness especially in test cycle relevant cold start conditions.

Due to these considerations, combustion process optimization not only toward low fuel consumption but also minimized pollutant emission is gaining in importance; at the same time, combustion noise has to remain as low as possible to satisfy tight NVH requirements. The fuel injection system is one of the main determining factors in achieving these targets: today, commercially available passenger car common rail systems employ pressures up to 2,500 bar, as presented by Matsumoto *et al.* [MAT 13], and sophisticated multiple injection patterns to optimize mixture formation for combustion processes utilizing high boosting pressures and exhaust gas recirculation (EGR) rates.

Injection rate shaping and its potential to further improve combustion development have been approached in numerous investigations in research and development, see Figure 9.1. Out of the need for low complexity and high cost-effectiveness of commercial systems, the main path in the automotive industry goes toward discontinuous processes using latest servo injector designs or directly controlled injectors which allow for a high number of single injection events at the shortest dwell times, thereby blurring the line between multiple injection and digital rate shaping. However, when used for continuous rate shaping, as shown by Predelli *et al.* [PRE 10], directly controlled injectors have to face the drawbacks of needle seat throttling on mixture formation. On the other hand, most experimental systems for pressure modulated continuous rate shaping have a high system complexity with additional high pressure components and/or control elements which may be difficult to integrate into an engine concept outside of the laboratory.

FEV and VKA's approach of fully integrating a pressure modulation system into a standard-sized passenger car injector has overcome these challenges. The new high pressure fast opening rate shaping (HiFORS) injector has been specifically designed for state-of-the-art injection pressures of up to 2,500 bar combined with completely flexible continuous rate shaping and shortest needle opening and closing times, thus avoiding needle seat throttling as far as possible.

9.2. HiFORS injector

Similar to the former FEV FIRST injector introduced by Laumen *et al.* [LAU 06], the HiFORS features a two-stage hydraulic servo system which transforms the actuation voltage of a single piezo actuator into a corresponding injection pressure. In this concept, the injector does not need any additional actuation elements or connections.

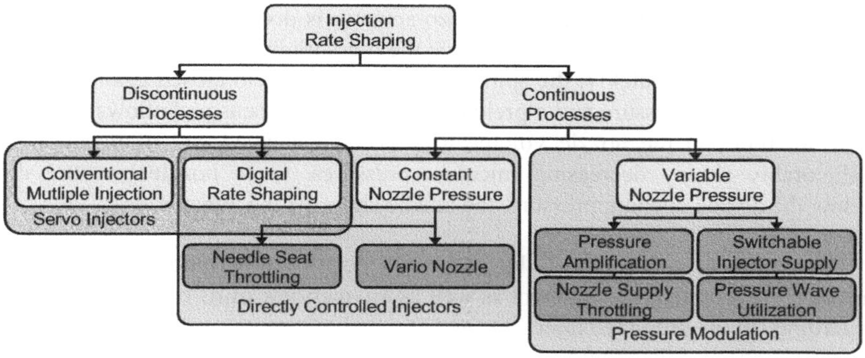

Figure 9.1. *Processes and used injection system technologies for injection rate shaping*

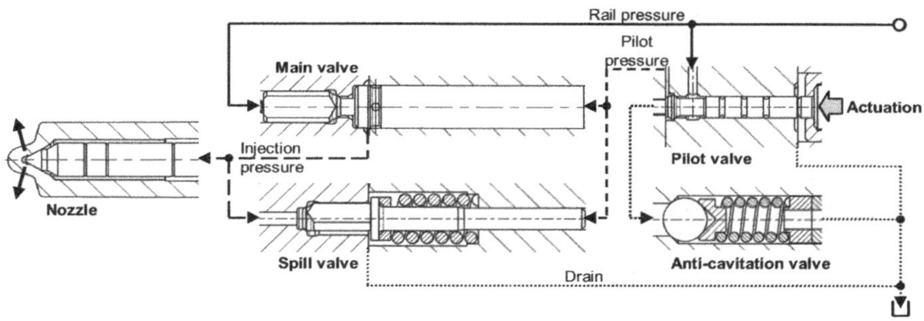

Figure 9.2. *HiFORS hydraulic scheme*

Figure 9.2 shows the injector's hydraulic scheme. The piezo actuator moves a pressure balanced, spring loaded 3/2-way pilot valve with a maximum travel in the range of few µm. The position of this valve governs the pressure inside a control volume which can be adjusted to any value between drain and rail pressure. An anti-cavitation valve in the drain path increases the back pressure to a level which prevents cavitation during high pressure drops at the pilot valve's control edges. The control pressure acts on the main valve piston and opens the main valve until the resultant forces of rail pressure, control pressure and injection pressure acting on the main valve assembly are at an equilibrium. Thus, the injection pressure is directly dependent on the control pressure and can be freely varied by adjusting the actuator voltage. The main valve itself is a 2/2-way valve without pressure balancing to minimize leakage. When injection pressure exceeds the nozzle needle's opening pressure preadjusted by spring tension, the nozzle opens very quickly and injection begins. For the end of injection, the piezo actuator is discharged and the pilot valve is pushed back into its rest position. Since the pressure in the relatively small control volume drops very quickly, the spill valve, which was held closed against injection pressure by pilot pressure and a preloaded spring, now opens and allows for a quick pressure relief in the nozzle volume to prevent a delayed end of injection with unfavorably slowly decreasing injection pressures. Once nozzle pressure drops below the nozzle opening pressure, the needle shuts off the injector.

Grzeschik *et al.* [GRZ 12] have given a detailed description of the HiFORS injector's layout and development as well as an overview of its operational behavior regarding high pressure injection and rate shaping capabilities in laboratory investigations and first engine operation.

9.3.Optical chamber rate shaping investigations

The optical investigation of the HiFORS spray and mixture formation and combustion properties were conducted in a high pressure, high temperature optically accessible combustion chamber at VKA, see Figure 9.3, using a 3-hole nozzle with a total hydraulic flow of 234 cm³/min and identical spray hole geometries to the nozzle used in the single cylinder investigations. The chamber's capabilities and the data processing approaches utilized are described in detail by Graziano *et al.*, [GRA 13]. The rate-shaping experiments were conducted in an air environment at a chamber pressure of 50 bar and temperature of 800 K, simulating diesel engine conditions at begin of injection during part load operation. Apart from square injection profiles, ramp and boot injection pressure shapes at rail pressure levels of 1,200 and 1,800 bar were used. The ramp profiles were shaped for a fixed ramp duration of 1 ms independent of the injection duration. The boot profiles were defined to hold approximately half rail pressure level for about half of the injection duration before quickly rising to maximum injection pressure.

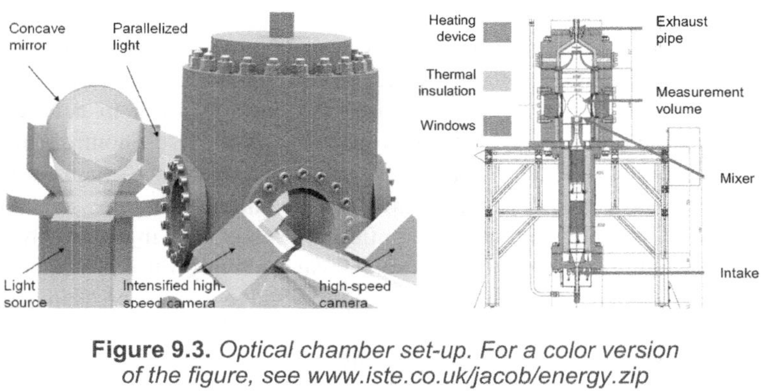

Figure 9.3. *Optical chamber set-up. For a color version of the figure, see www.iste.co.uk/jacob/energy.zip*

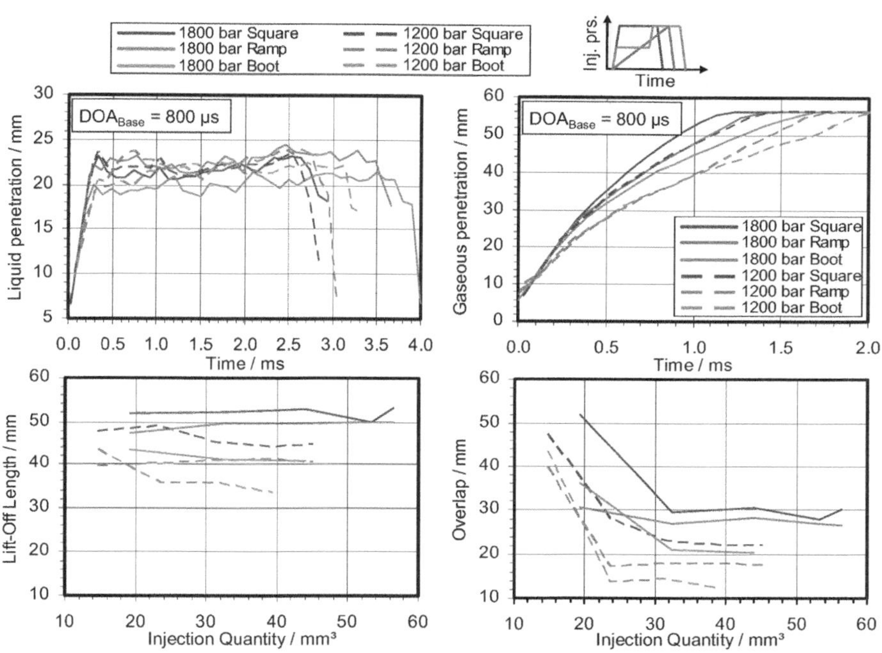

Figure 9.4. *Optical investigation results for different injection rate shapes and injection pressures. For a color version of the figure, see www.iste.co.uk/jacob/energy.zip*

The upper two diagrams in Figure 9.4 show spray formation behavior under the influence of different injection rate shapes. The individual injection durations have been chosen to yield the same injection quantities as a reference piezoelectrically actuated servo injector at a duration of actuation (DOA) of 800 µs. The liquid penetration length of the injection sprays shows no obvious dependency on neither rail pressure level nor rate shaping profile; this aligns with the literature, e.g. Siebers [SIE 99] who found that the liquid spray length is solely dependent on spray hole diameter, physical fuel properties and the thermodynamic state of the cylinder charge. The gaseous spray penetration, however, results from the impulse exchange between liquid and surrounding gas including the entrainment of ambient air and therefore shows a strong dependence on rail pressure and injection pressure profiles. As a result of combustion starting in the gaseous portion of the spray, both lift-off length (i.e. distance from nozzle tip to combustion zone) and overlap (distance from the tip of the liquid spray core to the combustion zone) are reduced for lower rail pressures as well as for rate shaping injection with reduced effective injection pressure at the beginning of injection. Since these values are indicators for the available distance and time for mixture preparation, they explain the tendency of increased soot formation due to non-sufficient oxygen availability when employing rate shaping in engine operation.

9.4. Single cylinder engine investigations

9.4.1. *Experimental design*

The single cylinder engine investigations were carried out on a 4-valve, 390 cm^3 engine with a compression ratio of 15. The part load point definition by engine speed and internal mean effective pressure (IMEP) covers the main operation range for relevant test cycles in a hypothetical C-class vehicle as introduced by Rottmann [ROT 11]; this includes fixed centers of combustion taken from the engine's reference calibration with a close-to-production piezoelectrically driven servo injector as well as target values for NO_x emissions for meeting EURO6 emission legislation, see Table 9.1.

Operating point	Engine speed	IMEP	p_{Rail}	p_{Intake}	Center of Combustion	Spec. NO_x target
	min^{-1}	Bar	Bar	bar	°CA ATDC	g/kWh
PL1	1500	4.3	720	1.07	5.8	0.2
PL2	1500	6.8	904	1.50	6.6	0.2
PL3	2280	9.4	1399	2.29	9.2	0.4
PL4	2400	14.8	1800	2.60	10.8	0.6
FL1	2000	27.0 (ref.)	1800 (ref.)	2.46	–	–
FL2	3000	29.0 (ref.)	2000 (ref.)	2.70	–	–
FL3	4000	28.6 (ref.)	2000 (ref.)	3.00	–	–

Table 9.1. *Single cylinder engine operating point definition*

The part load operating points are used for comparing the HiFORS injector's performance against the reference injector without the use of rate shaping as well as the potential analysis of rate shaping profiles in the higher load operating points PL3 and PL4. The HiFORS injector's high pressure capability is explored in three full load operating points at 2,000, 3,000 and 4,000 min^{-1} by increasing engine load as far as possible within given peak pressure, soot and exhaust gas temperature limits. All investigations were carried out using a close to production nozzle module with a hydraulic flow rate of 625 cm^3/min on both reference and HiFORS injectors.

9.4.2. *Base investigations*

The base investigations were carried out in the form of exhaust gas recirculation (EGR) sweeps in the defined part load operating points using a reference piezoelectrically actuated servo injector with and without pilot injection and the HiFORS injector without pilot injection. Due to the HiFORS injector's hydraulic layout, the needle opens while nozzle pressure is still being built up, resulting in a steep ramp-shaped injection pressure trace similar to former cam-driven UI systems even for rectangular actuation voltage profiles. As a result, the mass flow rate during ignition delay is reduced, with the effect of reduced maximum burning rate and a less steep cylinder pressure gradient very similar to the reference injection with pilot event, see Figure 9.5. An additional pilot injection for the HiFORS injector is, therefore, considered obsolete.

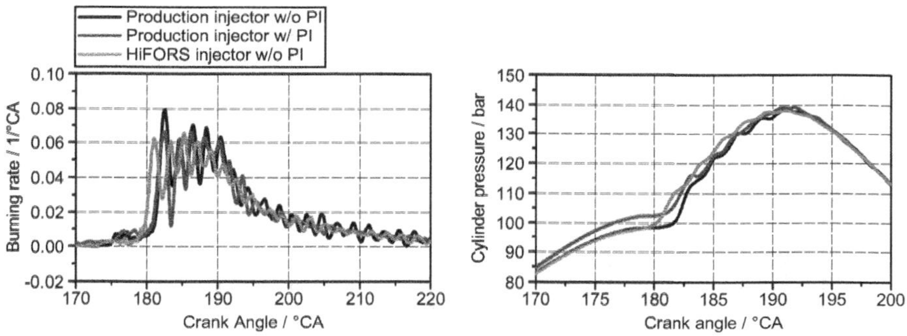

Figure 9.5. *Burning rate and cylinder pressure traces for HiFORS and production injectors at part load operating point PL4, indicated specific NOx emissions ≈ 0.5 g/kWh. For a color version of the figure, see www.iste.co.uk/jacob/energy.zip*

Figure 9.6 shows the overall results of the EGR sweeps when recalculated to the respective specific NO_x targets for EURO6 calibration. While point FL1 shows no reasonably measurable soot emissions at all for all injection patterns, in part load points FL2 to FL3 the filter smoke numbers of the HiFORS injector are even lower than those of the reference injector with single main injection; this is contributed to the relatively quick needle operation without long times under the effect of seat throttling, resulting in excellent mixture formation from the very beginning of injection. Only in high part load point PL4 are the HiFORS injector's soot emissions higher than both reference injections. This might be caused by higher internal pressure losses between fuel supply and nozzle which become evident at longer injection durations when nozzle opening and closing effects have less impact. The combustion sound level is in all operating points comparable to the reference injector with pilot injection, thus confirming the approach of omitting the pilot injection for the HiFORS injector. The indicated efficiency of the HiFORS injector also follows the reference injector with pilot injection in points PL2 to PL4, while in the low load point PL1, with short injection duration and therefore high importance of needle opening and closing behavior, the HiFORS injector has significant advantages.

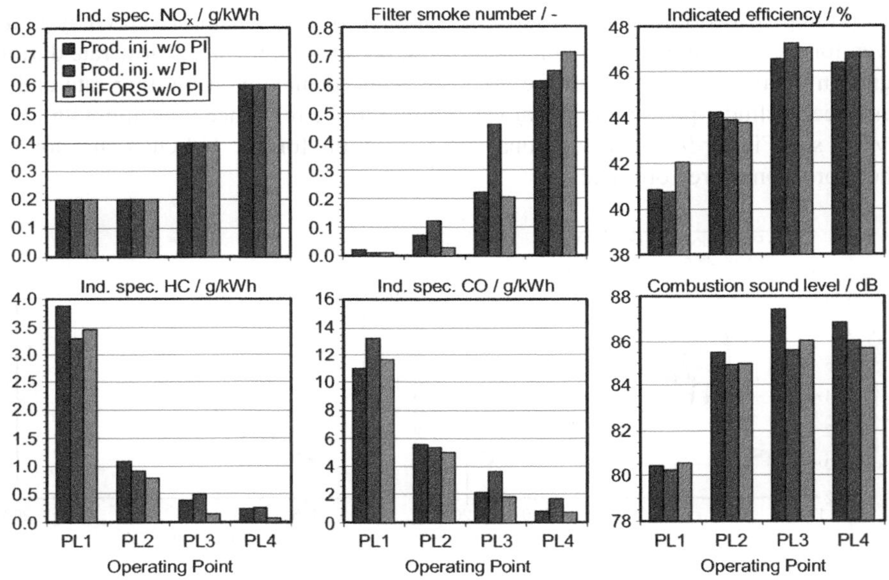

Figure 9.6. *Emissions, efficiency and combustion sound at part load operation for HiFORS and production injectors. For a color version of the figure, see www.iste.co.uk/jacob/energy.zip*

9.4.3. *Rate shaping*

Rate shaping investigations were carried out in higher part load operating points PL3 and PL4 where injection duration is sufficiently long to allow for significant alterations of the injection rate shape while, in contrast to full load conditions, combustion sound and NO_x reduction still play a role compared to purely performance-driven requirements which prefer rectangular injection profiles. In both points, variations of the ramp profile with decreasing slope steepness and low, medium and high boot profiles with increasing boot level durations were investigated. All values shown in Table 9.1 including the indicated NO_x emission level were fixed throughout the variations by adaptation of begin and duration of injection and EGR rate. Figure 9.7 shows the example of the low-level boot profile variation at PL4. The investigation starts with a square injection (blue curve) and the typical resultant cylinder pressure profile of a diesel engine combustion without pilot injection. When introducing and extending the boot level in the actuation voltage profile, the injector's hydraulic system will follow and form the according boot shape in the injection pressure profile. Due to the increasing period of fuel being injected at low injection rates, for constant engine load the total injection duration must increase; at the same time, the begin of injection is shifted to earlier points to keep the center of combustion constant. For increasing boot durations, the cylinder pressure rise caused by combustion starts earlier and tends to form a less steep pressure rise gradient. The profile variation is ended when rate shaping duration becomes so long that the injection pressure does not reach the maximum defined by the square injection anymore (red curve).

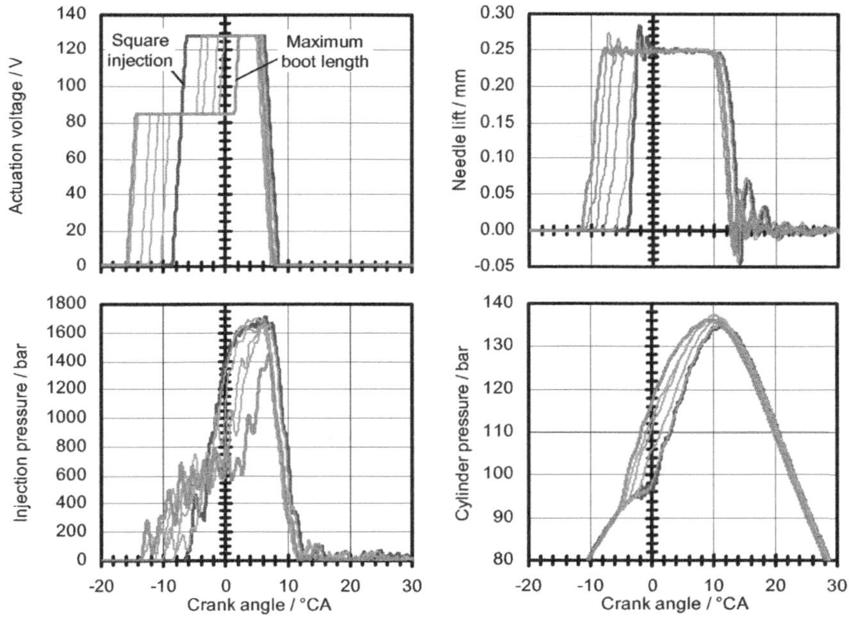

Figure 9.7. *Low boot rate shaping variation at part load operating point PL4. Blue: square injection profile; red: maximum boot length injection profile; gray: intermediate injection profiles. For a color version of the figure, see www.iste.co.uk/jacob/energy.zip*

Since combustion sound reduction at moderate drawbacks especially in soot emission is one of the main aims of injection rate shaping, the relation between these two parameters is shown in Figure 9.8 for both rate shaping part load operating points. In both cases, the variation starts with square injection profiles in the lower right-hand corner at low soot emissions and relatively high combustion sound levels. By applying rate shaping patterns with increasing shaping durations, the combustion yields reduced sound levels and increased soot emissions. All profiles follow very similar operating point-dependant paths within the measurement fluctuations. However, in both cases, the ramp variations only show limited capability in altering the engine's soot-sound behavior, the same applies for the high boot profiles in PL3.

The variation trajectories for PL3 and PL4 have quite different characteristics. At PL3, sound-level improvement can only be achieved by increasing soot emissions up to a point where further rate shaping only leads to more soot without any more sound benefit, with the critical parameters placed at about 83 dB and a filter smoke number of 1.0. At PL4, the potential for combustion sound reduction at limited additional soot emissions is smaller, but using the medium boot level there is the possibility to lower combustion sound by 1 dB without any drawbacks in soot formation.

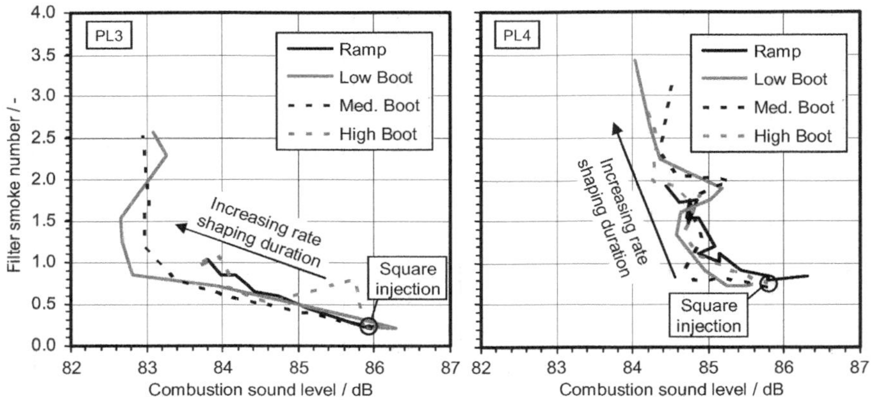

Figure 9.8. *Soot-sound trade-off for different injection rate shapes at part load operating points PL3 and PL4*

9.4.4. *Full load operation*

In the full load operating points, the potential of significantly increased rail pressures on engine performance is investigated. The operating points are defined by engine speed and upper limits for cylinder peak pressure, exhaust temperature and soot emissions, while engine load is increased in combination with adaptation of center of combustion as far as possible within these limits. For the exemplary full load operating point FL1 shown in Figure 9.9, the effective limits are peak cylinder pressure at 180 bar and a maximum filter smoke number of 1.0. The rail pressure for the HiFORS injector is varied from 1,600 to 2,500 bar, while the reference injector was operated at 1,800 bar rail pressure.

The results show that by using increased injection pressures, the maximum engine load can be increased from around 26.85 bar IMEP at 1,600 bar rail pressure to 27.5 bar at 2,500 bar rail pressure, thus improving the indicated engine full load performance by 2.4% without any need to improve the thermal or mechanical capabilities of engine components. To stay within peak pressure limits, the center of combustion (CoC) has to be slightly shifted from around 10.75 to 11.25°CA ATDC, resulting in an increase in exhaust temperature from 692.5 to 708°C and consequently a drop of indicated efficiency from 41.8 to 40.8%. Due to the shift of combustion center and therefore lower peak cylinder temperature, the indicated NO_x emissions are lowered from 8.6 to 7.65 g/kWh while higher cylinder pressure gradients during combustion at increased injection pressure lead to higher combustion sound level of 86.0 at 2,500 bar rail pressure compared to 85.25 at 1,600 bar.

Compared to the reference injector, the HiFORS injector's high injection pressure capability enables higher engine full load performance at better indicated efficiency although the combustion center has to be shifted to later points at the highest rail pressures. Given a fixed full load target, this opens up the possibility to optimize the combustion process by using nozzles with smaller hydraulic flow rates which improve operation behavior within the whole engine operation range. As another conclusion, for optimized utilization of highest injection pressures at full load conditions, the respective engine's peak pressure capability should be increased to avoid the need for combustion center shift with its impact on engine efficiency.

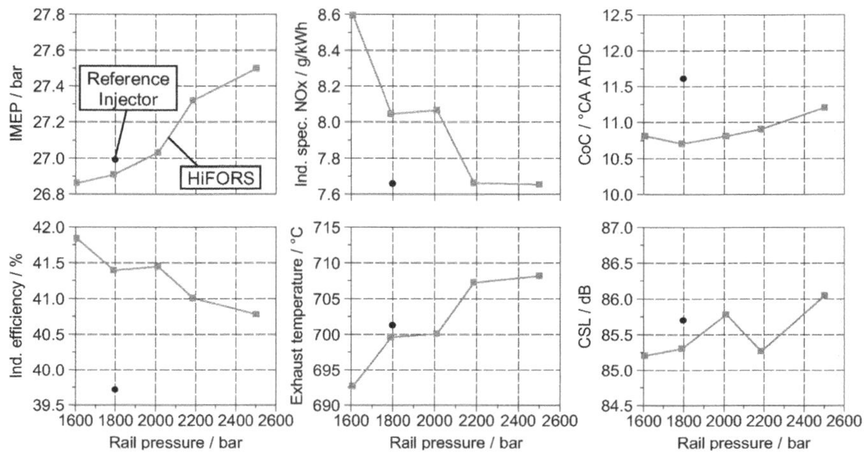

Figure 9.9. *Single cylinder engine behavior at full load operating point FL1*

9.5. Conclusion and outlook

The HiFORS injector is utilized for the optical and single cylinder engine investigation of advanced diesel combustion processes using a combination of high pressure injection up to 2,500 bar, quick needle opening and closing action and free injection rate shaping capability.

The influence of injection at different rail pressure levels with variable injection rate profiles on mixture formation and combustion in a laboratory environment is investigated by high pressure, high temperature optical chamber experiments. While the liquid spray core length remains constant for all variations, both reduced rail pressures and rate shaping approaches decrease gaseous spray penetration, resulting in combustion start closer to the nozzle tip with less spray development time and length needed for optimum mixture formation.

The new injector's impact on engine operation is analyzed using a single cylinder research engine. Injector performance is tested against a piezoelectrically actuated close to production servo injector using identical nozzles. Base EGR sweeps show significant advantages in NO_x-soot trade-off in low-to-medium part loads while maintaining combustion sound level and indicated efficiency of a standard injection pattern with pilot injection. The continuous rate shaping functionality opens additional potential for combustion sound reduction at moderately increased soot emissions. At full load operation, the injector's high pressure capability enables both increased engine performance and improved indicated efficiency compared to the reference injector.

The next steps will include the integration of the HiFORS injector into advanced closed-loop combustion control systems as described by Schnorbus [SCH 10] or Hinkelbein [HIN 10], thereby fully exploring the ability of continuous rate shaping to form an optimum diesel combustion process.

9.6. Acknowledgments

The research leading to these results has received funding from the European Union Seventh Framework Programme [FP7/2007-2011] under grant agreement No. 234032.

9.7. Bibliography

[GRA 13] GRAZIANO B., JAKOB M., KREMER F. *et al.*, "Investigation on the Ignition Sensitivity of 2-MTHF, Heptane and Di-n-butylether", *Proceedings of the European Combustion Meeting*, Lund, 2013.

[GRE 12] GRZESCHIK P., KÖRFER T., KOLBECK A. *et al.*, "FEV HiFORS: Neue Möglichkeiten zur simultanen Verringerung des Verbrauchs- und Emissionsverhaltens moderner Pkw-Dieselmotoren bei verbesserter Verbrennungsakustik", *8. Tagung: Diesel- und Benzindirekteinspritzung*, Berlin, 2012.

[HIN 10] HINKELBEIN J., Verbrennungs charakteristik regelung mittels Einspritzverlaufsmodulation bei direkteinspritzenden Dieselmotoren, PhD thesis, RWTH Aachen University, 2010.

[LAU 06] LAUMEN H.-J., PISCHINGER S., SCHMUCKER K.J. *et al.*, "Flexible Einspritzverlaufsformung – Auch bei PKW-Dieselmotoren ein wirkungsvolles Mittel zur Reduzierung von Verbrauch und Schadstoffemissionen?", *5. Tagung: Diesel- und Benzindirekteinspritzung*, Berlin, 2006.

[MAT 13] MATSUMOTO S., DATE K., TAEUCHI T. *et al.*, "Der neue Diesel-Magnetventil-Injektor von Denso", *MTZ – Motortechnische Zeitschrift,* vol. 74, no. 2, pp. 146–150, 2013.

[PRE 10] PREDELLI O., GRATZKE R., MAROHN R. *et al.*, "Kontinuierliche Einspritzverlaufsformung in Pkw-Dieselmotoren – Potenziale, Grenzen und Realisierungschancen", *31. Internationales Wiener Motorensymposium*, Vienna, 2010.

[ROT 11] ROTTMANN M.D., Einspritzratenformung am schnelllaufenden Dieselmotor, PhD thesis, RWTH Aachen University, 2011.

[SCH 10] SCHNORBUS T.G., Zylinderdruckgeführtes Einspritzmanagement beim Dieselmotor, PhD thesis, RWTH Aachen University, 2010.

[SIE 99] SIEBERS D.L., Scaling Liquid-Phase Fuel Penetration in Diesel Sprays based on Mixing-Limited Vaporization, SAE Technical Paper Series 1999-01-0528, 1999.

10

Development of Predictive Vehicle and Drivetrain Operating Strategies Based Upon Advanced Information and Communication Technologies

Environmental concerns and subsequent regulations for large CO_2 emission reductions have led to the development of innovative powertrain technologies for vehicles of all types. However, despite much development effort, CO_2 emissions have to be significantly further reduced. Thus, other complementary measures will be required, such as measures to drive the vehicle and control its powertrain more intelligently, by utilizing information from off-board, for the purpose of predictive control. Thus, in this chapter, an overview of various frontloaded development methods is given. These support the development of complex predictive control strategies that interact with real-world traffic systems, by using information and communication technologies, such as radar, video cameras, satellite navigation and v2x, to reduce energy consumption. These methods range from quasi-static vehicle simulation and analysis in the concept phase, to advanced co-simulation methods where the whole vehicle, its powertrain and connectivity, is accurately simulated, to the testing of actual prototypes which are able to process off-board information and use this information for energy optimized on-board management strategies.

10.1. Introduction

According to the European Commission, in 2010, the transport sector was responsible for 20% of greenhouse gas emissions in Europe, with the major part of that (more than 90%) being attributable to road transport.

Chapter written by Stephen JONES, Arno HUSS, Emre KURAL, Alexander MASSONER, Edoardo MORRA, Christa SIMON, Reinhard TATSCHL and Christian VOCK.

Consequently, various European authorities have introduced higher direct and indirect taxes, which penalize CO_2 emitting vehicles. These additional taxes, together with constantly increasing fuel prices, lead to ever increasing costs for vehicle operators and manufacturers. Specifically, the European Commission (European Commission, 2012) has regulated that the vehicle manufacturer fleet average to be achieved by essentially all new EU cars is 130 g of CO_2 per kilometer by 2015 and 95 g/km by 2020. Similar regulations are being planned in other key world markets. Hence, vehicle manufacturers, their suppliers and research institutes have very significantly accelerated their efforts to develop more energy efficient vehicles of all types, ranging from passenger cars to heavy duty vehicles, to meet these challenging requirements for far lower CO_2 emissions.

In the first instance, this has led to the development of more efficient conventional powertrain components, such as highly advanced and downsized internal combustion engines and more efficient multi-speed transmissions. Furthermore, more hybridized drivetrains (or powertrains), for example the Toyota Prius THS III powersplit, and even fully electrified vehicles were introduced and improved in terms of energy efficiency. However, even with these various powertrain improvements, vehicle manufacturers will find it hard to economically meet ever more demanding emission regulations.

Manufacturers and suppliers realize that even latest engine, transmission, electric machine (EM), battery and power electronics technologies may not in the longer term improve overall efficiency and CO_2 emissions enough. Hence, they are actively looking for complementary solutions to further improve the efficiency of vehicles by driving and controlling them more intelligently.

In this chapter, intelligent approaches, utilizing information and communication technologies (ICT) that allow predictive energy management in modern vehicles, are introduced. Specifically, this means the extensive usage of off-board information about the environment provided by technologies such as satellite navigation, radar, video and v2x, i.e. vehicle-to-infrastructure (v2i) and vehicle-to-vehicle (v2v).

The frontloaded approach to introduce ICT and predictive control in the powertrain and vehicle development process is subsequently demonstrated based on three project applications, showing different approaches, and corresponding degrees of engineering efforts. For simplicity, these examples are elaborated on passenger car examples, but can in principle be extrapolated to commercial vehicles and heavy duty applications.

First, the aim in the concept phase is to identify the potential of v2i technologies in terms of CO_2 emission reductions by specifying certain simulated scenarios or maneuvers. For example, by including information regarding upcoming traffic lights in an offline simulated driving maneuver, we can assess the emission reduction potential of predictive control, compared to the standard case where information about the traffic lights is not available. This simulative approach when intelligently and systematically applied provides quick and essential feedback of the potential ICT applications in the range of driving scenarios evaluated.

The second aim is to quantify the potential of ICT measures to reduce CO_2 emissions in typical urban driving areas. This method goes beyond the prior mentioned maneuver-based approach, as it tries to describe typical real-world urban traffic including a wide range of different maneuvers and types of vehicles and powertrains. For this purpose, certain ICT measures are defined in terms of their idealized eco-driving impact, where the driver gets recommendations, for example from a smartphone or via on-board displays that process data from the inter- and intranet to encourage energy-saving driving behavior. These measures are then investigated in simulation, with a representative number of different vehicle models, out of the actual fleet portfolios of the major vehicle manufactures. To identify the potential of ICT technologies in typical urban traffic scenarios, for the whole passenger car vehicle fleet, including state-of-the-art, as well as near future electrified vehicle technologies.

Third, by realistically developing and testing with the aid of simulation advanced predictive energy management control strategies, which directly act on the vehicle, such as a coasting assistant for the energy optimal speed approach to traffic lights or speed limits, or act on the powertrain, such as the optimal distribution of the e-torque provided by different EMs at each instant of time, we are able to control a real vehicle and the total powertrain system far more efficiently.

10.2. Energy potential analysis with offline simulation

Simplified offline simulations without any real-time data or traffic information are used to demonstrate the potential CO_2 emission improvements of ICT, via traffic control or intelligent driving strategies. A typical city driving situation, including numerous stop and go situations due to traffic lights and general congestion, represents a typical example of inefficient energy use. Clearly, the avoidance of frequent and sudden acceleration and deceleration maneuvers by intelligent application of ICT would drastically reduce the energy consumption in such urban traffic.

Figure 10.1 shows the two different driving behaviors with (the green line) and without (the red line) consideration of off-board traffic light information, for a purely electric vehicle (EV). For the given example, the change in driving behavior can raise the all-electric range by about 70%, under the assumption of a perfect driving "green wave". As a result, the application of ICT to traffic light dominated city driving can be seen as very promising.

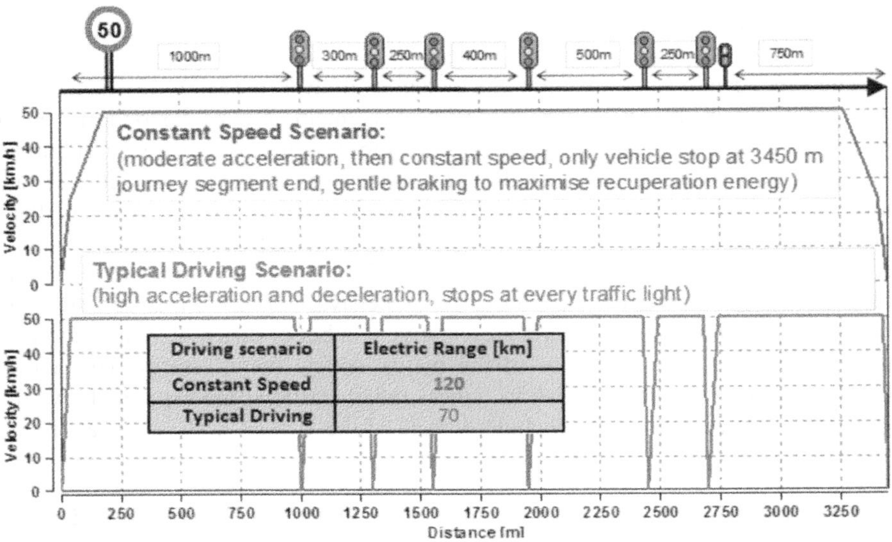

Figure 10.1. *Simulated course on urban road with short sequenced traffic lights). For a color version of the figure, see www.iste.co.uk/jacob/energy.zip*

One advantage of such offline simulation scenarios is that they can be created with relatively low efforts. Moreover, they provide an initial estimation of the potential of including infrastructure information into the efficiency analysis. However, the corresponding accuracy and validity is limited, since traffic, unexpected implementation problems or other issues will tend to prevent the wholly perfect driving shown in Figure 10.1. In other words, in reality, ICT traffic management may decrease the number of vehicle stops or avoid extreme velocity changes, but cannot fully avoid them. Thus, such offline results must be interpreted with caution by experts. Hence, the question naturally arises whether considering such simplified urban scenarios is able to give realistic estimations. One way to overcome this limitation is to consider not only different maneuvers separately, but also define a portfolio of scenarios typical for urban traffic and run them for a wide range of different vehicles. Such an approach is described in more detail in the next section.

10.3. Analysis of ICT measures for the reduction of CO_2 emissions in urban areas

ICT combined with advanced efficient vehicle and powertrain technologies has been identified as a potential breakthrough technology, for achieving further significant reductions in energy consumption and CO_2 emissions of vehicles in demanding real-world conditions. Hence, as a second approach, we aim to develop a novel methodology to reliably evaluate the impact of ICT-related measures on mobility, vehicle energy consumption and CO_2 emissions of vehicle fleets in realistic traffic scenarios. In this second approach, currently being elaborated in the frame of the European funded R&D project "ICT Emissions" (ICT Emissions, 2012), a wide range of different advanced vehicle models is being investigated by vehicle and powertrain simulation in virtual reality. These models cover different vehicle classes (A, B, C, D, E, F and J), and transmission types (MT, AMT, AT and CVT) as well as a wide range of electrification concepts, such as hybrid, plug-in hybrids and pure EVs.

To obtain a realistic aggregate estimation for the CO_2 emission levels of a typical urban vehicle fleet requires the weighting of the simulation results of the different vehicle and powertrain models with respect to their current or projected future numbers on public roads.

The aforementioned vehicle and powertrain models are simulated in one-dimensional (1D) (the longitudinal mode) over selected real-world driving cycles, which are generated through the offline usage of traffic flow simulation including real-world road data, in order to obtain more realistic driving scenarios.

Additionally, a collection of likely ICT measures in the corresponding driving maneuvers is identified. These ICT measures may include various predictive eco-driving modes, which are enabled using off-board data from satellite navigation, v2x, the Internet, a smart phone, etc. These ICT-based measures may include traffic management and control such as a "green wave strategy". Based upon these measures, the real-world driving cycles are appropriately manipulated to realistically represent the likely energy improved driving behavior in the investigated situations due to ICT application.

By comparing the overall fleet energy consumption of the simulations for the case, where the whole vehicle fleet runs through the defined driving cycles with and without various ICT measures, it is possible to estimate the effect of ICT measures on fleet CO_2 emissions in real traffic.

Such a combined approach, using various simulative methods and an overall fleet data analysis, allows a useful and early estimation of the benefits of ICT-based measures to reduce vehicle fleet CO_2 emissions, without the immediate need to commit to the very costly and time-consuming development and construction of connected prototype vehicles and their powertrains, and the near impossibility of conducting large-scale statistically valid experiments with large fleets of vehicles in real-life traffic. In this respect, the aim of the "ICT Emissions" project is to develop an integrated methodology that can be used to quantify the CO_2 emissions of alternative ICT solutions for road transport. The methodology will provide answers by integrating results from real traffic and emission models at the microscale (i.e. single vehicles and driving situations) and extrapolating them to the overall mixed fleet level (i.e. total road transport emissions) at a large regional level using a macrotraffic approach.

10.4. Development and implementation of predictive energy management systems for real-life application

In the previous section, diverse sources of ICT were investigated with respect to their theoretical CO_2 emission reduction potential over typical driving scenarios for single driving situations to assess the likely real-life fleet CO_2 emission impact. These investigations while quite involved are conducted as simply as is practically possible, using intelligent assumptions where required. In the current section, another complementary approach is described, which is required for use in the development, implementation, testing and validation of ICT-based predictive energy management strategies, in real prototype and eventually series production vehicles. Such a methodology must naturally follow the complete vehicle and powertrain development "V" process from concept simulation and requirements elicitation, to system design and function development and integration, through to application calibration on testbed and finally real-world driving and verification. This overall approach is demonstrated in the framework of the EU-funded R&D project "OpEneR" [OPE 12, JON 12a, JON 12b, JON 12c, JON 13a, JON 13b, LAV 13]. The "OpEneR" project represents cooperation between industry and research institutes. It envisages a networked system of on-board and off-board ICT systems, fusing data together in highly sophisticated energy-saving driving strategy. The main activities of this joint research project are described in detail in the following sections. More details can be found in the various recent publications listed above.

In "OpEneR", the ICT-driven predictive functionalities were initially developed with the aid of pure office PC co-simulation, combining various commercial tools in order to be able to make a three-dimensional (3D) simulation including simulation of powertrain and vehicle dynamics, control strategies and simulation of the environment. The functions were further tested on a powertrain testbed powered by similar simulation models with identical use cases as well as on various consortium members private test tracks and other facilities. Of course, the development of such advanced ICT functions in an international consortium requires iterative work, to find the best overall compromise among sometimes conflicting demands, or to simply manage unexpected development challenges. Thus, the development toolchain described is extremely flexible in that it enables rapid iterations, from office PC – to testbed – to track, and back again in a simultaneous or concurrent style of engineering. Hence, it supports not only the development of energy management functionalities but also their interaction with dynamic safety functionalities such as braking, TCS and ESP [LAV 13]. However, the "OpEneR" ICT-based predictive functions will finally be tested in 2014, on a real world and public driving route in Vigo, Spain which is equipped with various prototype ICT systems, such as v2x, for realistic development and testing of predictive energy management in the two already developed and highly advanced "OpEneR" 4WD pure EVs itself derived from the production PSA 3008 diesel hybrid.

The CTAG (Galician Automotive Technology Centre) test corridor in Vigo, Spain, is a test corridor of about 60 km with different kinds of public road types (motorway, highway and city scenarios). These facilities provide ICTs such as traffic flow and traffic light information and active traffic control management, and thus allow v2i and cooperative systems communication that supports up to 20 test vehicles.

10.4.1. *Optimal power-split strategy*

The "OpEneR" vehicle is a pure 4WD EV driven by two EMs, each mounted to one axle. In such a powertrain, an optimal torque-split or more accurately power-split algorithm for the two EMs can be used to increase the system's overall efficiency. Based primarily on the vehicle speed, overall mechanical power demand and battery supply voltage, a power-split algorithm was developed that determines an optimal power-split factor between the front and the rear EMs, see Figure 10.2. The calculated power distribution could vary from using only the front EM, to usage of both EMs to the use of only the rear one. In the "OpEneR" project, the resulting improvement in energy economy along the NEDC, FTP-75 and Vigo route was investigated in detailed simulation. For the NEDC, the improvement was 7.8%, for the FTP-75 5.5%, on the Vigo test corridor in direction of Vigo (A52) 4.0% and on the Vigo test corridor in direction of Ourense (A52) 4.9%. The use of ICT, in terms of both on-board and off-board information, allows the online generation of the projected vehicle speed and load profile, and thus the implementation of a more effective and robust predictive power-split algorithm.

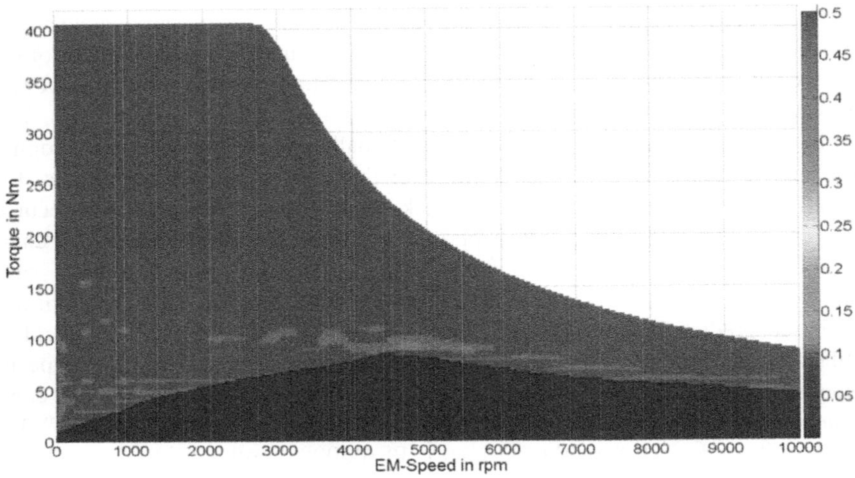

Figure 10.2. *Derived via offline calculation, the optimal values of the power-split factor for the "OpEneR" vehicle are obtained and stored in a look-up table for all possible driver torque demand values at each speed (TEM = 60°C and VEM = 320 V). For a color version of the figure, see www.iste.co.uk/jacob/energy.zip*

10.4.2. *Coasting assistant algorithm*

Another example from "OpEneR" relating to predictive energy management strategy is the introduction of a coasting assistant in a purely electric mode, for the energy optimal approach to legal speed limits and traffic flow speed restrictions or traffic lights. In this context, we typically differ between time-critical and time-uncritical approaches. This differentiation is necessary because it determines the optimization techniques to be used for the calculation of the optimal driving trajectory solution. We developed such a coasting assistance algorithm and optimized for application in real vehicles. This development was initially undertaken using an advanced office PC co-simulation environment, as introduced earlier and described further in the next section, but was also further developed, calibrated and validated using testbed and test track activities.

Figure 10.3 shows the simulated example of such an energy optimal speed profile to decelerate from 100 km/h to 50 km/h in 500 m, which is assumed to be the distance at which the information regarding the upcoming speed profile is transmitted to and received by the vehicle. The obtained speed profile is compared to a typical human behavior with a constant deceleration profile of approximately –1 m/s^2. The relative changes of the improved recuperated energy and the difference on maneuver time are depicted with bar plots.

For this specific maneuver, the human driver recuperates 25% less than in the energy optimized approach. However, this energy optimization comes at a price; namely, the time needed to finish the maneuver is increased by 6 s, see Figure 10.3. Clearly, the results highly depend on factors such as the starting point of the maneuver and the lower limit for the deceleration during the coasting maneuver which is also limited by the need to arrive at the end destination on time.

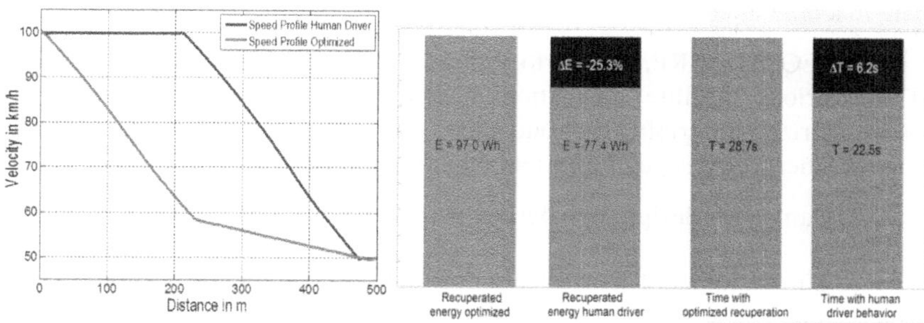

Figure 10.3. *Optimal speed profile for decelerating from 100 km/h to 50 km/h within 500 m a), energy consumption and time comparisons b). For a color version of the figure, see www.iste.co.uk/jacob/energy.zip*

10.4.3. *Introduction of a co-simulation platform*

To support the efficient frontloading (i.e. model-based development and precalibration of predictive energy management control strategies), and the subsequent migration to the complete vehicle or powertrain on the testbed (itself powered by simulation) and finally the test track, there is a need for a seamless and advanced vehicle and powertrain simulation toolchain. Such a seamless simulation toolchain supports the continuous reuse of simulation models, or parts thereof, throughout the development process and provides for fast test case migration throughout the development process from office PC – to HiL – to 4WD powertrain testbed – to track tests. Such techniques provide a highly effective, realistic, timely and also cost efficient platform for the development and refinement of ICT-based predictive strategies.

An example of such a co-simulation platform successfully introduced and used in the "OpEneR" project is the following:

– CRUISE provides a detailed description of the electrified powertrain including the EMs, the battery, as well as various subsystems such as on-board auxiliary systems and control interfaces;

– InMotion/CarMaker furnishes 3D vehicle dynamics, track and human driver functionality, enabling the virtual host vehicle to interact dynamically with traffic objects and its surroundings;

– Matlab/Simulink is used to include other subsystem models, such as the advanced Bosch cooperative regenerative braking system ESP*hev*®, Adaptive Cruise Control and various control strategies, e.g. predictive energy manager or dashboard human–machine interface);

– NAVTEQ ADASRP is used to virtually integrate actual digital map data, as used by various satellite navigation system providers, and delivers a virtual electronic horizon describing the road ahead, including the road gradient, curvature, number of lanes, fixed speed limits and other road infrastructure;

– AVL Cameo for design of experiment.

The co-simulation toolchain combining the above-mentioned simulation tools may be seen as one part of a much wider platform as depicted in Figure 10.4, which includes arbitrary simulation and testbed software to support complex vehicle and powertrain function development. This co-simulation toolchain is extendable to real-time on HiL or on the Powertrain Testbed. Such a co-simulation platform allows the representation of real-world driving in a realistic, but manageable way, where controllers can be virtually developed with the consideration of the real-world environment (e.g. curvatures, inclination and speed limits) from the early phase of project to its successful culmination.

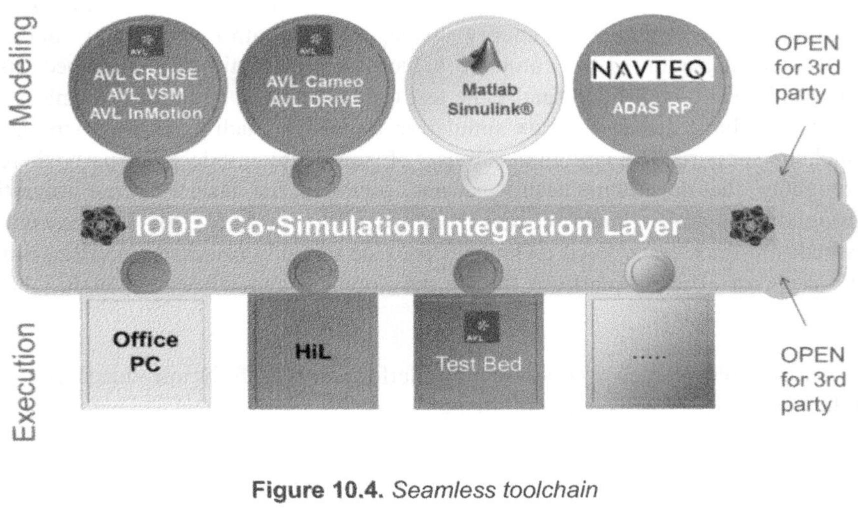

Figure 10.4. *Seamless toolchain*

10.4.4. *Testbed and real-world tests*

The predictive energy management strategies developed within the co-simulation platform can be further tested on the Powertrain Testbed using InMotion. Figure 10.5 shows how predictive energy management strategies developed in the "OpEneR" project are being tested on the AVL 4WD powertrain testbed. Of course, on such a testbed, the powertrain model, or parts thereof, is replaced by the physical vehicle and powertrain system of the prototype car simply by managing the related interfaces. The 3D vehicle dynamics model and the virtual real-world environment and the critically required maneuver-based use cases remain unchanged. This seamless approach helps to significantly reduce the time and cost of controller development and testing and thus supports concurrent engineering.

Figure 10.5. *Testing of predictive operating strategies with the "OpEneR" vehicle on InMotion 4WD Powertrain Testbed*

In a further step, the developed technologies were virtually pretested on simulated public roads (such as the real-world corridor in Spain), well before the release for real road testing is to be given. These public road models feature enhanced digital maps, and emulated v2v and v2i communications. As mentioned before, the detailed road information, e.g. curvatures, inclination and speed limits, must be included in the simulation toolchain for offline development and precalibration of predictive control strategies. The benefit of this approach is that the control algorithms can be developed and realistically tested in the co-simulation platform very early in the development process. The virtual reality extends the range of possibilities to evaluate developed functionalities. For example, the vehicle can be tested on various real-world driving routes all over the world without the need to ship the vehicle to other locations. Later, only after successful precalibration in the virtual environment in the office and at the testbed, the control algorithms will be actually tested on real public road driving routes for final calibration and improvement.

Another successful example of the application of such an innovative development process can also be seen in the "OpEneR" project. There, the innovative control strategies developed for the "OpEneR" prototype vehicles were tested on the Grossglockner Hochalpenstraße. The Grossglockner Hochalpenstraße is a well-known Austrian alpine road with several hairpin curves and high road gradients. Driving up the Grossglockner Hochalpenstraße is an ambitious test for all components of the prototype electrical powertrain. Important data such as temperatures of the cooling system, the inverters and the EMs were collected during the virtual Alpine test on the testbed. While driving up with an EV is a challenge for the electrical powertrain, driving down is a challenge for the recuperative brake system. With this virtually powered testbed, it was verified that both the cooling system and the regenerative braking system of the "OpEneR" prototypes were constructed properly to be able to drive on demanding road topologies such as the Grossglockner Hochalpenstraße.

10.5. Conclusions and outlook

Since the development of new technologies for engines, batteries or other parts of the modern powertrain may be insufficient (or simply too expensive for mass market adoption) to meet stringent forthcoming energy and consequently CO_2 emissions reduction targets, further innovative approaches are required to help lower overall vehicle and fleet emission levels.

In this chapter, we describe how through the integration of real-time off-board information, from diverse ICT sources, such as v2x, satellite navigation, radar, Internet to the vehicle and its powertrain subsystems, predictive control strategies can be used to significantly further improve the driving efficiency, and thus CO_2 emissions (as well as safety and convenience). Of course, the real efficiency and CO_2 improvements achievable depend strongly on the selection of an appropriate range of representative real-world driving cycles, which permits a simplified but still useful offline estimation of the improvement potential available.

The benefit of ICT on overall fleet emissions, including a mixture of vehicle and powertrain types, can be evaluated using a rather more complex combined approach. Here, the vehicle fleet mix, both as it is today and as it is projected to be in the future, is simulated in 1D, driving representative longitudinal driving cycles, both with and without ICT-based predictive solutions. The representative driving cycles are themselves generated separately from specialized urban traffic flow models. The aggregate estimation for the overall fleet CO_2 emissions requires the appropriate weighting of the various simulation results, according to the fleet mix assumed and proportion of time assumed to be spent driving each considered representative cycle. Such a combined approach is one aim of the joint "ICT Emissions" project.

In the case of the "OpEneR" project, it was shown how the electrical efficiency of an advanced 4WD EV may be significantly increased by utilizing off-board information for predictive powersplit and coast assistant functionalities. However, the successful selection and development of such ICT-based technologies into series production is extremely challenging, in both technical and organizational contexts. Thus, a key factor in the successful and rapid deployment of predictive strategies based on ICT will be the use of professional development processes, such as the "V" cycle, which integrate advanced methods and toolchains that seamlessly support the application of frontloaded approaches, concurrently running from the office PC – to the testbed – to the track. Such an advanced toolchain is being used today, in OpEneR and other projects, supporting predictive function development through simulation on the office PC and at the powertrain testbed, thereby reducing the need for expensive and difficult to repeat test cases on the track or the public road (when allowed).

To conclude, state-of-the-art system development requires new seamless simulation methods to allow realistic and highly reproducible testing of connected powertrain functions that process data from the vehicle environment in order to increase in general energy efficiency, safety and driving comfort. This necessarily results in a significantly increased system complexity that requires continuously improving engineering methods and processes.

10.6. Acknowledgments

The "OpEneR" project (285526) and the "ICE Emissions" project (288568) were both funded by the EC Seventh Framework Programme FP7-2011-ICT-GC. We would like to thank our partners in the "OpEneR" project, Robert Bosch GmbH, Peugeot Citroen Automobiles S.A., Robert Bosch Car Multimedia GmbH and Centro Tecnológico de Automoción de Galicia, and the Research Center for Information Technology of the University Karlsruhe. Moreover, we would also like to thank the Universidad Politecnica de Madrid (UPM) and Aristotle University of Thessaloniki for acting as project partners top AVL in the ICT Emissions project.

10.7. Bibliography

[EUR 13] EUROPEAN COMMISSION, Reducing CO_2 emissions from passenger cars – emissions – policies – climate action European Commission, available at: http://ec.europa.eu/clima/policies/transport/vehicles/cars/index_en.htm, September 2013.

[ICT 12] ICT EMISSIONS, Development of a methodology and tool to evaluate the impact of ICT measures on road transport emissions, grant agreement number 288568, funded by the EC Seventh Framework Programme FP7-ICT-2011-7, available at: http://www.ict-emissions.eu/, 2012.

[INT 12] INTERNATIONAL ENERGY AGENCY, CO_2 emissions from fuel combustion, IEA, Paris, 2012.

[JON 12a] JONES S. *et al.*, "Simulation environment development for the optimized energy management for fully electric vehicles", *Proc. DATE 12*, Dresden, Germany, 2012.

[JON 12b] JONES S. *et al.*, "Advanced 3D/1D co-simulation tool chain for the development of safe and efficient EV technologies", *IPG Technology Conference "Apply & Innovate 2012"*, Karlsruhe, Germany, 2012.

[JON 12c] JONES S. *et al.*, "Optimal electric vehicle energy efficiency & recovery in an intelligent transportation system", *Proc. 19th ITS World Congress*, Vienna, Austria, 2012.

[JON 13a] JONES S. *et al.*, "Optimal energy efficiency, vehicle stability and safety on the OpEneR EV with electrified front and rear axles", *Proc. 17th Advanced Microsystems for Automotive Application*, Berlin, Germany, 2013.

[JON 13b] JONES S. *et al.*, "Simulated development of safe and energy efficient driving of the 4WD OpEneR electric vehicle with an advanced cooperative regenerative braking system", *Proc. 9th ITS European Congress,* Dublin, Ireland, 2013.

[LAV 13] LAVERSANNE S. *et al.*, "Seamless development of predictive energy management, recuperation and safety systems: pure office simulation to the 4WD powertrain testbed", *15. MTZ Fachtagung "Virtual Powertrain Creation"*, Mainz, Germany, 2013.

[OPE 12] OPENER, Optimal Energy consumption and Recovery based on a system net-work, grant agreement number 285526, funded by the EC Seventh Framework Programme FP7-2011-ICT-GC, available at: http://www.fp7-opener.eu/, 2012.

PART 3

Energies and Fuels for Transports

Introduction to Part 3

The use of renewable energy sources and fuels in the transport sector is not yet widespread. In the past, the adoption of renewables in transport has been hampered by a number of barriers (high costs, misconceptions about their sustainability and a lack of supporting policies in many countries). However, the policy support that has been in place over the past decade and the gradual commercialization of renewables mean there is a realistic prospect that renewable solutions could be competitive with incumbent technologies and fuels by 2020, as long as policy support for deployment is reinforced and expanded. Part 3 addresses these issues. One chapter describes a system dynamics modeling approach simulating the prospective decision process of purchasers of cars for one of the competing fuel technologies in the EU, in order to determine feasible greenhouse gas reduction targets for each transport mode for the EU until 2050. Another chapter discusses the paradigm shift toward a new sustainable and innovative energy system of 2050 in the context of the road transport sector in the Nordic countries. The focus is on developing tools to understand, create and analyze prospective value chains up to the year 2050. A third chapter assesses the effect of rising fuel prices on energetic precocity, depending on income and zone of residence, whereas another chapter deals with the development of technologies for light-duty vehicles with methane-fuelled powertrains and specific aftertreatment systems, capable of achieving a 10% higher fuel conversion efficiency than that of a corresponding diesel engine technology, while complying with an emission level beyond Euro 6. The last chapter of part 3 considers the major technical issues concerning the integration of energy harvesting units in infrastructure elements, by the means of a preliminary model for an environmental and economic lifecycle assessment.

Introduction written by Zissis SAMARAS.

11

Measures to Promote the Diffusion of Alternative Fuel Vehicles in EU27

Passenger and freight transport, in terms of passenger-km and ton-km, grew continuously in the last decades in EU27, interrupted only by the economic crisis. The contribution of transport to total GHG emissions in 2007 reached a share of 20%. Facing fossil fuel scarcity in the coming decades, these developments demonstrate the responsibility of actors in the transport sector to stop this trend. Policy makers reacted by setting CO_2 targets for passenger car and light-duty vehicle fleets in EU27 until 2015 and 2020. Manufacturers already offer a range of efficient ICE vehicles, but even the 2020 targets can hardly be achieved by only selling vehicles with ICE. Alternative fuel technologies for passenger cars need to be pushed. The question raised is how to accelerate the diffusion of alternative fuel technologies efficiently. This chapter analyzes the potential and suitable incentives to accelerate the diffusion of alternative fuel technologies with the integrated assessment model ASTRA. It considers socioeconomic trends, such as aging societies, as well as rebound effects.

11.1. Introduction

Today, road transport activities are the major source for CO_2 emissions from transportation in Europe. According to the IEA [IEA 11], freight and passenger road transport were responsible for 93% of all transport-related CO_2 emissions within EU27 in 2009. About two-thirds of this emission can be allocated to passenger transport activities. Since 1990, passenger road CO_2 emissions have been steadily increasing.

In order to meet the CO_2 emission reduction targets of the EU for 2020 (–20% compared to 1990 levels) and long-term targets to remain on the 2 degree pathway according to the IPCC, significant changes in today's passenger road transport sector have been made. One way to achieve prospective CO_2 emission limits for transport is the reduction of transport demand by changing mobility behavior. Pricing

Chapter written by Michael KRAIL and Wolfgang SCHADE.

mechanisms and other incentives are powerful instruments to induce these changes, in combination with awareness of transport burdens for the climate. Policy makers are reluctant to implement incentives that afflict the population with additional financial burdens on passenger transport. Mobility is still an essential need for a functioning and growing economy. Hence, instruments leading toward less transport demand, modal shift toward rail and decreasing average trip lengths have to be accompanied by innovations reducing the fossil energy consumption of the rolling stock or an accelerated diffusion of alternative drives. The European Directive (EC) No. 443/2009 is one of the policy measures to enforce this development by setting CO_2 targets for the automotive industry.

In the long run, 95 g CO_2 per km is insufficient facing CO_2 emission reduction targets of up to 80% compared with the 1990 level. Alternative drive technologies such as battery electric vehicles (BEVs), plug-in hybrid vehicles (PHEVs) and fuel cell electric vehicles (FCEVs) need to substitute fossil fuel cars in the long-term. Studies on lifecycle emissions of these alternatives [TOR 06] demonstrate that implications on the transport systems need to be considered together with those on the energy system, as the carbon footprint of alternative fuel technologies depend strongly on the production pathway of the energy carrier. Furthermore, CO_2 emissions from vehicle production need to be taken into account, as the production of e.g. Lithium-Ion batteries is a carbon-intensive process, at least using current technologies. Nevertheless, in the long run there is currently no alternative to a shift toward electrified drives in passenger cars. The remaining question is: how can policy makers and the automotive industry accelerate the diffusion of alternative fuel technologies?

The chapter describes a system dynamics modeling approach simulating the prospective decision process of purchasers of cars for one of the competing fuel technologies in EU27. The approach was applied in the context of the EU FP7 project GHG-TransPoRD in order to determine feasible greenhouse gas reduction targets for each transport mode for the EU until 2050. The model takes into account the dynamics of the major constraints for a fast diffusion of alternative fuel technologies. Those are: currently high investment costs compared to conventional fossil fuel cars, limited filling station or recharging infrastructure, insecurity about calendar life of batteries and fuel cells, lower ranges (especially with BEVs) and lack of knowledge about the implications of a technological shift on individual mobility patterns. The implemented discrete choice approach incorporates the major drivers of car technology choice. An adapted learning curve approach is integrated in order to simulate the feedback between costs of major components in a car depending on cumulated number of car sales worldwide and the decision process. The vehicle fleet model is implemented as a module in the ASTRA (Assessment of

Transport Strategies) model. ASTRA is an integrated system dynamics model simulating the feedbacks between transport, economy and environment. Part 2 highlights the structure of the ASTRA model in which the vehicle fleet model is embedded. Part 3 presents the dynamics of the vehicle fleet model, its major causal chains and feedbacks. Results of some exemplary scenarios simulated with the model are illustrated and analyzed in Part 4. Finally, the chapter concludes with recommendations and an outlook on the need for further research.

11.2. The integrated assessment model ASTRA

Today, the demand for transport is closely interlinked with economic growth and the understanding of individual mobility as a major component of quality and freedom of life. Assessing the impacts of strategies to make transport greener requires considering transport not only as a closed system, but as a system interacting with others. On the one hand, social, economic, energy and environmental systems are driving dynamics in the transport system. On the other hand, transport strongly influences all other systems. A simple example is the breakdown of air transport in the North Atlantic area caused by the eruption of the Eyjafjallajökull volcano in Iceland in 2010. The high concentration of ash in the atmosphere caused, for example, economic losses for a number of airlines and for thousands of people prevented from coming home or to other destinations like the Hanover Fair, that revealed a number of empty booths that year. The shift from fossil fuels to renewable energy carriers like electricity or hydrogen is a result of the need to reduce transport-related CO_2 emissions, in order to reduce the degree of global warming and the depletion of fossil fuels. This shift means a revolution for individual motorized transport. It will influence mobility patterns set by the access to cheap oil for decades. It also impacts the economy which needs to provide the technical solutions to reduce CO_2 emissions, and the energy system which is required to enforce the use of renewable energy sources.

The ASTRA system dynamics model is a tool enabling integrated assessment of transport policy strategies. It links the systems of transport, society, economy and environment. Furthermore, the ASTRA model has been successfully linked to energy system models like the POLES world energy model. ASTRA has been developed and applied in a sequence of German, European and global research. The ASTRA model consists of the following modules interacting with each other: population (POP), macro-economic (MAC), foreign trade (FOT), transport (TRA), environment (ENV) and vehicle fleet (VFT) modules. ASTRA simulates the dynamics of the interaction of modules in the time period from 1995 to 2050. It covers all member states of EU27 plus Norway and Switzerland. In the following, a brief overview of the structure of the nine modules and their main interfaces is presented. Figure 11.1 highlights the main interactions between the modules. As the ASTRA model is composed of about 9,000 variables, there is a large number of feedback loops with different delays included. The following chapter specifying the design of the vehicle fleet model (VFT) describes the most important feedback loops. More detailed information about the other modules, impact chains and feedbacks can be found in Schade [SCH 05] and Krail [KRA 09].

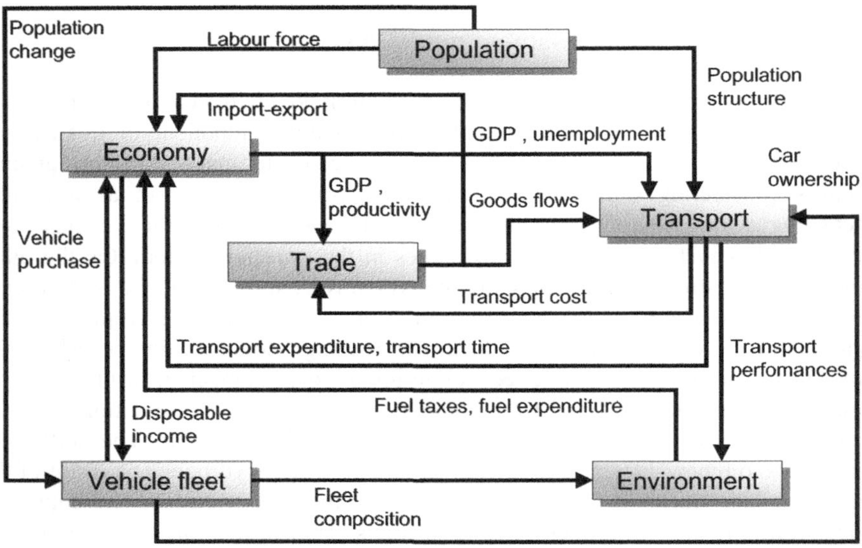

Figure 11.1. *Simplified structural diagram of the ASTRA integrated assessment model*

The population module provides the population development on a NUTS2 basis with 1-year age cohorts. Based on the age structure, given by the 1-year-age cohorts, important information is provided for other modules like the number of persons of working age, or the number of persons in age classes that permit us to acquire a driving license.

The economic module simulates the national economic framework, which embeds the other modules. The MAC incorporates elements of different economic theories. The model uses production functions of Cobb–Douglas type derived from neo-classical theory. Keynesian elements are considered, like the dependency of investments on consumption, which are extended by some further influences on investments like exports or government debt. Further elements of endogenous growth theory are incorporated, like the implementation of endogenous technical progress (e.g. depending on sectoral investment) as one important driver for overall economic development. In a nutshell, the demand side, the supply side, the sectoral interaction and the sector government are reflected by the MAC.

The trade module differentiates between trade within EU27 (INTRA-EU model) and trade between EU27 and the rest of the world (RoW) that is allocated into 15 world regions. Both models are differentiated into bilateral relationships by country pair by sector. Trade between EU27 member states is driven by world GDP growth, by GDP growth of the importing country of each country pair relation, by relative change of sectoral labor productivity between the countries and by averaged generalized cost of passenger and freight transport between the countries. The latter is chosen to represent an accessibility indicator for transport between the countries.

The transport model follows for both passenger and freight transport the classical four stage transport modeling approach: generation, distribution, modal split and a simplified assignment but not on a network basis. While transport demand is generated via a bottom-up approach using specific trip patterns of population groups, differentiated by country, age and income, the freight demand is derived from domestic production and monetary trade flows. Those are inputs from the MAC and the FOT module. Using transport cost and transport time matrices, the transport module performs the distribution and modal-split. Depending on the modal choices, transport expenditures are calculated and provided to the MAC module. Changes in transport times are transferred to the MAC module such that they influence total factor productivity. Considering load factors and occupancy rates, respectively, vehicle-km are calculated.

Major outputs of the TRA provided to the ENV module are vehicles-km travelled (VKT) per mode, per distance band and traffic situation, respectively. Based on these traffic flows and the information from the vehicle fleet model on total yearly fuel composition of the vehicle fleets and hence on the emission factors, the environmental module calculates the major greenhouse gas (GHG) and air pollutant emissions from transport: CO_2, NOx, CO, VOC and PM10. Besides emissions, fuel consumption and, based on this, fuel tax revenues from transport are estimated by the ENV.

The main objective of the vehicle fleet module (VFT) is the assessment of the structure of road vehicle fleets in terms of technological composition. All road vehicle fleets simulate, in addition to the diffusion of fuel technologies, the diffusion of emission standards and the age structure of fleets via cohort models. ASTRA differentiates between passenger cars, buses, light duty and heavy duty vehicles. The most comprehensive model, the passenger car model, is described in detail in the following chapter.

11.3. Diffusion of alternative fuel cars

Modeling the technological diffusion of vehicle fleets is crucial for a holistic assessment of climate and environmental policy impacts on the economy, transport and environmental systems. System dynamics suits perfectly as the methodology for this purpose. There are several feedbacks that can be modeled best in a system dynamics model which will be described in this section. Furthermore, it explains the chosen approach to simulate the diffusion of alternative and conventional drives in passenger cars.

The vehicle fleet module in ASTRA can be differentiated into three submodels which simulate: the ageing of the car stock, new cars registered and the choice of the appropriate fuel technology. The core of the model is a classical stock-flow model. New cars registered per time period constitute the inflow into the car stock which is differentiated by age cohorts. Cars are ageing within the stock. The outflow from the stock represents both scrapping of cars and export of cars outside the EU. With increasing age, the model supposes a growing probability of scrapping or exporting. Figure 11.2 highlights the stock-flow character of the vehicle fleet model and shows the interaction between the three submodels. The flow variables and the car stock variable are differentiated by age cohorts, emission standards (pre-Euro to Euro 6) and fuel technology. Several drivers determine in a linear way with varying significance the number of new registered cars per time period. The most important driver is the development of average disposable income per adult. Based on the principles of the national accounting system, ASTRA computes the disposable income of private households in real terms top-down from GDP for each country [KRA 09]. Other drivers with lower significance are the number of cars scrapped per year, the evolution of average car prices, average fuel prices and the number of persons above 18 years. The level of motorization plays a significant role as it closes a negative feedback loop dampening the number of new cars registered. It represents a saturation factor in the market. Another feedback is closed via the interrelation between new car registrations, investment and consumption, GDP and disposable income.

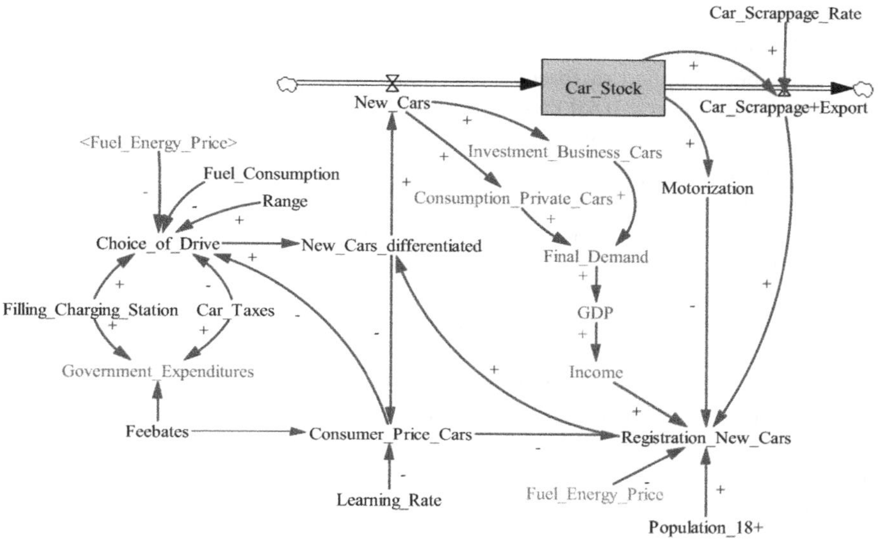

Figure 11.2. *Simplified causal diagram of the ASTRA VFT model*

Based on the time of the new car registrations, the total new cars registered are then allocated with a certain probability to emission standards. The second differentiation in this top-down process is the allocation to the available fuel technologies. This allocation is modeled in the fuel technology choice submodel. Today, the major alternatives to the conventional gasoline and diesel cars are: compressed natural gas (CNG) vehicles, liquefied petroleum gas (LPG) vehicles, hybrid electric vehicles (HEVs), plug-in hybrid electric vehicles (PHEVs), extended range electric vehicles (E-REVs), battery electric vehicles (BEVs), bioethanol (or flexi-fuel) vehicles and fuel cell electric vehicles (FCEVs). Some alternative drives are available and have been offered to consumers, at least by some OEM, for years like CNG cars, LPG, HEV and flexi-fuel cars. Other alternative drives like PHEV and BEV had their first market entry in the EU only recently. FCEVs are currently not available but some OEMs announced the introduction of the first FCEV in 2015 to the market (e.g. Daimler). ASTRA distinguishes between six main alternative fuel technologies and the conventional technologies gasoline and diesel. For reasons of simplification, PHEV and E-REV are assigned to the group of HEV.

In EU27, conventional gasoline and diesel cars will still dominate the vehicle fleet at least for the next decade according to experts, but the share of alternative drives is slowly increasing. According to the theory of diffusion, innovations diffuse with different speeds into the market. Common to all diffusion processes is an S-shaped pathway of diffusion over time. The diffusion of alternative drives can be simulated with the theory for analyzing discrete choice [MCF 01]. Originally, the theory has been applied to forecast the modal choice and transport demand in the context of planning the BART system in San Francisco. The choice for a transport mode out of a set of alternatives is similar to the choice of a suitable fuel technology in the car purchasing process. Each alternative has its consumer utilities which can be expressed by negative costs. Accessibility and availability of alternatives play a significant role for both. The accessibility of public transport is an example for a driver of modal choice, the density of the filling station infrastructure one for fuel technology choice. The possibility to also integrate non-quantitative impacts like individual preferences is very important as well.

Several US studies and a study from ARAL [ARA 05] elaborated via customer surveys potential factors influencing the decision of a car purchaser for a certain fuel technology. According to this study, customers assign a high value to economic efficiency for new cars. Price in combination with the provided performance of a car is the most significant factor, followed by the mileage of the car. Compared with older surveys the safety factor lost significance but, nevertheless, safety still plays an important role. Besides economic and technical factors influencing the car purchase decision, the study also included soft factors like design, image and prestige. The so-called residual disutility in logit functions can represent these soft factors influencing the acceptance of a fuel technology. Similar to the application of logit-functions in the modal-split stage, this model does not compute benefits but costs derived from the concept of total cost of ownership (TCO) that can be put into the logit function as negative benefits according to the following equation:

$$P_{cc,i} = \frac{\exp(-\lambda_i * C_{cc,i} + LC_{cc,i})}{\sum_{cc} \exp(-\lambda_i * C_{cc,i} + LC_{cc,i})} \qquad [11.1]$$

where: $P =$ share of purchased cars per car category cc and country i

$C =$ costs per vehicle-km per car category cc and country i

$\lambda =$ multiplier lambda per country i

$LC =$ logitconst per car category cc and country i representing the residual disutility

$cc =$ index for eleven car categories/fuel technologies

$i =$ index for EU27 countries plus Norway and Switzerland

The car fleet model calculates the required average costs per vehicle-km for each fuel technology in a bottom-up approach. First, the model computes variable costs per vehicle-km based on average fuel consumption factors for each technology and country-specific fuel prices provided by the POLES model described in Krail *et al.* [KRA 07]. The linkage to the world energy model POLES closes an important feedback loop. ASTRA calculates transport demand in terms of yearly fuel consumption, which is used as an input to simulate the fuel and energy price development. Finally, fuel prices are fed back and change the transport demand, influence the economy and the technology choice.

Fuel consumption factors for fossil fuel cars are derived from HBEFA [HBE 10]. Besides variable costs, the model also considers fixed costs for each car category. Fixed costs per car category and country are determined by vehicle circulation or registration taxes and purchase costs per country and car category, as well as country-specific average maintenance costs. All elements of fixed costs are transformed into costs per vehicle-km by the division of average yearly mileages per car category and country. As the conversion of purchase costs into costs per vehicle-km requires information on average lifetime per car category, this is derived from the car stock model via a feedback loop. Consumer prices of alternative fuel cars as well as of conventional fossil fuel cars develop dynamically via adapted one-factor learning curves (OFLCs):

$$pC_{cc,t} = mQ_{cc,t}^{-\varepsilon} \qquad [11.2]$$

where: $pC =$ consumer price for one unit (car)

$Q =$ cumulative production

$\varepsilon =$ elasticity of learning (learning rate)

$m =$ normalization parameter with respect to initial conditions

$cc =$ index for eleven car categories/fuel technologies

$t =$ year

ASTRA considers moderate learning rates derived from the literature for each alternative fuel technology [SCH 12]. Based on cumulated worldwide sale numbers, average car prices for these technologies decrease over time. It is assumed that the manufacturers will offer new alternative fuel cars without margin in the first 5 years. After 5 years, these cars are sold with the average margin. This effect is implemented as otherwise high costs at point of first market entry would prevent consumers from choosing the new technology and would slow down the cost decrease. Besides alternative fuel technologies, learning effects with low learning rates are also taken into account for implementing efficiency technologies in conventional gasoline and diesel cars.

Assuming completely rational purchase decision behavior based on all variables and fixed costs would disregard other important drivers like the distribution grid of filling or charging stations selling the required type of fuel. For conventional fuel types like gasoline and diesel, the distribution grid is characterized by a good quality in all EU27 countries. At present, owners or prospective customers of alternative fuel cars have to cope with the burden that the procurement of alternative fuels requires significantly longer and additional trips or is even not feasible due to a lack of filling stations. Janssen *et al.* [JAN 06] concluded in their paper on CNG market penetration that successful diffusion of new car technologies depends on a uniform development of technology and filling station infrastructure. Taking into account these significant impacts due to fuel supply differences, the model has to consider the quality of filling station grids as well. Hence, the four mentioned cost categories have to be completed by so-called fuel procurement costs. In order to generate these costs per vehicle-km for each car category and country, the model requires input in terms of approximated development of filling station numbers for each fuel type. The following equation describes the simulation of perceived total car costs per vehicle-km:

$$C_{cc,i} = \alpha_{cc,i} * pC_{cc,i} + taxC_{cc,i} + mC_i + vC_{cc,i} + procC_{cc,i} \qquad [11.3]$$

where: $C =$ perceived car cost per vehicle-km per car category cc and country i

$pC =$ purchase cost per vehicle-km per car category cc and country i

$taxC =$ taxation/registration cost per vehicle-km per car category cc and country i

$mC =$ maintenance cost per vehicle-km per country i

$vC =$ variable/fuel cost per vehicle-km per car category cc and country i

$procC =$ fuel procurement cost per vehicle-km per car category cc and country i

$\alpha =$ weighting factor representing the significance of purchasing costs

$cc =$ index for eleven car categories/technologies

$i =$ index for EU27 countries plus Norway and Switzerland

Finally, the logit function simulates the probability of the choice of a fuel technology based on the simulated perceived car costs.

11.4. Policies and scenarios

The model described above was applied in the EU FP7 project GHG-TransPoRD. In this context, the ASTRA model and its vehicle fleet module were prepared to assess feasible GHG reduction targets for transport as a whole as well as for each transport mode. This presumed to analyze the prospective CO_2 reduction potentials for passenger cars over the whole range of drives. The project identified a list of efficiency technologies for ICE as well as for alternative fuel cars. For each technology, the CO_2 emission reduction potential, an outlook on the expected first market entry, the additional costs compared to a similar conventional gasoline or diesel car and the expected learning rate were collected from different studies and discussed with experts from OEMs. Ten different scenarios were set to demonstrate the impacts of policies and technologies on the development of CO_2 emissions from transport. This section presents three selected scenarios and their implications:

– *MAX E&M*: the maximum efficiency under market conditions presumes that OEMs are forced to implement the maximum set of efficiency measures for both fossil and alternative fuel cars. Additional costs for each alternative are implemented and lead to increase in car prices. However, operation costs decrease due to optimized efficiency. Induced investment for R&D and adaption of production sites are also assumed. As opposed to the other scenarios, MAX E&M does not consider further policy measures promoting the choice of alternative fuel cars.

– *EV&HFC*: the electric vehicle and hydrogen fuel cell scenario assumes the same efficiency improvements as in the MAX E&M scenario. On top of that, it focuses on strong incentives for the purchase of alternative fuel cars. A feebate system is integrated which offers purchasers of alternative fuel cars up to €5,000 rebate and requests fees for ICE cars of up to €1,500 decreasing slightly over a decade. Filling and charging station infrastructure is supposed to adapt until 2050 to the density of fossil fuel filling stations. The average range of alternative fuel cars is presumed to increase up to 40% compared with the current status. Measures to promote alternative fuel technologies are implemented leading to an improved acceptance of alternative fuel cars. Finally, induced investments in R&D are also considered plus investments required to establish the production sites.

– *AMB REG*: the ambitious regulation scenario starts with the setting defined for MAX E&M. Additionally, alternative fuel technologies are subsidized like in the EV&HFC scenario. In order to compensate losses of fuel tax revenues, fossil fuel taxes increase in this scenario. Further policies like congestion charging were implemented to reduce the probability of rebound effects. Finally, the AMB REG scenario tests a phase-out of fossil fuel cars starting in the year 2035.

Figure 11.3(a) presents an overview of the estimated technological composition of EU27 car fleets. High efficiency conventional gasoline and diesel cars in *MAX E&M* have prevented alternative fuel technologies for a long time from being competitive. Hence, fossil fuel cars are supposed to dominate the EU27 car fleet with a share of 87% until 2050. Efficiency gains are able to compensate increasing fossil fuel prices such that the comparative cost advantage of fossil fuel-driven cars remains until at least 2040. The picture might look different when assuming a lower availability of oil resources until 2050, such that the applied price development constitutes a clearly optimistic assumption by the linked POLES world energy model.

The share of 62% alternative fuel cars in the EU27 car fleet in the *EV&HFC* scenario reflects the impact of the implemented feebate system in combination with accelerating the construction of the necessary filling station infrastructure. Despite the model's assumption that the costs for building the infrastructure are forwarded to consumers via higher alternative fuel or energy prices, alternative drives are the dominant technology in this scenario from 2030 to 2050. In the last two decades, particularly FCEVs, diffuse into the fleets, which is a result of the feebate system in combination with fewer barriers like a lack of filling station infrastructure (Figure 11.3(b)).

The most radical scenario is the *AMB REG* due to the complete phase out of fossil fuel cars from 2036 onwards. PHEV, BEV and FCEV account for about 89% of the whole EU27 fleet until 2050. The model assumes that about 34% of the fleet will be BEV. In this context, the ASTRA model does not consider constraints in terms of the limited number of persons that can satisfy their mobility needs via BEV. The ASTRA model implicitly assumes the possibility of a radical change of mobility patterns toward an increased use of new mobility concepts.

One of the main drivers of the diffusion of alternative fuel technologies is of course the dynamics in consumer prices. Figures 11.4(a) and (b) reflect the dynamics in the *MAX E&M* and *EV&HFC* scenario induced by the application of learning curves. In *MAX E&M,* alternative drives diffuse slowly into the EU27 vehicle fleet such that the technological learning is slowed down. Average prices of BEV are expected to remain over the whole simulation period above comparable alternatives. The projections show that FCEVs converge only after 2040. However, a general price increase is assumed for all cars due to a scarcity of natural resources used for the production of all types of cars.

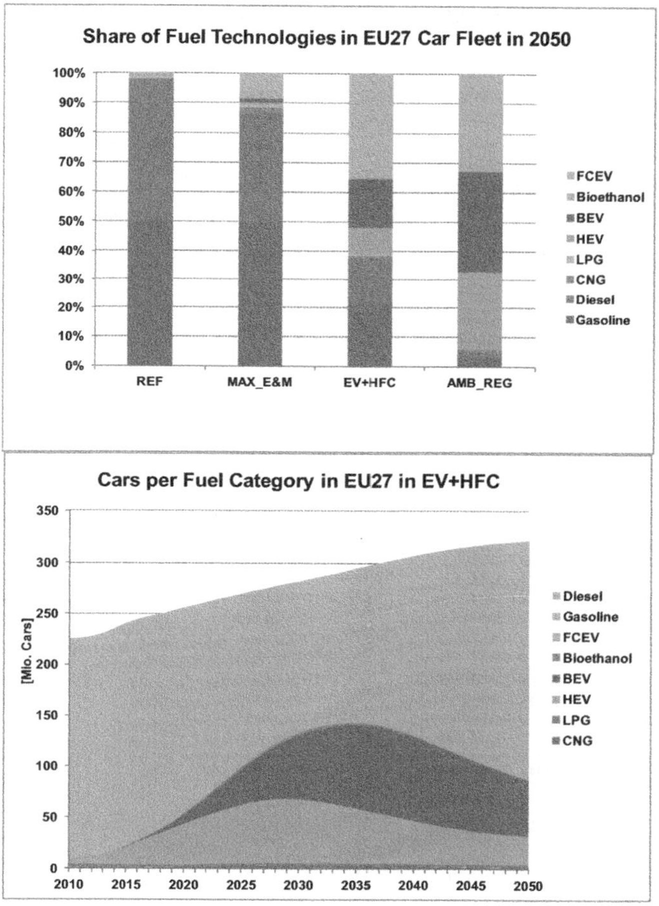

Figure 11.3. *a) Car fleet composition 2050 in all scenarios and b) diffusion pathway in EV+HFC scenario. For a color version of the figure, see www.iste.co.uk/jacob/energy.zip*

The difference to the price evolution in the *EV&HFC* scenario is obvious. The implementation of a feebate system accelerates the diffusion of alternative fuel technologies and reinforces the trend via faster technological learning. This leads to converging consumer price curves such that BEVs are expected to reach a similar price level as gasoline cars within the next 10 years. For FCEV, the trend is similar and the cost pathway decreases even more steeply. In the long run, the learning curve approach implemented in ASTRA leads to FCEV being the cheapest alternative.

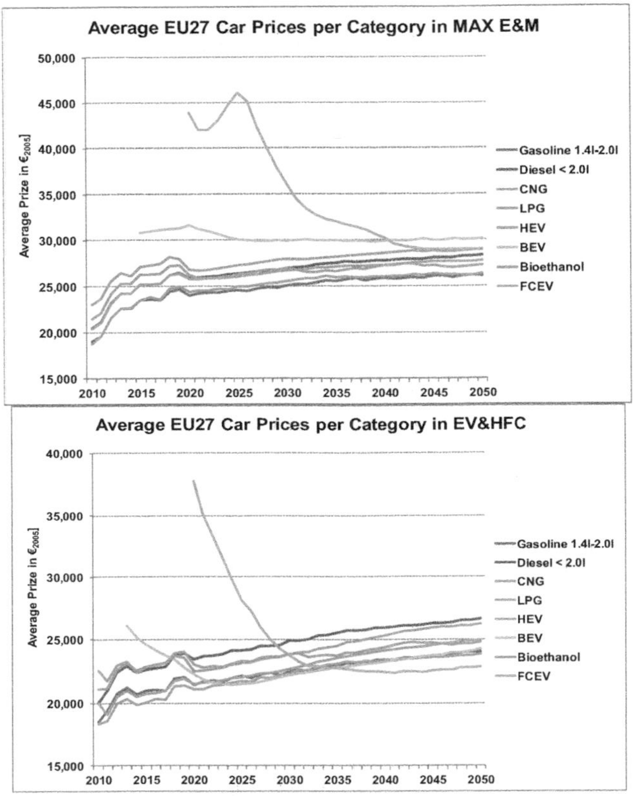

Figure 11.4. *a) Purchaser prices in MAX_E&M and b) purchaser prices in EV&HFC scenario. For a color version of the figure, see www.iste.co.uk/jacob/energy.zip*

The *EV&HFC* scenario in particular shows that higher prices can be accepted in the first years after market entry, such that the category of "Innovators" or "Early Adopters" is demanding these types of cars. The comparison of both scenarios demonstrates that there is a threshold and not all prices are accepted. New technologies need a powerful incentive pushing the technology. This happens in the *EV&HFC* as well as in the *AMB REG* scenario. The price developments in both scenarios are similar until 2035. The strong regulation policy assumed in the later scenario forbidding the purchase of fossil fuel cars after 2035 leads to an even stronger diffusion of alternative fuel technologies.

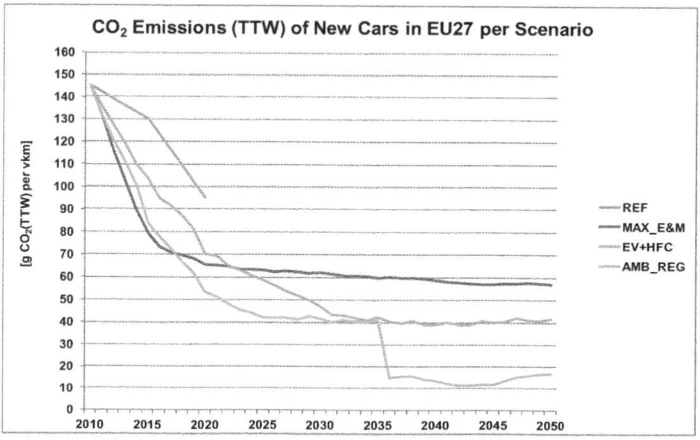

Figure 11.5. *Evolution of CO2 emissions per km from passenger cars for all scenarios*

Figure 11.5 presents the technological evolution of the efficiency of average new cars registered in EU27 according to the NEDC (New European Drive Cycle). It shows that the targets set by Directive (EC) No. 443/2009 of the European Parliament can be achieved or even be undersold. It also highlights that a rather quick decline of CO_2 emissions is feasible.

11.5. Conclusions

Simulating the impacts of climate policies in combination with technological development of passenger cars requires an integrated modeling approach. The vehicle fleet model integrated in the ASTRA system dynamics model fulfills the major requirements of such a strategic tool. It takes into account the various feedback loops that exist and determine the dynamics of such a complex system. In GHG-TransPoRD, the integrated transport, economy and environmental ASTRA model could be linked with the world energy model POLES. This is significant progress as the implications of the expected revolutionary shift from a fossil fuel-driven transport system to a renewable, electrified future transport system can only be assessed by an integrated transport and energy model. Transport demand drives the scarcity of fuel and energy and leads to increasing prices, which in turn change transport demand again. Innovations are boosted in such an environment but the scenarios set in GHG-TransPoRD showed that this is not a self-fulfilling prophecy.

Powerful policy measures and binding regulation need to be established in order to accelerate the diffusion of green alternatives. The CO_2 emission target set by the European Directive (EC) No. 443/2009 was a first step in the right direction. Nevertheless, efficiency improvements of cars with ICE are not sufficient facing CO_2 reduction targets of up to 80% compared with the level in 1990. Feebates or other mechanisms have the potential to reduce the comparative cost advantage of old technologies even in the short run. Technological learning leads to converging consumer prices over time for all fuel technologies.

As opposed to other models, the ASTRA model is able to reflect rebound effects. One of the most important effects is the increasing competitiveness of the car mode compared to other modes due to efforts of making passenger cars more efficient. Despite increasing purchase prices, the TCO over the whole lifetime of a car becomes lower compared with less efficient cars. As only perceived costs, or in other words the costs of operating a car, are the drivers of transport demand, the passenger car performance grows. This effect partly cannibalizes the technology-driven CO_2 savings achieved by improving efficiency or higher share of alternative fuel cars. An optimal set of policies needs to counteract such a rebound effect by setting further policies, e.g. increasing fuel taxes or introducing congestion charging, only two out of several alternatives.

11.6. Acknowledgments

The GHG-TransPoRD project has been funded under the FP7 framework by the EC. Further results of the project can be found on www.ghg-transpord.eu.

11.7. Bibliography

[ARA 05] ARAL, Trends beim Autokauf 2005, *Aral Study*, Aral Press, Bochum, 2005.

[HBE 10] HBEFA, *Handbook of Emission Factors HBEFA 3.1*, Infras, Basle, Switzerland, 2010.

[IEA 11] IEA – INTERNATIONAL ENERGY AGENCY, *CO_2 Emissions from Fuel Combustion*, International Energy Agency, Paris, 2011.

[JAN 06] JANSSEN A., LIENIN S.F., GASSMANN F. *et al.*, "Model aided policy development for the market penetration of natural gas vehicles in Switzerland", *Journal for Transportation Research Part A: Policy and active*, vol. 40, pp. 316–333, 2006.

[KRA 07] KRAIL M. *et al.*, *Outlook for Global Transport and Energy Demand*, Deliverable D3 of TRIAS project (Sustainability Impact Assessment of Strategies Integrating Transport, Technology and Energy Scenarios), Karlsruhe, 2007.

[KRA 09] KRAIL M., *System-based analysis of income distribution impacts on mobility behaviour*, Dissertation, University of Karlsruhe, 2009.

[MCF 01] MCFADDEN D., "Economic choices", *The American Economic Review*, vol. 91, pp. 351–378, 2001.

[SCH 05] SCHADE W., *Strategic Sustainability Analysis, Concept and Application for the Assessment of European Transport Policy*, NomosVerlag, 2005.

[SCH 12] SCHADE W. *et al.*, *Bottom-up quantifications of selected measures to reduce GHG emissions of transport for the time horizons 2020 and 2050: cost assessment of GHG mitigation measures of transport*, Deliverable D3.1 of GHG-TransPoRD, Karlsruhe, 2012.

[TOR 06] TORO F., HASENAUER U., WIETSCHEL M. *et al.*, *Technology trajectories for transport and its energy supply*, Deliverable 2 of TRIAS (Sustainability Impact Assessment of Strategies Integrating Transport, Technology and Energy Scenarios), Karlsruhe, 2006.

12

Creating Prospective Value Chains for Renewable Road Transport Energy Sources up to 2050 in Nordic Countries

If the Nordic energy and transport sectors are to meet the 2050 energy and climate policy targets, major systemic changes are necessary. Along with new technologies, changes are also required in other societal functions such as business models and consumer habits. The transition requires cooperation between public and private actors. This chapter discusses the paradigm change toward a 2050 Nordic road transport system based on renewable energy. More precisely, it proposes an approach for creation and analysis of prospective value chains up to the year 2050. The value networks arise from three alternative, but partly overlapping technology platforms, namely electricity, biofuels and hydrogen. The approach outlined in this chapter combines elements from the fields of system-level changes (transitions), value chain analysis and forward-looking policy design. It presents a novel, policy relevant application with a set of practical tools to support the development of implementation strategies and policy programmes in the fields of energy and transport.

12.1. Introduction

Sustainable energy technologies are driven especially by the climate change challenge, which necessitates a paradigm shift in global energy production and consumption structures. Currently, about 20% of the Nordic CO_2 emissions are due to the transport sector. If the Nordic energy and transport systems are to meet the 2050 energy and climate policy goals, a major transition is necessary. Along with new technologies, changes are also required in other societal sectors such as business models and consumer habits. The transition requires cooperation between

---

Chapter written by Anu TUOMINEN, Nina WESSBERG, Anna LEINONEN, Annele EEROLA and Simon BOLWIG.

public and private actors. Policy decisions should create potential for enterprises to provide renewable energy solutions in a way that they also attract consumers and transporters of goods. In order to be able to make informed policy decisions, foresight actions are needed to get an idea about the future trends and needs, and possible ways of shaping the future. Furthermore, greater knowledge and understanding is needed on the roles of various actors in the change process. In our understanding, actors are outlined in value chains and networks.

In this chapter, the paradigm change toward a new sustainable and innovative energy system of 2050 is discussed in the context of road transport sector in the Nordic countries. The focus is on developing tools to understand, create and analyze prospective value chains up to the year 2050. By 'value chain', we mean a network of activities needed in order to deliver a specific valuable product and service for the market, including activities related to energy sources or feedstock production; energy production; distribution and transportation; retail; consumption; regulation and governance; and research and development. In our case, the value chains arise from three alternative, but partly overlapping technology platforms, namely electricity, biofuels and hydrogen. In addition, the prerequisites for the feasibility of the resulting value chains are considered. The motivation for the chapter is to produce knowledge for future decision-making and policy support in order to create a grounds to enable sustainable energy solutions for the future transport sector. The proposed prospective value chain creation process seeks to identify the critical elements in the networks to be taken into consideration in strategic planning and decision-making.

Traditionally, value chains are considered in rather short-term business opportunity analyses. In our case, we need to outline the value chains, or networks, in the far future. Hence, our research questions are more specifically:

– what are the essential elements of the process for creating and understanding the key dimensions in prospective value chains for renewable energy and transport systems?

– what role can such a process have in assisting sustainable decisions and policy-making in the fields of energy and transport?

12.2. Theoretical background

The theoretical background of the chapter arises from integrating insights from three disciplines, long-term foresight, sociotechnical transitions and value chain analysis, in order to assist sustainable decisions and policy-making in the field of energy and transport.

12.2.1. *The functions of foresight in policy-making*

The impact of foresight on policy-making can be discussed through its three major functions: (1) informing, (2) facilitation and (3) guiding. The issue has been discussed widely recently among foresight experts and practitioners [GEO 06, DAC 08, WEB 09, KON 11]. Policymakers are dealing with increasingly complex issues that are highly interconnected and multidimensional. Extensive amounts of information are available, but the challenge is how to make sense of it in the given timescale of policy-making. The informing function of foresight refers to generations of insights regarding the dynamics of change, future challenges and policy options, along with new ideas, and transmitting them to policymakers as an input to policy conceptualization and design [DAC 08].

The second function, facilitation of policy implementation, gets its motivation from the changing nature of policy-making. There has been a shift in the conceptual understanding of policy process [WEB 09, DAC 08]. The shift has been from linear models of policy-making, consisting of successive phases such as formulation, implementation and evaluation phases, into cyclic models, where evaluations are supposed to feed back into the policy formation and implementation phases. In this function, foresight can be seen as an instrument which builds common awareness, networks and visions among stakeholders, and provides a forum for various actors for interaction, cooperation and learning in relation to specific change or transition. Foresight promotes the development of "future-oriented attitudes" among participants, which may ease policy implementation [DAC 08].

The third function, which is called policy guiding, refers to the capacities of foresight to support strategic planning or policy formulation. Weber *et al.* [WEB 12] describe this function as strategic counseling of the policy process. This function operates only if foresight process is carried out jointly with the policymakers in charge of the specific policy field. Foresight exercises may even bring to light the inadequacy of the current policy system to address the major challenges that society is facing [DAC 08]. Adaptive foresight is an example of a practical tool to support policy-making in this function [ERI 08]. The proposed process for analyzing prospective value chains presented in this chapter contributes all of these three functions and hence effectively supports strategic planning and policy design in the field of renewable energy for transport.

12.2.2. *Multi-level perspective to transition*

Change toward sustainable energy and transport systems in the climate change context is possible only through systemic innovations, for example in the fields of renewable energy production, energy efficiency, energy saving, behavioral changes in energy consumption, etc. These innovations require changes in all system functions. A dominating essence of a complex technological system is path dependence, which highlights that directions for future development are foreclosed or inhibited by directions in the past development.

Systemic change, i.e. transition in a path-dependent system, is a complex multidimensional societal change process, dealing with the co-evolution of technological, industrial, policy and social changes. A multi-level perspective (MLP) framework has been developed [GEE 05], in order to describe this complex process. The framework has also been developed and applied in the UK [FOX 10]. Geels has also applied the MLP framework to the transport system in his recent work as he has studied transitions toward low-carbon futures of automobile systems [GEE 12]. Three levels of change are abstracted in the MLP framework: landscape, regime and niche. *Landscape* forms an exogenous macro level environment that influences developments in niches and regimes. General developments in global operating environment, including, e.g. economic, cultural or environmental factors, compose the landscape level. *Regime* refers to the existing structures and actions of the system. In the context of this chapter, these structures and actions are related to e.g., nuclear power or mode-specific transport systems. The specific form of the regime is mainly shaped and maintained through the mutual adaptation and co-evolution of its actors and elements. The development is very path dependent. *Niches,* in turn, form the level where radical novelties emerge. Niches are local innovative solutions, experiments. Niches provide opportunities for the learning and incubation of alternative solutions that may gradually become strong enough to challenge the existing regime or adopt and transform the regime toward new directions (see e.g., [GEE 05, GEE 07, GEE 07]). The system transition, which in our case aims toward renewable energy solutions in the Nordic transport sector in 2050, is possible if the change processes in all these different levels are synergetic.

12.2.3. *Value chain analysis*

Global value chain (GVC) analysis has emerged since the early 1990s as a novel methodological tool for understanding the dynamics of economic globalization and international trade as well as particular industries such as automobile manufacturing [STU 08]. GVC is based on the analysis of discrete 'value chains' where input supply, production, trade and consumption and disposal are explicitly and (at least to some extent) coherently linked. In addition to the descriptive aspects of territoriality and input-output structure, much GVC discussion has revolved around two analytical issues: how GVCs are governed (in the context of a larger institutional framework) and how upgrading takes place along GVCs. Application of GVC into the context of technological innovation and renewable energy as in the TOP-NEST project is a novel field of research.

In traditional manufacturing network operations of suppliers, lead producers (such as OEMs – original equipment manufacturers) and customers are seen as independent sequential tasks, which form a value chain. Since the 1990s, however, this pattern has been changing and the theoretical discussion has also emphasized the transfer from value chains to value networks [NOR 94, PEP 06]. Value network perspective is considered to be more suited to those organizations where both the product and supply and demand chain is digitized, such as banking, insurance or telecommunication. In practice, co-operation of manufacturing networks is mostly limited to bilateral collaboration, e.g. vertical relationships between a customer and a supplier and the change toward network level decision-making and operations. As definitions, we may present the following [VAL 11]: (1) the value chain consists of the entire sequence of activities or parties that provide or receive value in the form of products or services, and (2) a value network consists of organizations (companies) co-operating with each other to benefit all network members. Recent literature also points out that external actors can have an important say in how a GVC is governed, e.g. governments, large NGOs, 'experts', certification bodies and service providers [PON 07, RII 09]. External actors are often important in emerging industries like those for renewables.

In the context of this chapter, we need to make a distinction between value chain analysis, in the sense of analyzing and developing of existing value chains, and the analysis of prospective value chains. As our approach is forward-looking and we want to anticipate value chains that may exist in the long-term future, we need slightly different viewpoints. The value network approach provides us with an interesting ground. Peppard and Rylander [PEP 06] present a procedure of network value analysis (NWA), which takes the network-oriented model as a starting point. The aim of NVA is to generate a comprehensive description of where value lies in a network and how value is created. The analysis is based on a step-wise procedure starting from definition of network objectives and identification of network participants to the analysis of value dimensions and influences, and shaping of the final value network. Another approach to future-oriented analysis of business models and business ecosystems is presented in Ahokangas *et al.* [AHO 12]. This approach aims at developing new business models for companies. The analysis starts from identification of the key actors and analysis of their needs and benefits in the business ecosystem. The second stage of the analysis is to generate scenarios for the identified ecosystem, so that new business models can be created. Our approach (see section 12.3.2) outlines the prospective value chains in the context of transport energy systems. It incorporates elements from the approaches mentioned above. The central aspects in our approach are value network thinking and future anticipation based on scenario approach.

12.3. Method

12.3.1. *Energy and transport in MLP framework*

Transport and energy systems, which we propose to analyze from the multi-level perspective (MLP), form the context for our method. Figure 12.1 presents the three basic components of the transport system: users, vehicles and transport infrastructure. In addition, in the middle of these components, all of which interact with one another, are a fourth and fifth component: transport services and transport system organization, governance and regulation. Each of these components is elaborated further into some key elements that characterize them. For example, transport vehicles and other means of transport rely on alternative technologies and materials, and besides the manufacturing processes, these also require maintenance. Different vehicle solutions make use of different fuels and other energy carriers, and they result in environmental impacts. Furthermore, the use of vehicles involves behavioral and business models, and different types of solutions are available concerning issues such as vehicle ownership (adapted from [AUV 12]). Figure 12.1 also presents the main elements of the energy system (primary energy sources, production and storage), which are linked to the transport system mainly through energy and transport infrastructures and are crucial for transport operations.

The transport and related energy system components and elements in Figure 12.1 can be analyzed from the multi-level perspective [GEE 04]). The three levels adjusted to the transport domain are landscape, transport system and technologies and solutions. The components and elements are positioned on the most appropriate levels to indicate their main application areas, but it should be acknowledged that here, there are no clear boundaries. This structuring is supported by the recent work by Geels [GEE 12], where similar early steps in exploring multi-level perspective in the study of transitions in the transport sector are taken. Geels suggests very similar definitions when drafting the automobility system in studying transitions toward low-carbon futures. Multi-level perspective in structuring the socio-technical system for land-based, hydrogen intensive road transport has also been used by van Bree *et al.* [BRE 10].

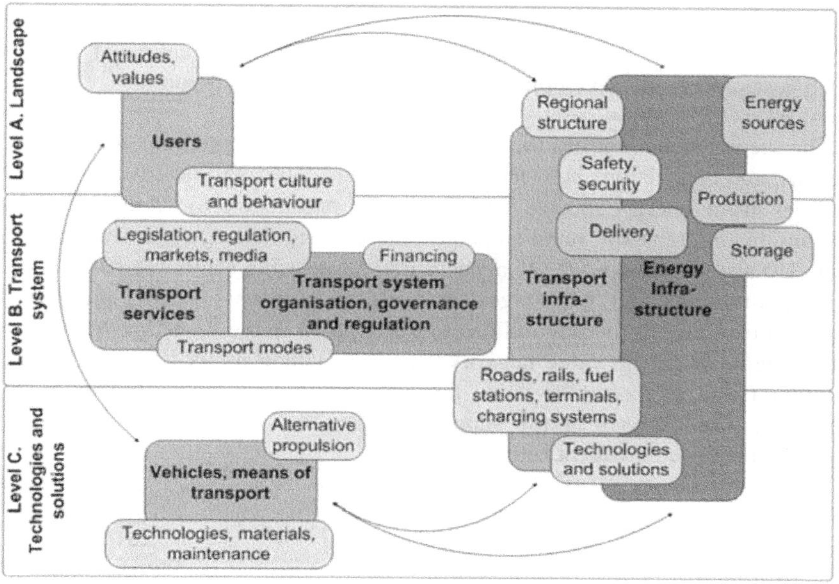

Figure 12.1. *The basic components of transport and related energy systems from the multi-level perspective*

12.3.2. *An approach for outlining prospective value chains for sustainable energy systems in road transport*

The proposed process consists of three stages (see Figure 11.2): step 1, building future context for the prospective value chains; step 2, identification of value network actors and; step 3, outlining of the prospective value networks.

The process starts with identification of the context where the prospective value networks will operate (step 1). For this purpose, various foresight methods, such as the Futures Wheel [GLE 94], and scenario methodology (e.g. [UNI 05]) can be used. The purpose of the Futures Wheel is to open future thinking and help in recognising key factors or drivers which affect the future development of energy and transport systems. Based on the Futures Wheel exercise, it is possible to create alternative futures using scenario methodology. The alternative futures create alternative possibilities for different technology platforms to develop. Next, to be able to anticipate technological development in different futures, the strengths, weaknesses and development possibilities of the different technology platforms need to be analyzed. The SWOT analysis of the technology platforms and evaluation of the scenarios from the perspective of various technologies present the tools for this.

The goal of the first step of the procedure is to recognize the ideal future for each technology platform. In other words, this enables the identification of the factors which best support the development of a single technology platform.

The second step identifies the value network actors and analyzes their individual interests, and connections between different actors. The analysis covers value chain activities from energy sources and feedstock production to energy production, distribution and transport, retail and consumption. Regulation, governance and R&D actors are also included in the analysis. First, all possible actors are listed and then their opportunities and advantages, as well as supportive needs, are analyzed. Opportunities refer to the possibilities to make profit in the value network (how the actor benefits from the value network?) and advantage refers to created value by the actor (what is the added value the actor provides to its customer or in the network?). The analysis of the supportive activities is needed to recognize the connection between different actors.

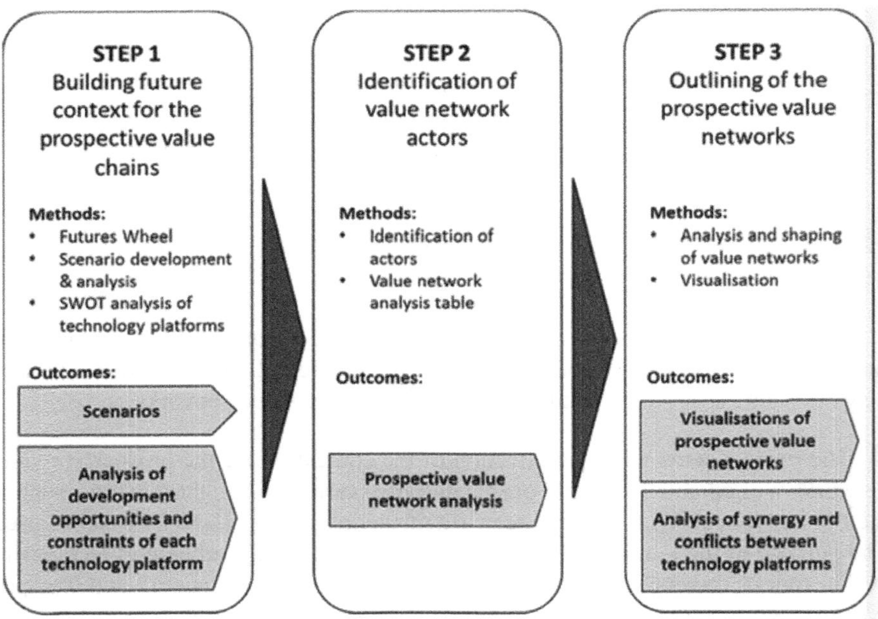

Figure 12.2. *A process for prospective value chain analysis*

The third step includes outlining the prospective value chains. Here, it is important to consider the following issues. Different technology platforms will co-exist in the future, but as mentioned above, different futures create different opportunities and development possibilities for different technology platforms. Therefore, we need to describe the level of technological development of the given technology platform in the outline of the value network. In other words, the outline of the value network works only in a selected scenario, and the level of technological development of a single technology platform varies across it. Furthermore, it is important to consider the different customers or transport segments of the value network, both in passenger and goods transport and also across in urban and long distance transport. The following sections present an example of the use of the method in the context of a 2050 Nordic road transport system based on renewable energy.

12.4. Building future contexts for the prospective value chains

Building future contexts started with a group brainstorming session using the Futures Wheel as an aid. An interdisciplinary group of researchers from Finland, Sweden, Denmark and Norway from the fields of technology, modeling and social sciences participated in the workshop. Based on the Futures Wheel exercise, the following three future directions were recognized: (1) "Green energy and transport": an environmentally sound transport system, resulting from strong environmental regulation and policy-making, (2) "Co-existence of technologies": various technologies co-exist in the transport system due to each of their limitations and advantages. A key driver in this line is technological development and (3) "Changing values as a base for new transport system": changing values are the main driver for change and new transport solutions are developed to meet new needs. From the exercise, two main dimensions as a basis for scenario building could be identified. The first one is the *policy coherence* factor: the level at which the whole society works toward transition. The second is the *societal structure* (centralized-decentralized). Using the dimensions presented in Figure 12.3, we formulated four different scenarios for 2050.

"Smart villages" present a decentralized community structure and energy system as well as coherent policies toward sustainability. Road transport is a dominant mode for both passenger and goods transport. Smart, low carbon public and goods transport solutions for small communities are deployed and attract more users. *"Urban beat"* describes centralized community structure and energy system together with coherent policies. Rail transport and non-motorized transport (cycling and walking) dominate in cities. For long distance travel and transportation of goods, high-speed rail is the main transport mode. The *"small steps"* scenario presents a centralized community structure and energy system, in which policies are incoherent. Various modes of transport, such as road, rail and NMT are used in cities. Road transport dominates long distance travel and transportation of goods. Finally, *"Prairie"* describes a decentralized community structure and energy system as well as incoherent policies. Road transport is the

dominant mode for both passenger and goods transport. However, the development and maintenance of the road network is poor.

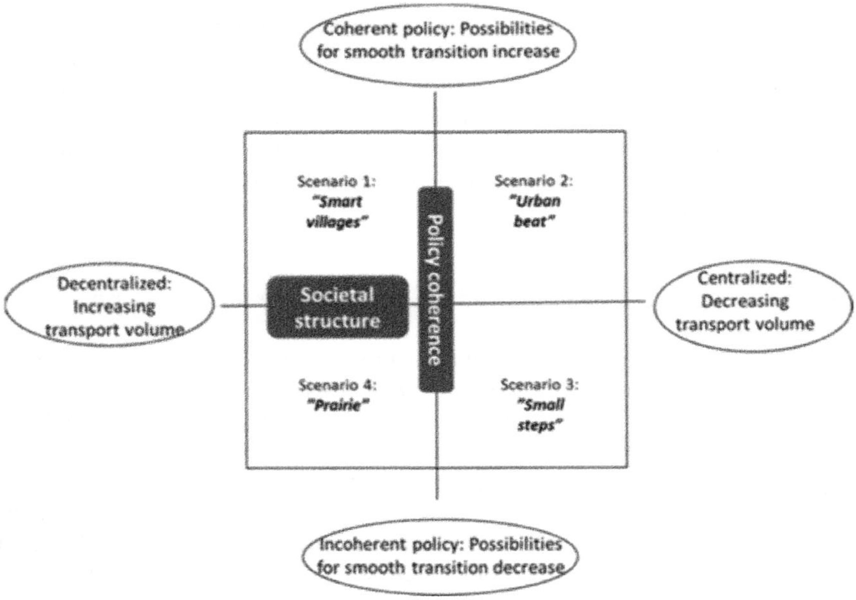

Figure 12.3. *The key dimensions for scenario building and the four transport scenarios formulated for 2050*

Next, alternative scenarios were analyzed against the three technology platforms. The analysis covered two issues. First, strengths, weaknesses, threats and opportunities of the technology platforms were identified by SWOT analysis. Second, the benefits and challenges that each scenario could provide for the three technology platforms were identified. Table 12.1 sums up the results of the scenario analysis.

The analysis reveals which scenarios are the most beneficial for each technology platform. For example, the characteristics of biofuels and hydrogen technology are in line with the decentralized structure and demand for local solutions of the "*Smart villages*" scenario. However, the same demands create a challenge for electricity, as they result in high need for energy storage and adjustment power. Biofuels are also strong in the "*Small steps*" and "*Prairie*" scenarios, because they are compatible with the existing energy and transport structures, and therefore they can evolve even if there is a shortage of R&D investments or shared objectives in the technology development.

Scenario	Dominant technologies *	Challenges for other technology platforms
Smart villages	Biofuel Hydrogen	Goods transport is based on road transport, which is challenging for electricity. High need for energy storage and adjustment power to guarantee the availability of electricity in the decentralized society.
Urban beat	Electricity	Small volumes for biofuels, because electrification of society is emphasized. Advancing bio economy creates competing uses for biomass (decreasing transportation volume may compensate this). Possible shortage of R&D investments. No demand for long distance electric passenger traffic (challenge for hydrogen).
Small steps	Biofuels	Contrasting interests hinder the development of public transport and its electrification. No coherent regulation or coordinated development activities. Companies are not willing to develop technology. Organization of heavy road transport is very challenging for electricity and hydrogen.
Prairie	Biofuels	Global economic crisis delays development, no money for implementation of new technologies. Disintegration of the infrastructure (challenge for electricity).

* benefitting from the circumstances

Table 12.1. *The results of the scenario analysis*

12.5. Analysis of value network actors: a biodiesel case study

The use of the prospective value network analysis process was tested with a second-generation biodiesel case study. In this case, a company has started to develop its biofuel business by building a hydro-treatment based biorefinery next to the pulp mill of the company. Crude tall oil is used as a primary raw material. The hydro-treatment biodiesel biorefinery investment is made without any public subsidy. It is the first market oriented hydrogen treatment biorefinery in the world. The chosen case study corresponds to the "*Small steps*" scenario, as it describes a situation where a forest industry company is refocusing its business. Drivers for this redirection are the dramatic decrease in the demand for traditional forest products and global monetary crisis. At the same time, there is a global need for increasing the use of renewable energy in road transport. In these circumstances, the production of biofuels is a tempting path for a forest company to follow. We could say that this case study is a strongly path-dependent example. The biorefinery company has long forest industry history in Finland. Pulp production started in 1880. An important strategic goal is to start to produce second-generation biofuels from wood material. Since the company is a traditional forest industry company, the production of second-generation biofuel nicely continues its history. Tall oil is produced from black liquor soap, a side flow of sulfur-chemical pulp mill process. In theory, other bio-oils and fats, Fischer-Tropsch wax, black liquor, lignin, turpentine and pyrolysis oil from forest biomass could also be used in the process. The same fuel distribution system can be used as used in fossil fuel distribution.

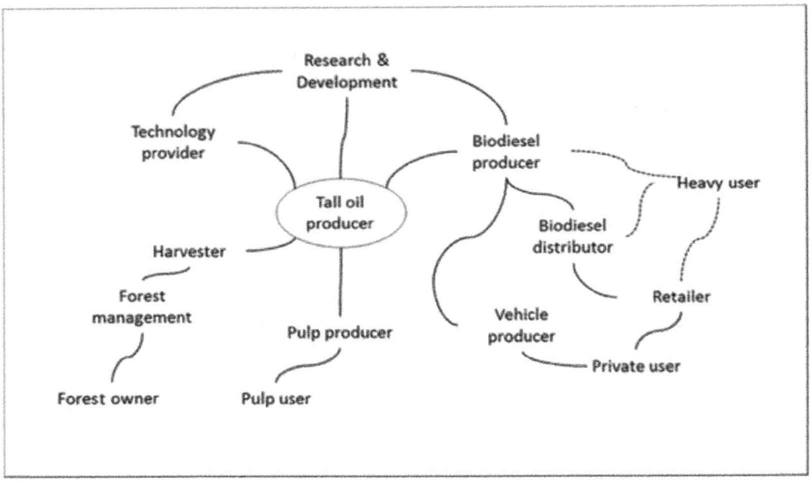

Figure 12.4. *The value network of a biodiesel example based on tall oil*

First, the actors which constitute the value chain network were identified. The network focal point is naturally the tall oil producer (see Figure 12.4). The network shows the relationships between different actors. After the network structure is constructed, value creation potential of the connections can be analyzed. The analysis may reveal new potential relationships or ways to organize the value chain structures in the future. For instance, the biorefinery company or the tall oil producer may be converted into a technology supplier in the future, as it owns the intellectual property rights to the process. Another example of the possible findings based on the value network chart is that there may be changes in the delivery part of the value network. For example, some heavy users of biofuels might get the fuel straight from a biodiesel producer instead of various intermediary actors. The network analysis also reveals that tall oil-based biodiesel production requires pulp production to be profitable. In addition, to be able to use this biodiesel option, we need a liquid fuel distribution system and the vehicle industry in favor of biodiesel as an energy source for road transport.

12.6. Discussion and conclusions

In this chapter, we have proposed an approach to create and analyze prospective value chains in the context of Nordic road transport for 2050 based on renewable energy. As an answer to our first research question, we have identified the three essential steps of the process for creating prospective value chains: (1) building future contexts for the prospective value chains, (2) identification of value network actors and (3) outlining the prospective value networks. The motivation behind our approach was that future value chains and future actors within have to be recognized in order to find out prerequisites for the future actions related to renewable energy sources for transport. Identification of key actors is important since they are responsible for the decisions toward low carbon futures. Hence, the process may provide an important foresight element to decision-making toward sustainable solutions. It contributes to all of the three functions of foresight in policy-making. The proposed approach may act as a checklist for the key issues to be covered in outlining prospective value chains in the road transport context. For instance, all the value chain activities should be discussed to find out the relevant points, from the perspective of actors and value creators in the network. The process integrates methods from different theoretical starting points, namely foresight, multi-level perspective and value chain theories. It also integrates energy and transport systems, and expands the context far into the future. The most challenging parts of outlining future actors are: first, to be able to imagine potential new actors as well as changing roles of the old ones, and second, to create potential new relationships between the actors. This task is especially difficult in a strongly path-dependent situation, as is the biodiesel case. We assume that, for instance, in testing this procedure in the hydrogen technology system the challenge may be slightly easier, because for instance in Finland the whole context is new and hydro technology actors are currently very few in number.

Getting relevant stakeholders to either take part in the workshops or give interviews is demanding. Stakeholders may be too busy to join workshops, which are quite usual methods used in today's research and business world. At least in this road transport context, the issue to be discussed is so large including energy, transport and transition policies, that the discussion would take time if done properly. There may also be confidentiality problems concerning new emerging technologies; stakeholders do not want to be part of public discussions. In the context of renewable energy, it is also challenging to look at different technological platforms, such as biofuels, hydrogen and electricity, simultaneously, in order to find synergies between them. It is not realistic to depend on only one option, but to also look for alternative technological solutions.

Despite the challenges, we believe that prospective value chain analysis helps various system actors to figure out landscape-level constraints, like values and global trends, as well as the needs which guide the society to change the existing regime in order to achieve goals such as renewable energy use. Value chain analysis may act as a tool to increase the knowledge of different actors in making decisions toward low carbon futures.

12.7. Acknowledgments

The research work presented in this chapter was carried out as a part of the TOP-NEST (Nordic Pathways for Sustainable Transport and Energy) research project, under the Nordic Energy Research Sustainable Energy Systems 2050 research funding programme, which is gratefully acknowledged. The authors also wish to thank all TOP-NEST project WP4 researchers: Lars Coenen (Circle Lund), Lars J. Nilsson and Alexandra Nikoleris (Lund University), Antje Klitkou, Dorothy Sutherland and Eric Iversen (NIFU), Juhani Laurikko and Tiina Koljonen (VTT) for their valuable contribution.

12.8. Bibliography

[AHO 12] AHOKANGAS P., MATINMIKKO M., MYLLYKOSKI J. *et al.*, "Future scenarios, ecosystems and business models for cognitive radio systems", *VTT Technology*, vol. 55, p. 54, 2012.

[AUV 12] AUVINEN H., TUOMINEN A., "Safe and secure transport system 2100", *VTT Technology*, vol. 5, 2012.

[DAC 08] DA COSTA O., WARNKE P., CAGNIN C. *et al.*, "The impact of foresight on policy-making: insights from the FORLEARN mutual learning process", *Technology Analysis & Strategic Management*, vol. 20, pp. 369–387, 2008.

[ERI 08] ERIKSSON A., WEBER M., "Adaptive foresight: navigating the complex landscape of policy strategies", *Technological Forecasting & Social Change*, vol. 75, pp. 462–482, 2008.

[FOX 10] FOXON T.J., HAMMOND G.P., PEARSON P.J.G., "Developing transition pathways for a low carbon electricity system in the UK", *Technological Forecasting and Social Change*, vol. 77, pp. 1203–1213, 2010.

[GEE 04] GEELS F.W., "From sectoral systems of innovation to socio-technical systems: insights about dynamics and change from sociology and institutional theory", *Res. Policy,* vol. 33, pp. 897–920, 2004.

[GEE 05] GEELS F.W., "Processes and patterns in transitions and system innovations: refining the coevolutionary multi-level perspective", *Technological Forecasting and Social Change*, vol. 72, pp. 681–696, 2005.

[GEE 07a] GEELS F.W., KEMP R., "Dynamics in socio-technical systems: typology of change processes and contrasting case studies", *Technology in Society*, vol. 29, pp. 441–455, 2007.

[GEE 07b] GEELS F.W., SCHOT J., "Typology of sociotechnical transition pathways", *Res. Policy*, vol. 36, pp. 399–417, 2007.

[GEE 12] GEELS F.W., "A socio-technical analysis of low-carbon transitions: introducing the multi-level perspective into transport studies", *J. Transp. Geography*, vol. 24, pp. 471–482, 2012.

[GEO 06] GEORGHIOU L., KEENAN M., "Evaluation of national foresight activities: assessing rationale, process and impact", *Technological Forecasting & Social Change*, vol. 73, pp. 761–777, 2006.

[GLE 04] GLENN J.C., The Futures Wheel, AC/UNU Millennium Project, Futures Research Methodology [CD-rom], 2004.

[GON 11] GÖNNÖLÄ T., SCAPOLO F., DESRUELLE P. *et al.*, "Foresight tackling societal challenges: impacts and implications on policy-making", *Futures*, vol. 43, pp. 252–264, 2011.

[NOR 94] NORMANN R., RAMIREZ R., *Designing Interactive Strategy: From the Value Chain to the Value Constellation*, John Wiley & Sons, 1994.

[PEP 06] PEPPARD J., RYLANDER A., "From value chain to value network: insights for mobile operators", *European Management Journal*, vol. 24, pp. 128–141, 2006.

[PON 07] PONTE S., Governance in the Value Chain for South African Wine, TRALAC Working Paper 9/2007, Trade Law Centre for Southern Africa, 2007.

[RIS 09] RIISGAARD L., "Global value chains, labour organization and private social standards: lessons from East African cut flower industries", *World Development,* vol. 37, pp. 326–340, 2009.

[STU 08] STURGEON T., BIESEBROECK J.V., GEREFFI G., "Value chains, networks and clusters: reframing the global automotive industry", *Journal of Economic Geography*, vol. 8, pp. 297–321, 2008.

[UNI 05] UNIDO, *Unido Technology Foresight Manual, Volume 1: Organization and methods*, United Nations Industrial Development Organization, Vienna, 2005.

[VAL 11] VALKOKARI K., VALJAKKA T., KANSOLA M., "Towards collaborative smart supply chains – capabilities for business development", *International Journal of Enterprise Network Management*, vol. 4, pp. 380–399, 2011.

[VAN 10] VANBREE B., VERBONG G.P.J., KRAMEr G.J., "A multi-level perspective on the introduction of hydrogen and battery-electric vehicles", *Technol. Forecast. Soc. Change*, vol. 77, pp. 529–540, 2010.

[WEB 09] WEBER M., KUBECZKO K., KAUFMANN A. *et al.*, "Trade-offs between policy impacts of future-oriented analysis: experiences from the innovation policy foresight and strategy process of the city of Vienna", *Technology Analysis & Strategic Management*, vol. 21, pp. 953–969, 2009.

13

The Consequences of Increasing Fuel Prices on Car Travel and Household Budgets

After decades of growth, average car use per adult suddenly leveled off during the 2000s. This break in trend resulted mainly from rising fuel prices, and was reinforced by the recessionary effects of the economic crisis and a beginning of the saturation of driving license and car ownership from the late 2000s, with the ending of a transition of the way of life toward a mobility achieved by car. From a basic model based on motorization rates and an indicator of energetic purchasing power, we show that the stagnation of car travel can be viewed as a lagged reaction to rising fuel prices, the speed of adjustment reflecting the ratio between short-run and long-run adaptability of household behaviors. As a result of increasing fuel prices, a higher proportion of households became financially vulnerable, with a higher risk of falling into energy poverty. This risk has increased most where situations of vulnerability are already concentrated, namely in areas of diffuse urbanization and among low-income households.

13.1. A break in trend in car travel: increasing financial constraints and the proximity of saturation

After decades of growth, average car travel has stagnated during the 2000s in France as in most developed countries. Reasons for such a break in trend are numerous and were discussed in detail in international studies [GOO 13]. The question was generally asked of future levels of car use, suggesting that car travel could have attained a peak to be followed by decline in the future [GOO 10]. In a previous paper [GRI 13], we found that such a stagnation was not limited to the largest conurbations, but could also be seen in low-density areas.

Chapter written by Richard GRIMAL.

Long-term trends in car travel can be viewed as an interaction between changing travel needs, attitudes and preferences, and restrictive constraints. Until the mid-1990s, the combination of increasing demand, a changing way of life and fewer constraints stimulated car travel growth. First, restrictive constraints hindering the expression of latent demand have been declining: higher driving license rates, decreasing motorization costs, increasing incomes and moderate fuel prices. Simultaneously, travel needs have been increasing with a changing way of life: a greater participation of women to the workforce – generating additional home-to-work trips – and the democratization of leisure travel. The changing way of life also modified the expression of travel needs: in particular, the individualization of driving among households, through its extension to women, making car ownership an increasingly individualized good [PAP 08] and resulting in increasing vehicle*kilometers traveled (VMT). The changing spatial environment – an evolution that was partially permitted by the diffusion of car ownership, as it released constraints on localization – was also an important factor of car travel growth. As a result, for given household needs – related, for example, to home-to-work trips – there was an increasing amount of car travel, resulting in greater car dependency [NEW 89]. Combined with the long term improvement of socioeconomic conditions, increasing travel needs can be seen as the main drivers of growth for car travel, leading to a relative decline of car travel inequalities depending on income over time [COL 13a].

However, from the 2000s, average car travel per adult leveled off in all income groups and areas of residence. It declined in urban areas, while slowing down in diffuse areas, where average car travel per adult peaked later – around 2008 – and at a higher level [COL 13a]: the difference in trend could possibly be related to a greater resilience of urban areas, whose inhabitants can turn toward efficient alternatives – transit, walking and cycling – when fuel prices are rising. This break in trend corresponds to a changing context, now characterized by increasing financial constraints and possibly a reversal starting in the way of life, related to changing attitudes and living conditions of the upcoming generations. But it also results from the exhaustion of the diffusion process of driving license and car ownership, which is now seemingly converging toward saturation. Paradoxically, harder economic conditions combined with the effects of the economic crisis since 2008 could have resulted in increasing travel needs and a greater car dependence, an interpretation that is consistent with other findings from the literature [CAS 14]. These changing conditions are shown in Table 13.1, where the period 1977–2013 was divided into seven subperiods, the first five being 6 years each, the last two (2007–2010 and 2011–2013) being shorter to focus on recent changes. Average values per subperiod were reported for some key determinants of car travel:

– the average motorization rate of adults (number of cars divided by the number of adults);

– the average simulated driving license rate. As driving license rates were only available in French Car Fleet Surveys – for the period 1993 to 2010 – they were simulated for French Household Continuous Surveys, using generational rates observed in French Car Fleet Surveys for adults over 30 and applying them to the generational structure of households in French Household Continuous Surveys. Reference generations are themselves defined as 5-year brackets, the first one being born between 1885 and 1890, and the last one after 1990;

– the average activity rate of adults in households with a head of household between 25 and 65. This indicator reflects the increasing participation of women in the workforce;

– an indicator of spatial distribution for active adults – at the place of residence – being the rate of active adults living in low-density areas, corresponding to outer suburbs and communities from the rural space;

– the average income per consumption unit (corrected from inflation, in constant €s 2013), using the Oxford scale for reasons of continuity with ancient data – 1 for the first adult, 0.7 for other adults and 0.5 for children below 14;

– an index of average fuel price (basis 100 in 1974), based on fuel prices per engine type (gas/diesel/GPL) weighted by the structure of the fleet, also corrected from inflation;

– an indicator of average energy purchasing power, defined as the average real income per consumption unit divided by the fuel price index.

	1977–1982	**1983–1988**	**1989–1994**	**1995–2000**	**2001–2006**	**2007–2010**	**2011–2013**
Average motorization rate of adults	44.3%	50.6%	55. 2%	61.3%	66.8%	71.6%	73.0%
Average simulated driving license rate	71.3%	76.7%	81.3%	84.3%	86.3%	88.1%	89.5%
Average activity rate of adults in households with head of household between 25 and 65	62.0%	62.0%	63.8%	63.0%	66.6%	71.3%	77.3%
Rate of active adults living in low-density areas (outer suburbs + rural areas)	36.6%	37.5%	39.9%	38.4%	41.1%	43.7%	43.3%

Average real income per consumption unit (in constant €'s 2013)	15 629	15 753	17 257	18 827	19 425	19 900	19 892
Indicator of average real fuel price (base 100 in 1974)	104	101	90	99	112	126	142
Indicator of energy purchasing power (average real income per consumption unit divided by the real fuel price index)	15 028	15 597	19 174	19 017	17 344	15 794	14 008

Source: French Household Continuous Surveys (1977–1992) and French Car Fleet Surveys (1993–2010)

Table 13.1. *Average values of determinants of car travel by subperiod*

Over a long-term period, constraints hindering access to a high level of automobility have been decreasing. However, from the evolution of financial constraints associated with car travel, three distinct periods can be highlighted. From the mid-1970s to the early 1990s, the average energy purchasing power has been increasing, because of rising incomes and, from the mid-1980s, the oil counter-shock, resulting in lower fuel prices. Then, during the 1990s, the level of energy purchasing power remained fairly stable, as an increasing average income was still balancing the effects of higher fuel prices from the mid-1990s. Finally, it has been falling from the downturn of the millennium, firstly because of continued increasing fuel prices, which attained their historical peak around 2012, and also later on because of a stagnating average income from the economic crisis. Therefore, higher fuel prices have certainly played a decisive role in the break in trend in car travel starting from the early 2000s, which could have been reinforced by the recessionary effects of the so-called "subprime" crisis from the late 2000s.

However, increasing financial constraints for households may also have come from other budget factors, especially housing prices that have been strongly rising from the late 1990s and through the 2000s [ROU 10], with ambiguous effects on subjective wealth perception, depending on whether households have finished paying off their loans or not. As these expenditures are quite inelastic, they may have contributed to the increased pressure on transport budgets and represented an incentive to reduce unnecessary travel among new generations that are still paying off their loans, though there is some evidence from cross-sectional views that trade-offs between transport and housing expenditures could be limited [COU 13].

In parallel, another reason for increasing average car travel per adult was the increasing proportion of adults holding a driving license and having a personal vehicle at their disposal, as these are necessary conditions to driving. For instance, the simulated definitive proportion of driving license holders among adults rose from 71.3 to 89.5% between periods 1977–1982 and 2011–2013. Similarly, the proportion of motorized adults increased from 44.3 to 73.0% over the same period. However, diffusion of driving license and car ownership is now slowing down as levels of disposal are approaching saturation: the driving license rate of adults only increased from three points over a decade, between periods 2001–2006 and 2011–2013, and the motorization rate of adults, after a steady rise until the mid-2000s, also levels off from 2006. This could be an indication that the diffusion process of driving license and car ownership is now nearly approaching its end. This evolution is nonetheless progressive rather than sudden, as no clear break in trend can be seen in driving license and car ownership growth. Therefore, proximity to saturation should be regarded as an additional factor of curbing for car travel rather than the primary cause of the break in trend of the 2000s.

Saturation levels of car ownership, however, remain a matter of discussion. The saturation level of car ownership was first thought of as being a vehicle per household, before moving toward a vehicle per driving adult. However, Collet *et al.* [COL 13b], using logistic curves to simulate long-term thresholds of car ownership, found a lower saturation level of 0.76 cars per adult at fuel price levels of 2010, which could be explained by other reasons limiting the diffusion of car ownership; for instance a localization in dense metropolitan cities characterized either by the lack of need for a car or by strong parking difficulties and congestion.

When it comes to travel needs, we can see that they have always been increasing over time, even during the recent period characterized by stronger financial constraints. For instance, with the increasing participation of women in the workforce, the rate of activity within the 26–65 age bracket has always been increasing. We can even notice an accelerated growth during the 2000s, from 63.0%

in the 1995–2000 period to 77.3% between 2011 and 2013, a paradoxical evolution that could result from increasing financial constraints following the economic crisis, inciting more adults – especially women – to find a job to increase the household income. The proportion of active adults living in low-density areas has also always been increasing over time, resulting in higher trip distances – especially for home-to-work trips – and more car dependency. Here again, this tendency accelerated from the 2000s, which could be a consequence of increasing housing prices all over the 2000s, pushing middle-class households toward the outer suburbs. Therefore, the period starting from the 2000s could be paradoxically a moment of increasing travel needs, stimulated by pressures on housing markets and the economic crisis from the late 2000s.

According to this analysis, the main driving force behind the stagnation was the long-term trend toward increasing fuel prices altogether with a greater volatility from the late 1990s until now. This trend was recently reinforced by other factors, in particular clearer signs of saturation in driving license and car ownership from 2006, and the recession from 2009. Indeed, the transition of the way of life toward motorized travel [WIE 99] will soon be complete with the full replacement of generations born before World War II. As a result of higher financial pressure, households have rationalized car travel in order to maintain their fuel budget shares. Consequences of this break in trend should also be considered, as the downturn of the millennium could be characterized by a contradiction between increasing travel needs, paradoxically resulting from greater economic difficulties, and tougher financial constraints hindering the expression of these needs.

13.2. A lagged reaction to rising fuel prices

13.2.1. *Model specification*

In this section, we implement a basic model of average car travel per adult – VMT traveled – based on motorization rates of adults and energetic purchasing power, as defined in section 13.1. Three areas of residence are considered: core cities, inner suburbs (in the continuity of core cities, within built-up areas surrounding them) and low-density areas, including outer suburbs (outside built-up areas, defined from a percentage of households whose head is working in the built-up area) and rural areas. In order to assess the predictability of the break in trend in car travel, the model is first estimated on a restricted period (1974–1995) preceding it, and then used to forecast average car travel until 2010. Three variants of the model are compared:

– the first one is based solely on motorization rates (a);

– in the second one, average car travel also depends on energy purchasing power (b);

– in the last model, a lagged dependent variable is included to test the assumption of a delay in household's reaction to changing travel costs (c). An example of similar specification – usually referred to as the "partial adjustment" model – was, for instance, implemented in Dargay [DAR 02].

We now display respective model specifications:

$$-\ (a)\ Yt = \beta_0 + \beta_M\, Mt + \varepsilon_t \qquad [13.1]$$

$$-\ (b)\ Yt = \beta_0 + \beta_M\, Mt + \beta_{PP}\, PPt + \varepsilon_t \qquad [13.2]$$

$$-\ (c)\ Yt = \theta\beta_0 + (1\text{-}\theta)\ Yt\text{-}1 + \theta\beta_M\, Mt + \theta\beta_{PP}\, PPt + \theta\varepsilon_{t-1} \qquad [13.3]$$

Y_t Average car travel per adult at time t

Y_{t-1} Average car travel per adult at time $t\text{-}1$

M_t Motorization rate of adults at time t

PP_t Indicator of energy purchasing power a time t

ε_t Disturbance of the model at time t

β_0 Intercept

β_M Parameter associated with motorization rate of adults

β_{PP} Parameter associated with average energy purchasing power

θ Partial adjustment parameter

Instead of direct exogenous and endogenous variables, the model is estimated on their logarithm, in order to allow a direct interpretation of the parameters as elasticities. The formula [13.3] is derived from the following formulas:

$$-\ Yt - Yt\text{-}1 = \theta\ (Yt\text{-}1^* \text{-}\ Yt\text{-}1) \qquad [13.4]$$

$$-\ Yt\text{-}1^* = \beta 0 + \beta_M\, Mt\text{-}1 + \beta_{PP}\, PPt\text{-}1 + \varepsilon_{t-1} \qquad [13.5]$$

Y_{t-1}^* Desired or potential demand at time $t\text{-}1$

θ can be interpreted as the yearly rate of adjustment to changes in external conditions. It is also the ratio between short-run and long-run elasticities. Distinct estimates are produced for every time-series. Quartiles of income are ordered from low-income to high-income households.

13.2.2. *Model estimation*

Parameter estimates are shown in Table 13.2, along with R^2 statistics. Model (a) already describes quite well car travel growth in all areas and income groups, though with a better goodness-of-fit in low-density areas and medium-income groups. The goodness-of-fit is improved through the introduction of an indicator of energy purchasing power, better improvements being obtained for core cities and extreme quartiles of income, where the fluctuations of energy purchasing power have a greater influence on car travel. However, apparent differences between areas for short-run elasticities are also caused by different speeds of adjustment to economic conditions. In the long-term, the sensitivity of car travel to the variations of purchasing power is quite independent from the area – a result in line with other findings from the literature – for instance, Calvet et Marical [MAR 11] – which could be explained by the willingness to maintain fuel budget shares at their current level – but with a different speed of adjustment, which happens to be much faster for low-density areas, with 70% of the long-run adjustment achieved within a single year, against 42% in inner suburbs and 56% in core cities. The speed of adjustment also happens to be slower in the 4th quartile of income and best improvements obtained through the introduction of a lag are for high-income groups. Within other income groups, there is no significant difference between short-run and long-run sensitivity, low-income groups being the most sensitive to financial constraints on the short-run.

Variable interpretations can be given for the value of this parameter. The first one is to consider it as an illustration of the degree of need and emergency for adapting behaviors, either through increasing car travel, favored by good economic conditions, given a strong latent demand, or alternatively through the rationalization of car travel, given the level of transport budgets in a context of rising fuel prices. But as it represents the relation between short-run and long-run elasticities, it could also reflect to some extent the variable strategic ability of households to adapt their behaviors through long-run decisions such as a residential relocation, modal shift or buying more efficient vehicles, beyond short-run emergency measures which can be taken to rationalize travel behaviors and control travel budgets.

	Core cities	Inner suburbs	Low-density areas	1st quartile	2nd quartile	3rd quartile	4th quartile	All included
Motorization alone								
R2	0.92	0.91	0.97	0.93	0.97	0.96	0.92	0.97
βM	0.98	0.95	1.25	1.15	1.20	1.06	1.11	1.09
Motorization and purchasing power, no lagged dependent variable								
R2	0.96	0.93	0.99	0.96	0.98	0.98	0.95	0.99
βM	0.74	0.84	0.94	0.96	1.01	0.82	0.87	0.90
βPP	0.31	0.19	0.40	0.41	0.25	0.31	0.30	0.29
Motorization and purchasing power, with a lagged dependent variable								
R2	0.96	0.94	0.99	0.95	0.99	0.98	0.97	0.99
θ	0.56	0.42	0.70	0.57	0.63	0.80	0.24	0.32
βM	0.70	0.70	0.97	0.89	1.03	0.84	0.65	0.75
βPP	0.36	0.37	0.37	0.45	0.23	0.31	0.63	0.36

Source. French Household Continuous Surveys (1977–1992) and French Car Fleet Surveys (1993–2010)

Table 13.2. *Goodness-of-fit and parameter estimates per quartile of income and zone of residence*

Keeping in mind these interpretations, we can explain differences in sensitivities to financial constraints and speeds of adjustment between areas and income groups. Part of the explanation is related to a variable car dependency in relation with localization. For instance, the gap between short-run and long-run elasticities is greater within conurbations, as inhabitants of these areas face a lower degree of necessity for car travel and the emergency of adapting behaviors in case of rising fuel prices is weak, given low transport budgets. Simultaneously, they have greater opportunities to move toward alternatives to the car on the long-run. On the opposite, inhabitants of low-density areas have a stronger necessity for car travel and fewer opportunities to adapt their behaviors beyond emergency measures, but the incentive to reduce car travel is also stronger in case of rising fuel prices, given higher fuel budget shares. Differences between income groups are also partly explained by effects of localization, as high-income groups often live in densely populated, transit-oriented areas, while middle-class households often live in car-dependent areas. The degree of necessity of additional car travel is also weaker in high-income groups as they are closer to saturation. However, they may have a greater strategic ability on the long-run, for instance through professional and residential mobility, or the removal of superfluous leisure trips, while middle-class households show a lower ability to adjust their behaviors on the long-run, beyond immediate measures.

13.2.3. *Forecasting car travel*

Models, after being estimated, are used to forecast car travel, in particular on the period 1995–2010, characterized by strongly increasing fuel prices. Simulation results are shown in Figure 13.1 for all areas and for the 4th quartile of income. The model based on motorization alone reasonably simulates long-term trends until 1994. Predictions are very good in low-density areas where car travel growth is almost linear and quite insensitive to fuel price fluctuations (the oil shocks from the late 1970s and the oil counter-shock from the mid-1980s). Then, from the mid-1990s to the early 2000s, predictions remain quite good in low-density areas, where the slowdown of car travel partially corresponds to a temporary slowdown in motorization rates. On the contrary, in urban areas, real car travel growth is increasing faster than predicted, in particular in core cities, despite an important slowdown of motorization rates. But the most important disconnection between car travel and motorization begins from the 2000s, where despite a strong increase in motorization rates in all areas, average car travel stagnates or is even decreasing. Obviously, this model neither reproduces fluctuations of car travel related to financial constraints, which happen to be of greater amplitude in urban areas than in low-density areas, nor explains the break in trend in behaviors starting from the 2000s.

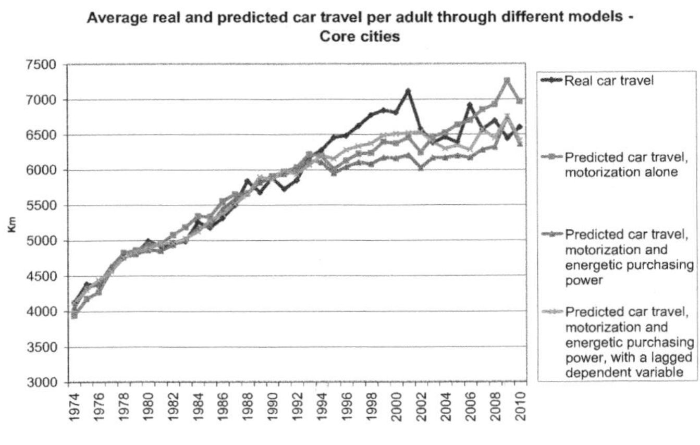

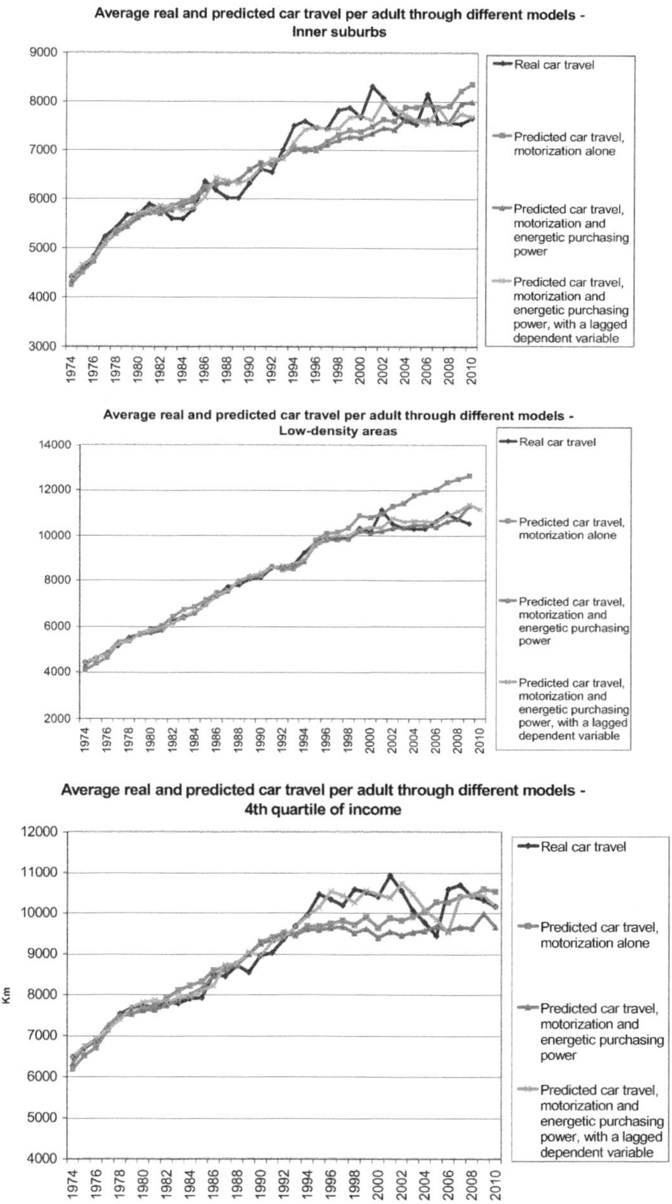

a) core cities; b) inner suburbs; c) low-density areas d) 4th quartile

Figure 13.1. *Average real and simulated car travel through models (1), (2) and (3). For a color version of the figure, see www.iste.co.uk/jacob/energy.zip*

The model taking into account energetic purchasing power is anticipating a slowdown in car travel growth in all areas. Predictions become almost perfect in low-density areas, where the date of the breakdown perfectly corresponds to the reality, as well as average values of car travel. In core cities, the model predicts a slowdown from 1993, while the reality corresponds to a break in trend from 2001: it looks like behaviors have not changed continuously in reaction to increasing fuel prices, but rather after the peak of 2000 which must have represented some psychological threshold. The same phenomenon is observable within inner suburbs. The introduction of a lag does not have an important effect on predictions relative to low-density areas. This result is consistent with what we found about the speed of adjustment to external conditions, which is almost instantaneous in those areas. On the contrary, within conurbations, predictions are significantly improved through the introduction of a lag. Within core cities, predictions become almost perfect from 2002, with a predicted date of the break in trend now almost corresponding to the truth. Within inner suburbs, predictions are almost perfect from the mid-1990s, with a slowdown followed by a slight decline from 2002. Finally, predictions relative to high-income groups project some light on phenomena of saturation. The break-in-trend happens sooner, from 1995. From then on, car travel has known important fluctuations. The model based on motorization alone already correctly anticipates the stagnation of car travel from the mid-1990s, illustrating a greater proximity to saturation. Predictions are not much improved through the inclusion of an indicator of purchasing power, but a very substantial improvement is brought through the introduction of a lag, the period starting from the mid-1990s being now almost perfectly described.

To conclude this section, the assumption of a lagged reaction to rising fuel prices seems to hold: almost instantaneous in low-density areas, this reaction was rather progressive in urban areas, suggesting that while this reaction may be temporary and limited in low-density areas, it may recover strategic long-run adjustments of behaviors within conurbations. In core cities and among high-income groups, it was amplified by quite a long pause in motorization rates, which may now have become close to saturation.

13.3. The rise of fuel poverty

13.3.1. *Data and methodology*

In this section, we assess the effect of rising fuel prices on energy poverty, depending on income and zone of residence. A normative approach is designed to determine a threshold of vulnerability, inspired by Nicolas *et al.* [NIC 12]. However, in this study, the calculation of this threshold is based only on fuel expenditures rather than on total travel costs. For reasons of data availability, fuel budget shares were calculated only on French Car Fleet Surveys, from 1993 to 2010. Household fuel consumptions are obtained as the sum of fuel consumptions of all household vehicles, which are themselves calculated from unitary vehicle consumptions (in l/100 km) combined with travelled distances reported for every vehicle. Household fuel budgets are then estimated as the product of vehicle fuel consumptions per fuel prices by type of motorization (gas/diesel/GPL). Fuel budget shares are finally obtained by dividing household fuel budgets by household incomes. Indexes of inflation are used to take into account the erosion of fuel budgets in real terms.

13.3.2. *Impacts of rising fuel prices on household budgets and the rationalization of car travel*

The real dynamic of car travel is then compared with a theoretical view into which average car travel per adult is supposed to derive from several factors:

– the dynamic of fuel prices in current €s;

– changing types of motorization, mainly the gradual replacement of gas engines by diesel engines;

– the increasing energetic efficiency of vehicles, for a given type of motorization, expressed in km/l;

– inflation;

– the standard of living, expressed in constant €s per consumption unit.

Theoretical average car travel for a given year is calculated under the assumption of a strategy into which households would try to maintain their fuel budget share at a constant. This strategy looks quite realistic when considering average fuel budget shares, which remained pretty stable at a level roughly 3–4% of from the mid-1980s, despite intermediary fluctuations related to the volatility of fuel prices, according to National Transport Accounts [CGD 14]. Under this assumption, factors mentioned above can be considered as theoretical growth components of car travel, which are shown in Figure 13.2(a). Technical and economic factors, income included, represent a potential growth of 3.1% between 1993 and 2010, a result not far from the real evolution which was of 7%. Therefore, the assumption of a fuel-budget-share-controlling strategy holds quite well when considering long-run trends. However, fuel price fluctuations were very important during this period, with two peaks in 2000 and 2006 and another even higher one in 2012, corresponding each time to the threshold of an average 4.6–4.8% budget share. As strategies to rationalize car travel are starting from these peaks, we can assume this threshold to stand either for objective financial constraints which have caused a feeling of greater vulnerability among households, or for some psychological barrier in the willingness-to-pay of car users. We can also see from Figure 13.2(b) that budget shares have followed the fluctuations of fuel prices, meaning that households did not manage to control their budget shares, which have for instance increased from 3.6 to 4.9% between 2003 and 2006. Indeed, fuel price fluctuations were too fast to allow households to adapt on the short-run, only allowing them to limit the increase in fuel budget shares. Over the whole decade, the benefit obtained through car travel rationalization can be estimated to a – 0.46 pts in fuel budget shares.

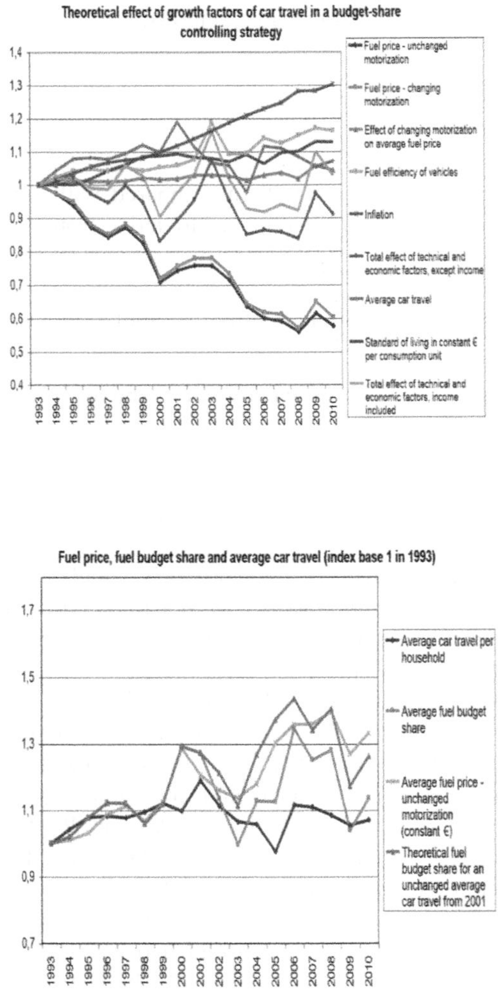

Figure 13.2. *a) Theoretical effect of growth factors on car travel in a budget-share controlling strategy and b) fuel price, fuel budget share and average car travel (index base 1 in 1993). For a color version of the figure, see www.iste.co.uk/jacob/energy.zip*

13.3.3. *An increasing vulnerability, especially in areas of diffuse urbanization and among low-income households*

Finally, we examine the effects of increasing fuel prices on household vulnerability, and whether these effects have been homogeneous or not between zones of residence and income groups. In the first definition, vulnerability is supposed to depend only on budget shares. Inspired by Nicolas *et al.* [NIC 12], a threshold of vulnerability was empirically determined at 7%, corresponding to a drop out in the distribution of budget shares. Using this threshold, approximately corresponding to the double average share, 18.7% of households can be considered vulnerable. In the second definition, budget shares are crossed with a poverty threshold – which is estimated as half median income – as a given budget share does not recover the same risk of falling into fuel poverty depending on household income:

– if fuel budget share is inferior to 2%, the household is not considered vulnerable, whatever its income;

– if it stands between 2 and 10%, vulnerability depends on income. The household is then considered vulnerable only if its remaining income after deduction of fuel expenditures is inferior to the poverty threshold;

– if it is higher than 10%, the household is considered vulnerable, whatever its income.

The second definition is more restrictive and focused on low-income households which are identified as representing a greater risk of energetic poverty, with lower rates of vulnerability as a consequence. It could be used to prioritize public help. Of course, it is also open to criticism, as it is omitting households that have limited fuel expenditures because they do not have access to a car, either for financial reasons or because they never had a driving license, which could nonetheless be socially and economically threatened through this deficiency [POC 07]. However, these households will be considered as already excluded from mobility, vulnerability being defined as the risk of falling into a precarious mobility or difficult financial situation consecutive to high travel costs.

Depending on fuel prices, the proportion of vulnerable households has fluctuated from 10 points, with two peaks at 27.2 and 26.4%, respectively, in 2000 and 2006, compared to 16.7% in 1993 and 18.3% 2003. Though temporary, the impact of fuel price volatility on energetic poverty was thus far from being negligible. With the more restricted definition, it fluctuated between 10 and 15%. Fuel budget shares have always been higher in low-density areas – between 5 and 6.5% – than in conurbations, where they fluctuated between 3 and 4.5%, a result which is consistent with the literature [ORF 99], and explained by higher transport costs combined with a lower income on average. It is also in low-density areas, which were already the most vulnerable areas in 1993, that the proportion of vulnerable households has increased the most, for instance from 24 to 37.4% between 1993 and 2001 – according to the first definition of vulnerability. In urban areas, it rose from 12.7% to more than 20% during the same period. Therefore, though vulnerability has increased everywhere, the gap between urban areas and low-density areas increased.

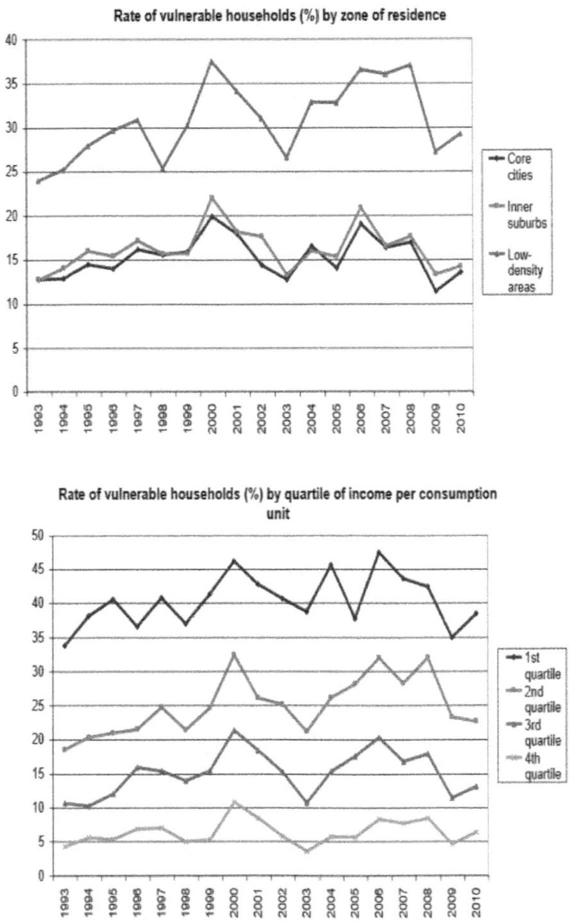

Figure 13.3. *Rate of vulnerability of households according to the unrestricted definition a) by zone of residence and b) by quartile of income. For a color version of the figure, see www.iste.co.uk/jacob/energy.zip*

The impact of rising fuel prices also differed depending on income. Obviously, fuel budget shares have always been much higher among low-income households: from 6.4 to 2.8% in 1993 from the 1st to the 4th quartile of income, and from 7.6 to 2.9% in 2010. Besides, the difference in budget shares between income groups is increasing in periods of higher fuel prices, as fuel shares have remained quite stable in the 4th income group – fluctuating between 2.8 and 3.4% – while fluctuating between 6.4 and 9.3% in the 1st quartile. Reasons for a lower volatility among high-income households are a higher income denominator, as well as a higher income growth, compensating the increase in fuel prices. As a result, the rate of vulnerability was much more volatile among low-income households, from 33.8 to 47.4% – with the large definition – than among high-income households – from 4.2 to 10.8%.

Therefore, the risk of falling into energy poverty increased most where it was already the highest, within low-density areas and among low-income households, as is illustrated by Figures 13.3(a) and 13.3(b). However, public policies aiming at preventing or fighting energy poverty should not be limited to areas of diffuse urbanization: given urbanization rates, almost half of vulnerable households live within conurbations, distributed equally between core cities and inner suburbs. Public help should in fact primarily target low-income households, where most situations of vulnerability are concentrated: the 1st quartile gathers 66% of vulnerable households, among which 34.3% are living in low-density areas, 17.3% in core cities and 14.3% in inner suburbs. Another important group would be households from the 2nd quartile living in low-density areas, constituting 11.5% of vulnerable households.

However, this approach does not take into account the probably variable resilience of populations and territories, depending on transport alternatives and land use patterns, which may finally provide a relative benefit to dense, functionally mixed and central areas. We can expect this better resilience to reinforce social competition to access core cities and their close suburbs. High-income households will probably be in a better position to adapt to this transition through residential mobility and the acquisition of more efficient vehicles. Considering policy implications, a central issue will be the ability of public decision makers to offer additional transit supply, pedestrian and cycling facilities and to manage the overcrowding of public transport which has already become a reality in several big cities. Adaptations of transport supply need time, while the effects of an energy shock can be very sudden. In this scenario, the pressure on fuel budgets will represent a strong incentive for the renewal of the car fleet toward more efficient vehicles, stimulating the research and technical progress. But, on the other hand, substantial social difficulties will rise and the quality of life could be degraded.

13.4. Bibliography

[CAL 11] CALVET L., MARICAL F., "Consommation de carburant : effet des prix à court et à long terme par type de population", *Economie et Statistiques*, vol. 446, pp. 25–44, 2011.

[CAS 14] CASCAJO R., DIAZ OLVERA L., MONFORT SALVADOR V. *et al.*, "Situation de crise et mobilité quotidienne. Le cas de l'Espagne", *51e colloque de l'Association de Science Régionale de Langue Française (ASRDLF)*, Marne-la-Vallée University, p. 213, France, 2014.

[COL 13a] COLLET R., MADRE J.-L., HIVERT L., "Vers quels plafonds pour la motorisation et l'usage de la voiture?", *Economie et Statistiques*, vol. 457–458, pp. 129–139, 2013.

[COL 13b] COULOMBEL N., LEURENT F., "Les ménages arbitrent-ils entre coût du logement et coût du transport? Une réponse dans le cas francilien", *Economie et Statistiques*, vol. 457–458, pp. 53–75, 2013.

[COM 14] COMMISSARIAT GÉNÉRAL AU DÉVELOPPEMENT DURABLE, 51e rapport à la Commission des Comptes Transport de la Nation (CCTN), *Références*, July 2014.

[DAR 02] DARGAY J., "Determinants of car ownership in rural and urban areas: a pseudo-panel analysis", *Transportation Research Part E: Logistics and Transportation Review*, vol. 38, no. 5, pp. 351–366, 2002.

[GOO 10] GOODWIN Ph., "Peak car – series of five articles", *Local Transport Today*, London, June 2010–June 2011.

[GOO 13] GOODWIN PH., VAN DENDER K., "Peak car – themes and issues", *Transport Reviews, special issue on peak car*, vol. 33, no. 3, pp. 243–254.

[GRI 13] GRIMAL R., COLLET J.L., MADRE J.L., "Is the stagnation of individual car travel a general phenomenon in France? A time-series analysis by zone of residence and standard of living", *Transport Reviews, special issue on peak car*, vol. 33, no. 3, pp. 291–309, 2013.

[HIV 12] HIVERT L., MADRE J.-L., "L'adaptation des ménages aux hausses du prix des carburants", in *La mobilité urbaine des années 2000*, éditions CERTU, Troisième partie "Mobilité, société et économie", pp. 76–77, 2012.

[KUH 12] KUHNIMOF T., ARMOOGUM J., BUEHLER R. *et al.*, "Men shape a downward trend in car use among young adults – Evidence from six industrialized countries", *Transport Reviews*, vol. 32, no. 6, pp. 761–779, 2012.

[NEW 89] NEWMAN P.W.G., KENWORTHY J.R., "Cities and automobile dependence. An international sourcebook", *Complete Gower Technical*, Gower Publishing, p. 388, 1989.

[NIC 12] NICOLAS J.-P., VERRY D., VANCO F., "Utiliser la voiture pour se déplacer au quotidien: taux d'effort et vulnérabilité des ménages face à l'augmentation du prix des carburants", *Revue d'Economie Régionale et Urbaine*, vol. 1, pp. 5–30, 2012.

[ORF 99] ORFEUIL J.P., POLACCHINI A., "Les dépenses des ménages franciliens pour le logement et les transports", *Recherche Transports Sécurité*, vol. 63, pp. 31–46, 1999.

[PAP 08] PAPON F., HIVERT L., "Adulterous behaviour within the car-owner couple: some analyses from French panel data on car rental and car sharing within households", *IATSS Research*, vol. 32, no. 2, pp. 6–15, 2008.

[POC 07] POCHET P., NICOLAS J.-P., "Social inequalities in access to cars and daily mobility: the French experience", *International Conference on Public Transport and Urban Citizenship: Making Dublin the capital of Ireland*, The Policy Institute, Trinity College, Dublin, 2007.

[ROU 10] ROUGERIE C., FRIGGIT J., Prix des logements anciens – Dans les années 2000, ils ont augmenté beaucoup plus vite que les loyers et les revenus, *INSEE Première*, 1297, 2010.

[WIE 99] WIEL M., *La Transition urbaine ou le passage de la ville pédestre à la ville motorisée*, Mardaga, 1999.

14

The Development of an Innovative On-board CNG Storage System for Methane-Fuelled Cars Conducted Within the FP7 EU Project 'InGAS'

This chapter outlines part of the activities conducted within the context of the InGAS "Integrated GAS powertrain" (www.ingas-eu.org) collaborative project conducted within the context of the EU 7th Framework Programme and coordinated by Centro Ricerche Fiat. The overall scope of this pre-competitive research project was the development of the technologies for light-duty vehicles with methane-fuelled powertrains and specific after-treatment systems, capable of achieving a 10% higher fuel conversion efficiency than that of a corresponding diesel engine technology (with respect to the state-of-the-art at the outset of the project) while complying with an emission level beyond Euro 6. Within this general context, the specific aim of the subproject, the results of which are outlined in this chapter, was the design and physical realization of an advanced, on-board CNG storage system for methane-fuelled cars, which included the development of specific components such as advanced, light-weight and affordable Type IV pressure vessels, novel "in-tank" valves for the vessels and an innovative electronic pressure regulator.

14.1. Development of innovative, low-cost Type IV CNG vessels

One of the main objectives of the project was the development, realization and testing of innovative high-pressure vessels (working pressure of 200 bar) which can

Chapter written by David Mark STORER, Giorgio MENZATO, Michael KLESCHINSKI, Volker STRUBEL, Bernhard KIENER, Jerzy KALETA, Pawel GASIOR, Wojciech BLAZEJEWSKI, Georg MAIR and Christian GREGOR.

offer both low-weight (i.e. significantly less than conventional steel Type I vessels) and affordability (with respect to Type IV vessels made of carbon-fiber) (Figure 14.1).

The aim was to achieve an optimal design solution based on a thermoset glass and carbon hybrid composite structure to result in both a lightweight and reasonably low cost product. During the project, more than 50 design solutions were investigated by *xperion*, and a total of 125 cylinders were produced; most of these vessels were needed for the extensive series of tests, conducted within the project according to the regulations ECE-R110 by the partners *BAM* and *WRUT*, specialists in CNG-vessel testing.

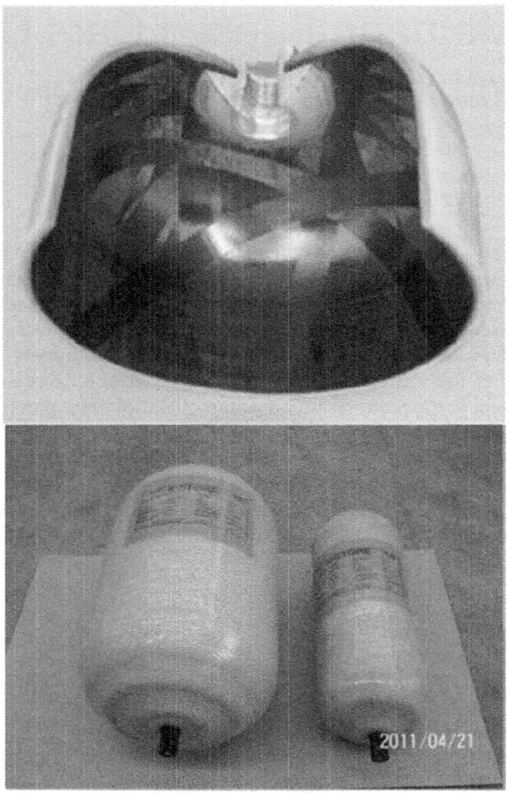

Figure 14.1. *The new lightweight, low-cost hybrid-composite vessels CNG vessels developed and tested during the project. For a color version of the figure, see www.iste.co.uk/jacob/energy.zip*

Of the various tests conducted (Figure 14.2), it was possible to conclude that the most safety critical was the vertical drop test.

In general, the vessels had been optimized for weight rather than robustness; nevertheless, by appropriately redesigning the area around the neck of the vessels, it was possible to respect all of the ECE-R110 tests without the need to add a shock absorbing protector onto the domes of the cylinder. The final design was deemed to offer the best compromise and balance between weight saving, cost and capacity.

Figure 14.2. *Extensive testing of the new vessels was performed during the project (clockwise from top left: burst tests, fire engulfment, bullet penetration and acoustic-emission testing) . For a color version of the figure, see www.iste.co.uk/jacob/energy.zip*

When the final composition of the vessel construction had been identified, and the ECE-R110 tests completed, two different vessel dimensions were designed and manufactured with, respectively, 21 and 60.5 l water capacity for integration within the demonstrator vehicle.

As regards the issue of mass production of the vessels in an automotive high-volume manufacturing process, which was also investigated during the project, the cycle time of a wet winding process (more than 30 min for an automotive CNG vessel) was considered to be rather long in comparison to other regular manufacturing processes in the automotive industry.

To overcome this hurdle, the recommendation would be to split the production chain into several different production cells, which can be combined into high-volume production lines. In particular, several production cells for wet winding can feed one curing oven. By multiplying the production cells, a production volume of up to 500.000 units per annum would be possible, involving a reasonable investment and by applying well-known, robust production technologies.

One main advantage of this high-volume manufacturing concept is that several vessel dimensions can be produced in parallel. Furthermore, by introducing modularity into the vessel manufacturing process, redundancy can also be achieved for maintenance or in case of a breakdown of one production cell, while enabling a high degree of flexibility to cater for varying customer demands.

14.2. Development of advanced components for the on-board CNG storage and supply system

During the project, *Ventrex* developed a new generation of an electronic pressure regulator in order to achieve functional optimization and size reduction (Figure 14.3).

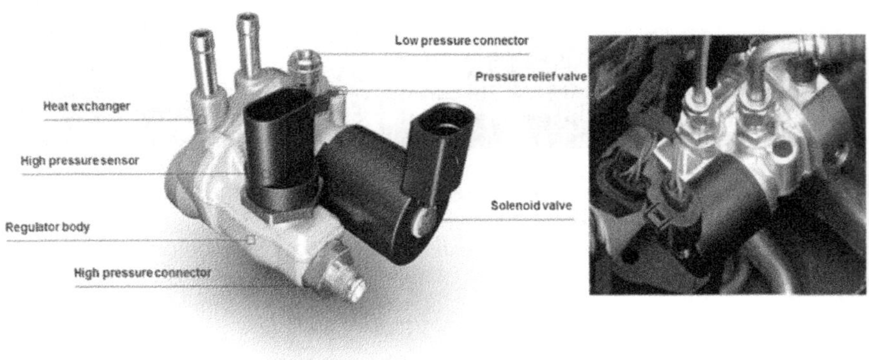

Figure 14.3. *The electronic pressure regulator developed and tested during the project. For a color version of the figure, see www.iste.co.uk/jacob/energy.zip*

In parallel with the design work, many validation tests were performed in the laboratory before incorporating the pressure regulator and control unit in the demonstrator vehicle.

An advanced in-tank shut-off valve was also developed by *Ventrex* with a series of different concepts tested in order to produce an ECE-R110 conforming valve which could be installed into the pressure vessel (Figure 14.4).

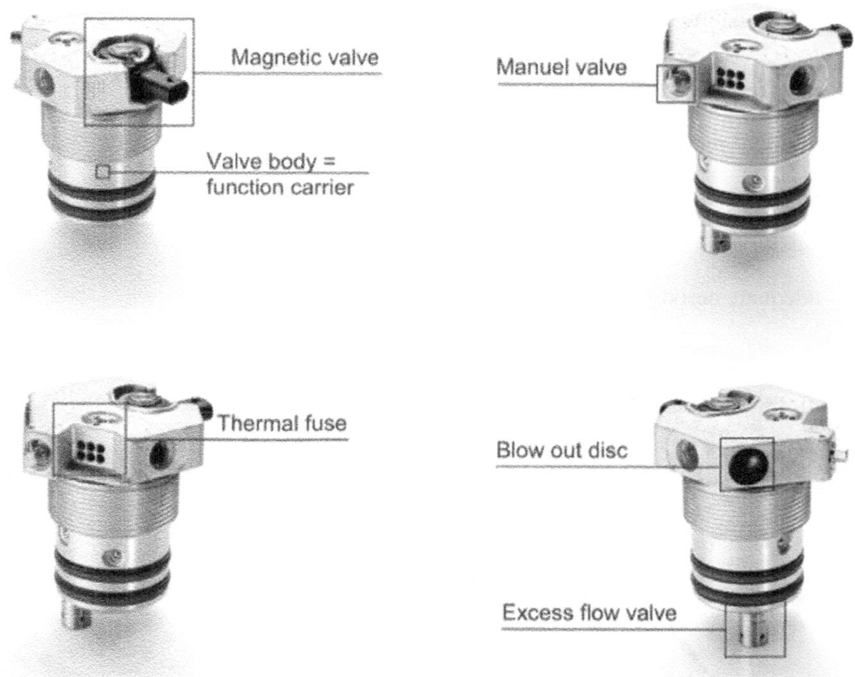

Figure 14.4. *The advanced in-tank valve developed and tested during the project*

Following the successful completion of a series of functional and misuse tests, a set of valves was supplied for mounting in the vessels to be installed on the test vehicle. Having installed these newly-developed components on the test vehicles, a series of functional tests was performed on the vehicle to test and verify the highly satisfactory performance of these components, which were found to meet the specifications and project targets completely.

14.3. Design of CNG storage module

During the project, an innovative CNG storage module for an economy-segment car was designed and validated using numerical simulation: this CNG storage module (Figure 14.5) integrated both the advanced CNG cylinders developed within the project and a redesigned rear suspension system, enabling the technical specifications defined at the outset of the project to be respected, in particular:

– tank capacity of approximately 130 l;

– minimal impact on the space within the passenger compartment and in particular no reduction in the habitability on the rear passengers;

– minimal reduction (–10%) of the space in the luggage compartment;

– minimized the variations of the vehicle settings so as to reduce to a minimum the impact of the vehicle performance characteristics, also with respect to overall efficiency;

– adequate aerodynamic performance.

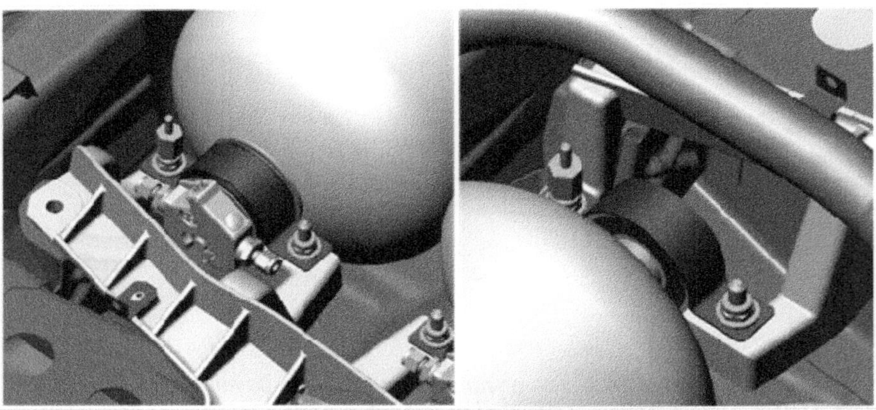

Figure 14.5. *Details of the neck mounting system of the vessels in the advanced CNG storage module developed during the project*

On this basis, the solution proposed should offer adequate CNG capacity to achieve a driving range in the order of 500 km with less additional weight when compared to other more traditional solutions (e.g. Type I CNG vessels) while providing good intrinsic safety and a high level of compatibility with the automotive assembly process (as described subsequently).

14.4. Integration of CNG storage module in vehicle

In order to be able to integrate the advanced CNG storage module developed in the project, the under-body of the vehicle was completely redesigned in order to accommodate the module while respecting the following specifications:

– develop the under-body for the integration of the advanced CNG storage module;

– ensure a high level of intrinsic safety of the vehicle;

– enable straightforward assembly/disassembly of the module on the vehicle;

– limit the required variations on the weight, including additional weight, so as to reduce to a minimum the impact of the vehicle performance characteristics.

The realization of the advanced CNG storage module and its successful installation on the demonstrator vehicle (Figure 14.6) served to validate the solution in terms of technical feasibility and technological viability, respecting fully the specifications and overall requirements of the project.

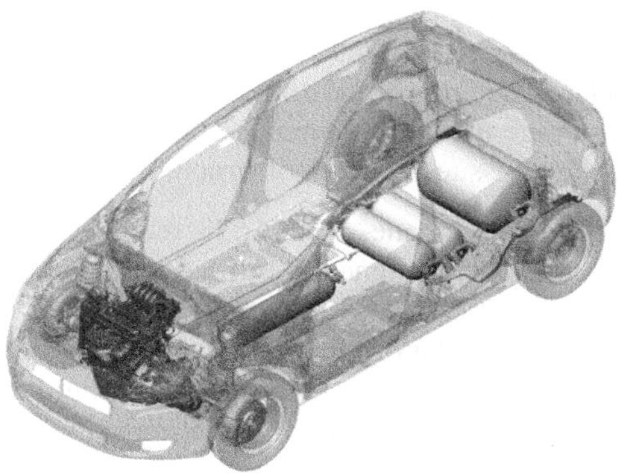

Figure 14.6. *Image of the advanced CNG storage module integrated into the FIAT Grande Punto vehicle during the project*

14.5. Performance validation

A comprehensive virtual performance assessment was conducted to validate the proposed CNG storage solution in terms of stiffness, crashworthiness and vibro-acoustic performance:

– Using a detailed FE model, it was possible to verify that both the bending and torsional stiffness, and the first bending and torsional eigen-frequencies of the Body-in-White, fulfilled the predefined specifications. Furthermore, by conducting static and dynamic simulations with and without the vessels, it was ascertained that the presence of the vessels does not influence to any significant extent the salient static and dynamic characteristics of the vehicle body due to the presence of bushings in the connection between the vessels and the vehicle.

– A detailed numerical analysis of the vehicle complete with the advanced storage system, which was further supported by an experimental analysis on the demonstrator vehicle, revealed that the dynamic behaviour of the vessels does not adversely influence the dynamics of the vehicle when excited by road-surface-induced oscillations due to the filtering effect through the attachments of the vessels to the vehicle.

– The detailed analysis of the structural resistance of the storage module served to validate the selection of the aluminum alloy (AlSi9MgMn) as an appropriate material for the frames of the CNG storage module. However, the results of the analysis also indicated that before freezing such a design solution for mass production, further detailed optimization of the vessel supporting brackets may be appropriate to reinforce these potentially critical regions.

– The crashworthiness analysis served to validate the technical feasibility of the modular advanced CNG storage system. With respect to the European frontal, lateral and rear impact standards, the numerical simulations demonstrated that the level of forces acting on the storage module in general, and the vessels in particular, would not give rise to any specific issues in terms of structural integrity.

– Instead the more severe Japanese TRIAS33 rear impact test standard, which Fiat vehicles are conventionally designed to fulfill, would give rise to relatively high levels of forces acting specifically on the larger, rear vessel due to the relatively confined deformation space in the original configuration. Correspondingly, a series of design variations was identified and analyzed, including the repositioning of the rear exhaust muffler in order to avoid any possible impact with the rear vessels, and a relatively small design variation in the rear floor section of the vehicle to avoid contact in this area during a rear impact crash scenario.

14.6. Considerations regarding production issues

Overall, the results of the virtual performance analysis successfully demonstrated the validity of the advanced CNG module concept while highlighting areas for final design optimization, with a view to proposing this solution for mass production of a vehicle with this layout.

In terms of the respective investments and variable costs related to manufacturing and the production process, an analysis was performed focusing on the CNG system components, the under-body transformation and the assembly of the vehicle. Since projected production volumes affect not only proportional investment costs, but also the selection of the most appropriate and cost-effective production technologies, comparisons were made with respect to a range of typical production volumes which, depending on the extent of market penetration, might be expected for a vehicle of this segment a medium level of between 20,000 and 50,000 vehicles per annum and a high level of 100,000 vehicles per annum.

From the results of this comparison, it would appear that the option of using neck mounting as opposed to independent strap mounting would be more expensive overall. However, it is important to note that this analysis has been conducted specifically with respect to the economy-segment vehicle selected for this activity and its transformation in order to achieve the predefined CNG storage capacity. In practical terms, for this vehicle, the independent strap mounting would not be feasible since more space would be required for the single production line to enable the independent mounting of three vessels, which is typically unavailable in an existing production plant; instead, the modular configuration using neck-mounted vessels enables the possibility of secondary lines running in parallel with the main production line. This represents an important practical advantage which tends to offset the slightly higher costs estimated for this specific case.

14.7. Conclusions

Within the InGAS subproject which specifically addressed the development of an innovative, on-board CNG storage system for methane-fuelled cars, it was possible to achieve all the main specified performance targets in terms of:

– proposing a new modular design and construction concepts for advanced CNG vehicles;

– enabling extended vehicle range by virtue of the advanced, highly-integrated CNG storage module;

– developing and validating advanced CNG storage components (valves and electronic pressure regulator);

– ensuring full compatibility of the components of the system;

– assessing the ECE-R110 regulations via scientific validation.

The additional cost of current-production CNG vehicles compared to the gasoline/diesel derivate results mainly from costs of the storage components and from higher vehicle assembly costs: instead, the integration into the assembly lines of the conventional gasoline and diesel derivatives should be feasible by virtue of the modular CNG storage system concepts, which can be preassembled and tested outside the main vehicle assembly lines.

Overall, the project has delivered technical solutions of considerable interest and relevance from an industrial perspective being both innovative and economically viable, particularly regarding:

– the advanced CNG thermoset vessel design based on a glass and carbon hybrid composite structure developed, validated and realized within the project, and tested extensively, which offers a technically valid and sustainable solution for the innovative lightweight, low-cost composite vessels for the on-board storage of CNG;

– the new generation of an electronic pressure regulator and advanced in-tank shut off valves designed and realized which offer significant advantages in terms of dimensions, weight, number of components and cost with respect to previous versions of these innovative, high-precision components;

– the advanced CNG storage module concept using hybrid composite Type IV vessels attached to the vehicle via neck mounting developed in collaboration.

In general, these innovations could be exploited without significant further development for next-generation, mass-produced, CNG vehicles, thus overcoming the main obstacles for mass-market entry of CNG vehicles related to the storage system, namely:

– the driving range, which is typically limited to less than 350 km in the CNG mode;

– significantly higher vehicle weight of CNG vehicles compared with gasoline or diesel vehicles, due to the weight of the CNG storage system;

– inferior vehicle performance, particularly in terms of handling, acceleration and braking, due to the higher mass of the CNG storage system;

– higher vehicle costs and hence longer payback time for the higher initial investment when purchasing the vehicle.

Overall, the developments made within this project, particularly with regard to the development of a low-cost, lightweight CNG storage system, could serve to enable the significant benefits of methane-fuelled vehicles in terms of reduced CO_2 emissions and environmental impact to be exploited more over the coming years through increased market penetration.

14.8. Acknowledgments

The activities described in this chapter were undertaken within the context of the EU FP7 Integrated Project "InGAS: Integrated GAS powertrain – Low emissions, CO_2 optimized and efficient CNG engines for passenger cars (PC) and light-duty vehicles (LDV)" (SST-2007-RTD) which started 1st October 2008 and was coordinated by Massimo Ferrera of *CRF*.

The authors gratefully acknowledge the support of the EU Commission in this regard and in particular the advice and assistance of the EC Project Officer Maurizio Maggiore.

15

Sustainability Assessment of Infrastructure Elements with Integrated Energy Harvesting Technologies

The possibility of integrating energy harvesting devices into the bridge/tunnel structures along Coastal Highway Route E39 has been investigated in a feasibility study by the Norwegian Public Roads Administration (NPRA). The main advantage of integrating energy conversion devices in a structural element is the reduction of costs compared to stand-alone devices. The construction could be used as a foundation, a mooring point and provide a dry environment for electrical devices. Easy access to the production site could also reduce the cost for installation, operation and maintenance.

Two important challenges related to harvesting renewable energy by infrastructure elements, without concerning about the energy source, are to store it or feed the energy to the grid. In the second case, tailoring generation to demand is of critical importance. Tasks such as supply and demand management, for instance, peak hour management, what kind of storage should be used – electrical or thermal – need be solved. Furthermore, integrating energy production devices in a structure might cause negative environmental impacts and affect the life expectancy and maintenance costs of such structures. The potential environmental impacts associated with renewable technologies are the consequences for bird life or marine fauna at the fjord crossing locations, as well as noise and visual impact. Thus, a sustainability assessment should be performed in order to quantify the ecological, economical and societal impacts of the suggested alternatives.

15.1. Background

Road transport is crucial for the territorial cohesion within the European Union and to open new business opportunities abroad. In terms of infrastructure (with more

---

Chapter written by Bijan ADL-ZARRABI, Mohammad HOSEINI, York OSTERMEYER and Holger WALLBAUM.

than 5,000,000 km of paved roads) and transport of both goods and passengers, it could be considered as the most important transport network in EU-27.

The "EU-2020 strategy" [EUR 10a] includes the "Flagship Initiative Resource efficient Europe" [EUR 11a], where European Commission presents proposals aiming at cleaner, more efficient and more sustainable transport through the adoption of measures such as research and innovation, setting current standards and developing the necessary infrastructure support as well as regulatory measures such as pricing. The White Paper on Transport [EUR 11b], takes into account major policy initiatives for a competitive and resource efficient transport system under sustainable developments. Furthermore, European Road Transport Research Advisory Group (ERTRAC) sets out the following ambition: "Towards a 50% more efficient road transport system by 2030" [ERT 10].

However, these targets have challenges and threats to be tackled such as the deterioration of existing infrastructures, their susceptibility to climate change and the expected increase in road freight transport (between 40 and 50% compared to today) [ERT 09], maintaining competitive strength of EU transportation while addressing environmental goals, without forgetting the limited financing due to the economic stringency. Therefore, there is an ambitious goal to fulfill these needs but it is clear that it should be achieved in the most cost-effective way.

Sustainable construction has been launched as a lead market initiative. Recommendations stated say that sustainable construction as a lead market should address mainly legislation, standardization, labeling and public procurement [EUR 07]. Regarding buildings, certification systems such as the European BREEAM and DG 2010/31/EC [EUR 10b]: "Performance of buildings" and LEED (Leadership in Energy and Environmental Design) in the US could be considered as recognized measures of sustainability performance. Similar approaches for road infrastructures could be The Greenroads Rating System in the US to promote sustainability best practices [GRE 11], the Environmental Product Declaration (EPD©) for road based on lifecycle analysis (LCA) currently in progress, and other local, but relevant approaches like the French methodology: ROUTE-DURABLE for "sustainable roads" or the Dutch Environmental LCA tool: DuboCalc.

Current best practices for road products and infrastructures such as the widely implemented CE Marking established by the EU Construction Products Regulation No 305/2011 take into consideration a set of technical requirements for their established uses, information on the content of hazardous, and in terms of environmental impact it is only remarked that all construction materials should not have an exceedingly high impact over their entire lifecycle, on the environmental quality or on the climate during their construction, use and demolition.

A complete environmental approach including socioeconomic aspects and new constraints such as the resilience to climate change and the implementation of freight corridors and associated loads and of promising technologies such as energy harvesting, greener drainage systems among others is not even mentioned by current best practices for road products and infrastructures, therefore there is a clear gap for improvement and harmonization to cover all the mentioned aspects.

The selection and assessment of the key performance indicators suitable for these analysis of road infrastructure is being addressed by recent or ongoing European research projects, but each one is focused on different parameters and indicators: the ERA-NET EVITA: environmental indicators; SBAKPI socioeconomic aspects, COST Action 354 technical performance including safety) or different parts of roads: COST Action 354 pavements; PANTURA bridge. The conclusion is again that a holistic and harmonized approach is still necessary.

This is the right moment to integrate all this generated knowledge (due in large part to the efforts of the EC) and create new European Directives and a widely recognized tool to move from the research to the expected implementation of a more sustainable road transport network.

The development of a harmonized and holistic assessment tool for road infrastructures that include environmental, technical and socioeconomic performance proposed within this chapter could be suitable for motivating green and smart procurement, thus strengthening eco-innovations and cost-effectiveness throughout Europe with the final goal of moving to safer, greener and cost-effective road infrastructures. Lifecycle engineering (LCE) is an engineering discipline that focuses on a systematic approach to design the entire lifecycle of a product by incorporating the environmental indicators along with the economic, technical and social aspects during the product development. It could be considered as a decision-making methodology that integrates LCA, lifecycle cost analysis (LCCA), social lifecycle analysis (SLCA) and technical performance indicators. This integrated discipline could be the most accurate to develop the holistic and advanced sustainable infrastructure design and assessment tool.

The purpose of this chapter is to identify the major technical issues concerning the integration of energy production units in infrastructure elements by the means of a preliminary model for an environmental and economic lifecycle assessment.

15.2. Highway E39 in Norway

Highway E39 on the west coast of Norway connects Kristiansand in the south to Trondheim in central Norway. The highway has eight ferry links (see

Figure 15.1). The Ministry of Transport and Communications has given the project "Coastal Highway Route E39" a mandate to assess technological solutions for fjord crossings. The project has also been mandated to investigate how the bridge/tunnel structure can be exploited to produce energy from wind, sun, waves and tidal currents. The project consists of four components namely, social impacts, fjord crossings, energy and implementation strategies and types of contracts.

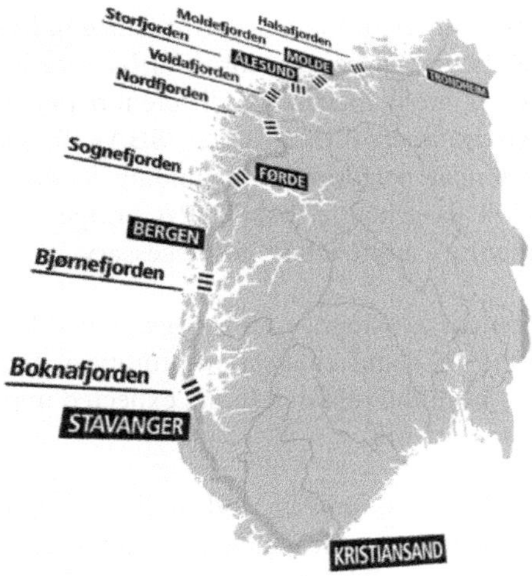

Figure 15.1. *The E39 highway with eight ferry links*

Norway is a substantial producer of renewable energy, due to abundant hydropower resources (more than 99% of mainland Norway's electricity production is from hydropower plants). Even so, there is still vast untapped potential in wind, tidal and wave power. Producing electricity from additional renewable energy sources will reduce the Norwegian carbon footprint and contribute to the Norwegian government's goal of carbon neutrality by 2050.

One of the challenges for renewable energy is to develop technologies that are cost competitive with non-renewable energy. In addition, there are technical challenges associated with large-scale exploitation of energy sources. Using the bridge/tunnel structure as part of the energy-producing facility can probably reduce the establishment costs and thus increase the potential for the development of renewable power plants.

A feasibility study was performed. The feasibility study was carried out by Rambøll, Norconsult and the SP Technical Research Institute of Sweden. The purpose of these studies was to investigate the potential for renewable energy production (solar, wind, tidal water and wave energy) by integrating devices into the structure on fjord crossings and to explore the technical challenges and opportunities. The conclusions of the study indicated pros and cons of each energy source related to location and, to some extent, type of structure.

The general conclusion of the study is as follows: "Power production in connection with bridge constructions poses many challenges. The greatest challenge, and the one that is more or less impossible to overcome, is the limitation imposed by the selection of bridge site location. With this in mind, it is crucial to choose energy production devices which are optimized for the existing conditions at the site. It is also very important to plan the integration of devices in an early design stage of the structure in order to achieve the desired added value".

15.2.1. *A summary of feasibility study*

15.2.1.1. *Wind power*

The physical overall wind power potential in Norway is estimated to be thousands of TWh/year, but a great deal of the potential is not available for exploitation because of environmental and economic circumstances [VIN 12]. A good commercial wind site should have an average wind speed in the region of 7–11 m/s [AME 12]. It is common for wind turbines to start producing power at 3–4 m/s [WIN 12]. Full load hours vary from almost 1,000 h in the Nordfjord to more than 3,000 h in the Boknafjord, making these two fjords the most suitable locations for wind power [SAE 12]. The average wind speed at 50 m height for the different crossings varies from 8 m/s above the Boknafjord to 5 m/s above the Nordfjord [VIN 09].

Onshore wind technology is relatively mature and well developed, while offshore technology is still under development. The offshore technologies are more of interest for the E39 project, since bridges located in the fjords would face similar environmental challenges, such as bottom-fixed foundations due to the great depths. There are three main types of wind turbines that could be of interest for the E39 project namely; horizontal axis wind turbines (HAWT), vertical axis wind turbines (VAWT) and ducted augmented wind turbines (DAWT).

The improvement of DAWT technology will continue to be in focus, as there are many advantages that could be expected from this type of turbine compared to the traditional horizontal turbines. A number of ideas have been launched for how to combine wind turbines and bridge constructions, for example placing the turbines on the extended pontoons of a floating bridge [CHR 12] or integrating DAWTs in the roadway [ÅKE 12].

The estimated power production for the crossing locations varies from 0.3 to 2.6 GWh/km/year for suspension bridges and 0.3–30 GWh/km/year for floating bridges. The greatest potential for power production is at the Boknafjord crossing. Table 15.1 shows the estimated energy production at the Boknafjord crossing (assuming 8.4 km length [ÅKE 12]).

HAWT (small) [1] [1]	2.7	23
HAWT (large) [3] [3]	30	252
VAWT [3] [3]	–	–
DAWT [5][5]	1.4	12

Table 15.1. *Estimated energy production from wind power*

The highest value for floating bridges is based on the installation of HAWTs with a diameter of 101 m placed every 250 m bridge [SAE 12]. This alternative will subject the bridge to extreme additional loads, which is considered unrealistic at this stage. For suspension bridges, DAWTs integrated in the roadway seem to be the best alternative [ÅKE 12]. However, a suspension bridge is not possible at the Boknafjord crossing. For detailed calculations, see [CHR 12, ÅKE 12], and [HOS 12].

For most of the suggested technologies, the added value and synergies gained by integrating wind turbines in the construction are expected to be low. The costs of integration will most likely be increased due to necessary modification and increased loads. However, there is an exception. Using DAWTs integrated into the construction could be more cost-effective and has great potential for power production [ÅKE 12]. Other options that might be viable are to attach turbines as shovels around the pillars or installing VAWTs in the zone near the shore [CHR 12].

15.2.1.2. *Solar energy*

Typical solar irradiance in Norway is 600–1,000 kWh/m^2/year [SOL 12], which corresponds to about 1,500 times the total annual energy consumption. The potential for exploiting this energy is vast, but little direct sunlight limits the actual power production potential. However, there is a lot of indirect, diffuse sunlight that might lead to some production.

Photovoltaic (PV) panels would be more suitable than solar thermal installations, since these fjord crossings are located in sparsely settled areas; the heat generated would have to transport long distances, which is not a viable option. Electricity, on the other hand, can easily be connected to the grid and transferred to a remote point of use. Thus, such restrictions do not apply for PV panels. However, solar water heating could be used for tap water heating, defrosting and snow melting that may be needed on or near the bridge [KOV 12].

Table 15.2 presents the estimated power production for different mounting alternatives of the solar panels. The calculations have been carried out by SP and are based on at 45 tilt angle for optimal utilization of the irradiance. The assumption is that there is no large variance for the estimated power production at the different fjord crossing locations.

Side mounted	0.6	0.6
Roof mounted	1.9	1.7
Mounted on wires	8.4	–
Mounted on pylons	0.3 per pylon	–

Table 15.2. *Estimated energy production solar power*

Solar panels mounted on the wires of a suspension bridge give the greatest potential for power production. However, this alternative would most likely increase the additional wind loads on the bridge structure and therefore is not recommended. A solar power plant could be easily installed on a bridge with relatively small added value. The solar panels could be mounted directly on the construction without using any other mounting systems if this is planned in the design phase. This would reduce the total costs. For detailed calculations, see [CHR 12, KOV 12, SAE 12] and [HOS 12].

Solar panels on the bridges along the E39 are unlikely to be feasible as production systems for delivery to the grid based on the today's solar power technologies and market. If the panels are mounted on the sides, as a roof, on the pylons and on the wires, the energy production would be in the acceptable range. The panels mounted on wires would account for around 75% of the total production [KOV 12], but since this is the mounting alternative that gives the greatest additional load, all-round mounting is not an optimal solution. However, pilot installations could be of interest in order to gain new knowledge through experience and to contribute to the further development of solar panels.

An advantage of solar power generation in favor of wind power is the higher reliability of energy availability with sun shine being more heterogeneous than wind in the assessed case. However, solar radiation is relatively low in Norway, limiting the economic performance even of tailored cells.

15.2.1.3. *Tidal power*

Waves have a very high energy concentration and are a natural store of kinetic energy that can travel thousands of kilometers with little energy loss [CAL 06]. It is estimated that the contribution of wave energy to the Norwegian energy portfolio could reach between 12 and 30 TWh per year [SAN 07].

Waves vary with time and location. Wave energy scatter diagrams typically contain information about annual distribution of the significant wave heights (H_s in meters) and peak wave period (T_p in seconds). Data from at least a year are required to ensure that seasonal differences in the energy flux are accounted for.

It is impossible to make exact calculations of the expected power production, as there is not sufficient wave statistic data from the fjord crossing locations. However, rough calculations have been carried out by SP based on the rated power and the number of installed WEC devices. The results presented in Table 15.3 are based on the most suitable and applicable devices, listed in [VEN 12], which could start producing power when the significant wave height is below 1 m. These calculations are not adequate, but they do give an indication of the expected power production (if the significant wave heights are high and long enough, $H_s > 1$ m and $T_p > 3$–4 s).

The results show that the highest potential for power production is found in the Moldefjord, the Boknafjord and the Bjørnafjord (long crossing). Assuming that the crossing with the greatest wind speed also will have the most powerful waves, the Storfjord crossing could also be a crossing location with a potential for extracting wave energy. It is most likely that the power production would be in the middle or lower range of these intervals. However, it should be expected that the significant wave heights would be more than 1 m in the Boknafjord. The results show that there is significant potential for producing energy from waves if the significant wave height is more than 1 m. For detailed calculations, see [CHR 12, VEN 12] and [HOS 12].

The costs of installing WEC devices depend on the device chosen. There are many different types that are already on the market or being developed. A production cost of 1.5–3 NOK/kWh is expected, based on experience with the market [CHR 12]. The costs for WEC devices would be reduced if they were integrated into the bridge construction. The cost reduction could be expected to be about 40%, which corresponds to a production cost of 0.9–1.8 NOK/kwh [CAL 06].

Crossing	Suspension/ floating bridge [GWh/km/year]	Submerged tunnel [GWh/km/year]	Suspension/ floating bridge [GWh/year]	Submerged tunnel [GWh/year]
Halsafjord	15–193	13–163	27–347	23–293
Moldefjord	20–235	17–200	160–1880	136–1,600
Storfjord	18–218	16–185	61–741	54–629
Voldafjord	16–198	13–168	32–396	26–336
Nordfjord	15–188	13–160	26–320	22–272
Sognefjord	18–221	16–188	67–843	59–717
Bjørnafjord (short)	15–184	13–157	24–295	21–251
Bjørnafjord (long)	20–230	16–196	114–1,320	91–1,122
Boknafjord	20–236	17–201	168–1,986	143–1,688

Table 15.3. *Estimated potential for WEC devices producing energy at Hs<1 m and T_p<3–4 s*

There are advantages when combining WEC devices and bridge constructions, especially in terms of cost reductions. The main challenges are the reliability of the energy resource and the additional loads on the structure. It is crucial that the integration of WEC devices be included in an early phase of the bridge design.

15.3. Identification of environmental and economic lifecycle challenges

The built environment plays a significant role in sustainable development and it is a major focus of attention within the environmental movement. In addition to indirect impacts, such as deforestation and the degradation by consuming 40% of the world's energy, 40% of the world's materials, 55% of the wood cut for non-fuel use and 12.2% of the total water used [HOF 08, US 08, UNE 06, ROO 95]. Since buildings represent about 40% of all CO_2 emissions, while building construction and maintenance represent about 10% of the total building construction processes (production of materials, construction, maintenance, operation, deconstruction), it means that it will be equivalent to (40%x10%=4%) of total CO_2 emissions are associated with construction processes. If, as we said before, road construction is about 17% of total construction sector, total CO_2 emissions due to construction processes (production of materials, construction, maintenance and operation) will be (40%x10%x17%=0,68% of total emissions). Moreover, road transport represents 80% of total energy consumption of transport sector and 86% of total freight transport.

The greening of infrastructure and the built environment may provide one of the major opportunities to achieve sustainability for urban environment, offering drinking, storm and wastewater management, temperature attenuation, carbon sequestration, increased biodiversity and reductions in air pollution. However, since economic and ecological resources are limited and trade-offs in decision-making processes are ubiquitous, a concept is needed that supports balancing the different interests, expectations and opportunities. Such a concept is explained in the following sections in the example of the E39 that has been introduced earlier.

15.4. Pareto optimized concepts balancing economic and environmental aspects

As better performing infrastructure is in most cases more expensive than a less well-performing one, the justification for the additional expenses normally follows a lifetime-based argumentation. In the case of an economic argumentation, the higher initial costs are justified by lower running costs for maintenance and usage (lifecycle costs = LCC). In the case of an environmental argumentation, more environmental impact in the construction phase (embodied impact) is justified by lower impact in the usage phase, for example due to less treatment for maintenance or even the generation of energy that can substitute other energy mixes.

Models that evaluate the trade-off between environmental and economic performance of built structures show, however, that a hoped for synergy (win-win) of economic and environmental optimization only exists to a certain point after which there is a characteristic trade-off between better performance in one dimension and a decreasing performance in the other (Figure 15.2). Even when Pareto optimal concepts are chosen (by tailoring the individual components toward optimal performance regarding the applied indicators), the idea of strong sustainability (a synergetic maximal performance in both dimensions for a solution) cannot be achieved. It has to be noted that a lot can be achieved with comparably little money when moving a small step away from economically optimal concepts toward slightly more environmentally friendly ones.

Though this has to be assessed individually for each solution, several examples show that projects that simply aim for extreme environmental performance (in Figure 15.2, for example based on the LCA indicator RECIPE) by passive measures of efficiency fail to be economically feasible. Energy generation by infrastructures can be a turning point in this perspective as it is simply a matter of lifetime when the ecological investment pays economically off. As infrastructures exceed buildings in terms of their lifetime, they have huge potential in this regard.

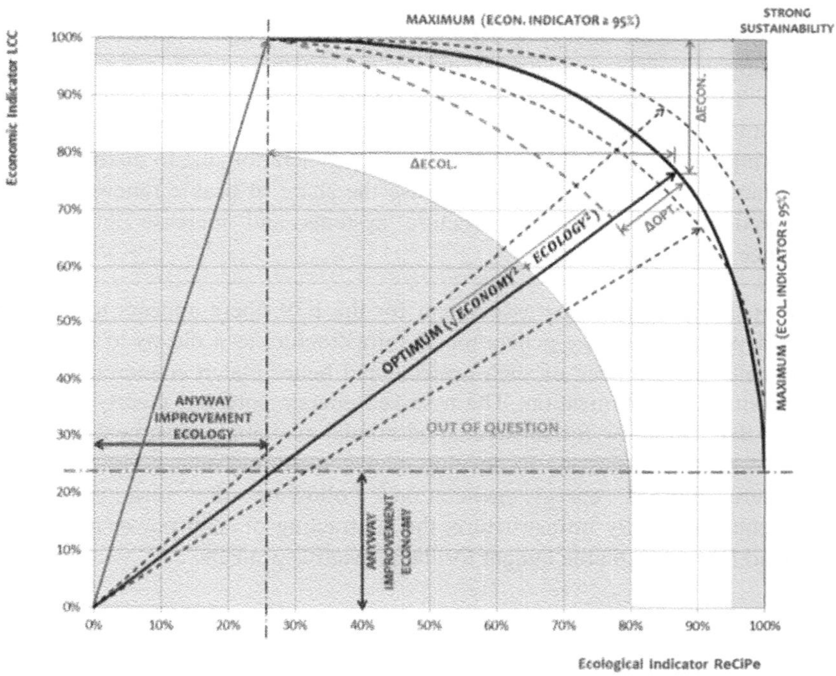

Figure 15.2. *An example of results obtained by Pareto optimization*

In terms of feasibility, the different options for generating energy via infrastructures have to be assessed individually.

15.5. Conclusions

Power production in connection with bridge constructions poses many challenges. The greatest challenge, and the one that is more or less impossible to overcome, is the limitation imposed by the selection of bridge site location. With this in mind, it is crucial to choose energy production devices which are optimized for the existing conditions at the site.

When considering power production from wind and waves, the Norwegian climate has a great advantage. The strongest winds and most powerful waves occur in the winter, when the national energy demand is at its highest. The opposite applies for solar energy, which means that the energy generated needs to be stored or additional energy sources are needed during the winter months.

The tailoring of the energy generation to the energy consumption and bridging the remaining gap via energy storage remains the big challenge of renewable energy generation. This probably cannot be solved on a regional level and is challenging in many settings on a countrywide level. Solutions on an EU level are necessary to allow making better use of local peaks in energy generation due to distribution over larger distances. This in return will increase the share of usable renewable energy also created from merged infrastructure systems and increase the economic performance of such projects.

As a positive result of the assessment for the E39, there appears to be a solid basis for hope to use energy generation in infrastructure as a means to improve the environmental performance of such projects and better match economic feasibility and environmental optimization. The assessed energy potential clearly indicates a dimension that can allow for substantial additional available investment. For this to happen, it would be necessary to develop management and ownership concepts for infrastructure merged energy generation. Following the idea of a more environmentally friendly infrastructure, the contracting of such power plants might create an additional available budget for the respective projects.

15.6. Bibliography

[ÅKE 12] ÅKESSON L., Wind energy technology survey for ferry free E39 project, SP Technical research institute of Sweden, 2012.

[AME 12] AMERICAN WIND ENERGY ASSOCIATION, available at: http://www.awea.org/learnabout/utility/transmission/index.cfm. [Accessed 4 July 2012], 2012.

[CAL 06] CALLAGHAN J., "Future marine energy – results of the marine energy challenge: cost competitiveness and growth of wave and tidal stream energy", *Carbon Trust,* 2006.

[CHR 12] CHRISTOPHERSEN E.B., "Technology survey for renewable energy integrated to bridge constructions", *Rambøll,* 2012.

[ERT 09] ERTRAC – EUROPEAN ROAD TRANSPORT RESEARCH ADVISORY COUNCIL, ERTRAC, Road Transport Scenario 2030+ "Road to Implementation", available at: http://www.ertrac.org/pictures/downloadmanager/download.php?id=8, 2009.

[ERT 10] ERTRAC – EUROPEAN ROAD TRANSPORT RESEARCH ADVISORY COUNCIL, ERTRAC Strategic Research Agenda 2010 – Towards a 50% more efficient road transport system by 2030, available at: [ftp://ftp.cordis.europa.eu/pub/technology-platforms/docs/ertrac-sra-2010_en.pdf], 2010.

[EUR 07] EUROPEAN COMMISSION, A lead market initiative for Europe, available at: [http://eur-lex.europa.eu/LexUriServ/LexUriServ.do?uri=COM:2007:0860:FIN:en:PDF], 2007.

[EUR 10a] EUROPEAN COMMISSION, EUROPE 2020 A strategy for smart, sustainable and inclusive growth, available at: [http://eur-lex.europa.eu/LexUriServ/LexUriServ.do?uri=COM:2010:2020:FIN:EN:PDF], 2010.

[EUR 10b] EUROPEAN COMMISSION, Energy performance of buildings, available at: http://eur-lex.europa.eu/LexUriServ/LexUriServ.do?uri=OJ:L:2010:153:0013:0035:EN:PDF [Accessed 23 September 2013], 2010.

[EUR 11a] EUROPEAN COMMISSION, A resource-efficient Europe – Flagship initiative of the Europe 2020 Strategy, available at: [http://ec.europa.eu/resource-efficient-europe/pdf/resource_efficient_europe_en.pdf], 2011.

[EUR 11b] EUROPEAN COMMISSION, Roadmap to a Single European Transport Area – Towards a competitive and resource efficient transport system, available at: http://eur-lex.europa.eu/LexUriServ/LexUriServ.do?uri=COM:2011:0144:FIN:EN:PDF, 2011.

[GRE 11] GREENROADS FOUNDATION, The Greenroads Rating System. United States, available at: https://www.greenroads.org/files/235.pdf, 2011.

[HOF 08] HOFFMAN A.J., HENN R., "Overcoming the social and psychological barriers to green building", *Organization & Environment*, vol. 21, no. 4, pp. 390–419, 2008.

[HOS 12] HOSEINI M., "Ferjefri E39, Subproject Energy", available at: [http://www.vegvesen.no/_attachment/435449/binary/731805], December 2012.

[KAR 12] KARTVERKET, VANNSTAND., available at: http://vannstand.no/index.php/nb/statistikk/midler, 2012.

[KOV 12] KOVACS P., WAHLGREN P., Solar energy technology survey for ferry free E39 project, SP Technical research institute of Sweden, 2012.

[ROO 95] ROODMAN D.M., LENSSEN N.K., PETERSON J.A., *A Building Revolution: How Ecology and Health Concerns Are Transforming Construction*, Worldwatch, Washington, 1995.

[SAE 12] SÆTA, SIMONSEN, SOLLI *et al.*, Technology survey for renewable energy integrated to bridge structures, Norconsult, 2012.

[SAN 07] SANDGREN J., HJORTL E., MIRANDA W. *et al.*, Potensialstudie av havenergi i Norge, SWECO Grøner for Enova SF, 2007.

[SOL 12] SOLENERGIRESSURSER I NORGE, available at: http://fornybar.no/sitepageview.aspx?sitePageID=1648, 2012.

[UNE 06] UNEP SUSTAINABLE BUILDINGS & CLIMATE INITIATIVE (SBCI), Sustainable Building and Construction Initiative, Information note, Paris, UNEP, Division of Technology, Industry and Economics, 2006.

[US 06] U.S. GREEN BUILDING COUNCIL RESEARCH COMMITTEE, A National Green Building Research Agenda, Washington, 2008.

[VEN 12] VENNETTI D., Wave and tidal energy technology survey for ferry free E39 project, SP Technical research institute of Sweden, 2012.

[VIN 09] VINDKART FOR NORGE, Kartbok 1c: Årsmiddelvind i 50 m høyde, Kjeller Vindteknikk, 2009.

[VIN 12] VINDRESSURSEN I NORGE, available at: http://fornybar.no/sitepageview.aspx?sitePageID=1740, 2012.

[WIN 12] WIND POWER PROGRAM, available at: http://www.wind-power-program.com/turbine_characteristics.htm, 2012.

PART 4

Greenhouse Gas Mitigation

Introduction to Part 4

The transports sector is responsible for around 25% of global CO_2 emission due to fuel combustion (30% in OECD countries), and 15% of overall greenhouse gas emissions. Road transport, and particularly passenger cars, represent a high share of these CO_2 emissions, but road freight could account for up to 30–40% of the road sector. Furthermore, due to a continuous growth of motilities (passengers and goods) and despite significant improvements of vehicle technologies and increase in non-fossil fuels, transport has recorded a high increase in its contribution over the past few years; it is the only sector that did not significantly reduce its GHG in recent years. To reach the GHG reduction targets of around 80 to 95% until 2050 (compared to 1990 levels), the transport sector should deeply evolve or even restructure, toward new technologies and non-fossil energies, and new organizations and services.

Three main questions are thus raised in this part: which overall mitigation strategies should be followed for reaching these targets, to what extent can costs or incentive measures support these strategies and, finally, what is the contribution of freight and how can it be assessed and regulated in view of better organizations?

Mitigation strategies and pricing/incentive measures

In the first chapter, a transport strategy analysis is proposed that studied reduction technologies for all modes and different scenarios, through a quantitative model to assess economic and environmental impacts. An ambitious policy strategy would guarantee reductions of 60–70% of GHG in 2050. This strategy would support R&D for biofuels for air transport and for cross-modality, and it would focus first on conventional engines and then on alternative energy (i.e. electricity and hydrogen from renewables). Pricing measures and ambitious regulations such as a ban of fossil fuel cars in 2035 were also suggested.

Introduction written by Michel ANDRÉ.

The second chapter raises the question of pricing measures and particularly of fuel costs and analyzes heavy goods transport and its evolution in Sweden. Despite higher fuel prices and the availability of cost-effective engine technologies, a quasi-constant increase in CO_2 emission is observed, linked to the road traffic transport volumes. Technical, institutional and financial barriers, as well as contract structures and ownership patterns, thus reduce the incentives for transport firms to implement these technologies. Finally, when fuel savings are realized, rebound effects can cancel out improved energy efficiency.

The third chapter analyzes the incidence of super-credits (or incentives) to promote low-carbon vehicles (electric, plug-in hybrids and electric vehicles with range extenders). Different electric vehicle penetration scenarios are examined for the EU-27 region, up to 2050. The simulation demonstrates that the influence on the actual GHG emissions is highly dependent on the EV penetration, while the incentive itself would have a limited effect.

Finally, the final chapter demonstrates the feasibility of using GPS-based mobility data for characterizing spatial distribution of the mobility and assesses the potential of modal shift to public transport according to its availability within different geographical areas. This potential and GHG reduction in fact seem quite limited for city centers. Technological scenarios are also tested for public and private transport (electric and plug-in vehicles) that demonstrate higher potential.

The case of freight transport chains

The freight transport sector is also concerned by GHG reduction, with a particular challenge due to the strong increase in goods transport linked to future economic growth as well as to the particular complexity and length of the supply chains across various transport modes.

The COFRET European research project studied appropriate methodologies for assessing GHG emissions or carbon footprint from freight transport chains. Harmonized and agreed methodologies are indeed required for the estimation of potential gains, identification of best practices and comparison of different approaches, while considering the accuracy of the calculations, especially in the case of complex multi-modal supply chains. After a review of the existing methodologies and their limits, supply chain elements (SCEs) are defined as modular operations of the transport within the logistics chain, and vehicle operation systems and allocation principles are also discussed as a basis toward a standardization of the assessment. These concepts are highlighted through real-life supply chains.

Accordingly, the last chapter analyzes the carbon intensity of French shippers from a large survey (around 3,000 shippers and the following of three shipments for each one). The variability of the carbon intensity is characterized according to a typology of shippers, and the specific carbon intensity (per tonne or tonne x km) can vary from 1 to 15 between the different groups of shippers. An in-depth knowledge of the determinants of freight transport demand and the factors influencing logistical choices would explain that variability and make freight GHG mitigation easier.

16

GHG Mitigation Strategy in the European Transport Sector

Transport accounts for roughly one quarter of all greenhouse gas emissions (GHG) in Europe. Transport is the only sector that has not reduced its GHG in recent years. To meet the European and global targets to reduce GHG in industrialized countries by 80 to 95% until 2050 compared with 1990 requires that the transport sector is put on a pathway to drastically reduce its emissions. The chapter demonstrates that with a combined R&D and transport policy strategy, reductions of 60 to 70% of GHG until 2050 will be achievable for the European transport sector.

The analysis to develop the transport strategy combines an analysis of the innovation system of the transport sector with respect to developing GHG reduction technologies of all modes and a model-based quantitative scenario exercise to assess the transport, economic and environmental impact of the scenarios.

We intend to support R&D for biofuels for air transport, and R&D for cross-modal transport. The policy strategy suggests to focus on road transport (i.e. efficiency of ICE) in the short term and on alternative energy technologies (i.e. electricity and hydrogen both from renewables) in the medium term. Additionally, the transport strategy must include pricing measures and ambitious regulations (e.g. ban of fossil fuel cars by 2035).

16.1. Introduction

Transport accounts for roughly one quarter of all greenhouse gas emissions (GHG) in Europe. Transport is the only sector that has not reduced its GHG in recent years. However, European and global targets to reduce GHG in industrialized countries by 80 to 95% until 2050 compared with 1990 requires that the

Chapter written by Wolfgang SCHADE and Michael KRAIL.

transport sector is put on a pathway to drastically reduce its emissions. To achieve such reductions will be a challenge for industry and policymakers. This challenge can only be tackled by a consistent strategy that considers both R&D for transport innovations and transport policy, also taking into account the potential changes of behavior and production patterns.

Such a transport strategy is the outcome of the European FP7 co-funded research project GHG-TransPoRD (http://www.ghg-transpord.eu). The development of the GHG reduction strategy of transport consisted of five main steps undertaken by GHG-TransPoRD and explained in the following sections of this chapter:

– analysis of R&D funding and the innovation systems in all transport modes and for alternative fuels;

– scoping of GHG reduction measures and technical reduction potentials;

– techno-economic assessment of cost of GHG reduction measures;

– stepwise scenario building and model-based assessment of scenarios and their GHG reduction pathways;

– summary and conclusions on a suitable GHG reduction strategy for transport.

First, it is important to understand the status quo i.e. today's structure of GHG emissions of the transport sector in Europe (see Figure 16.1). At the top, the distribution of CO_2 emissions for transport originating in EU27 is presented. This includes international bunkers for planes and ships leaving Europe, which then can be named the comprehensive figure. Looking at the comprehensive numbers, it is obvious that in particular three modes must contribute to the significant GHG reductions: road, air and maritime shipping. Together, these modes account for 96% of all EU transport GHG emissions arising from transport activities.

The bottom of the figure shows only the domestic transport emissions, i.e. those GHG emissions arising from transport activities within European countries (see Figure 16.1 right-hand side). Domestic road transport accounts for 94% of EU27 domestic emissions, which is roughly split into one-third arising from road freight transport (29% of total domestic) and two-thirds coming from road passenger transport (65% of total domestic). This analysis of status quo reveals that priorities for GHG mitigation of transport should have the following order: (1) passenger road transport, (2) freight road transport and (3) maritime and air transport. Concerning the latter, it should be mentioned that these two modes over the last years, except the crisis years 2008/2009, revealed by far the highest growth rates of transport demand and thus emissions.

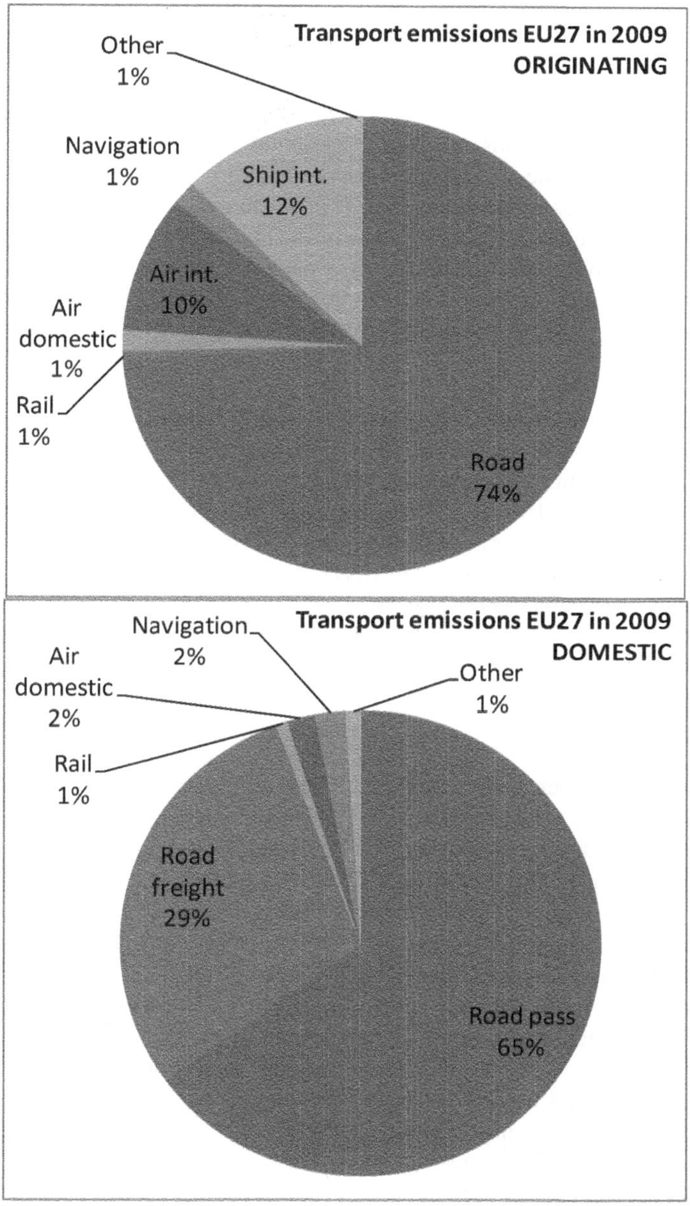

Figure 16.1. *Structure of GHG emissions of transport in EU27 in 2009 ([SCH 12 using own estimates and EEA 2011). For a color version of the figure, see www.iste.co.uk/jacob/energy.zip*

16.2. Analysis of R&D funding and the innovation systems of the transport modes

Three different approaches have been combined by GHG-TransPoRD for the analysis of how innovations are emerging in the transport sector and how well positioned would be the European transport industry to develop GHG reduction technologies: (1) a quantitative assessment of R&D investments; (2) the analysis of patents; and (3) the qualitative description of the innovation system transport (ISyT). Each of these activities would be worth a paper on its own (also see [LED 10], [WIE 12]). This section provides a summary of findings to show the strength of the European transport industry in terms of global R&D and innovation efforts.

Table 16.1 reveals the strength of the European R&D in the transport sector. The transport sector is the largest industrial R&D investor in the EU. In terms of global sales, the share of European transport industry accounted in 2008 for roughly one-third 774 €bn (34.2%). However, €40.3 bn spent on R&D represents more than two-fifths (42.4%) of global transport R&D by EU transport industry. So, the EU industry spends overproportionaly on R&D compared to other global regions. The other main innovative parties in the past have been companies with headquarters in Japan and the USA [LED 10]. In recent years, significant R&D capacity in the automotive sector has been built-up in China, and this trend is expected to be continued
[OLI 12].

Herewithin, research efforts of the automotive industry are clearly dominating, followed by those of the aviation sector. R&D investments of the automotive sector have been further disaggregated into road passenger and road freight transport and supplier components. We found significantly higher levels of R&D investment volumes and a higher R&D intensity of car manufacturers compared to manufacturers of commercial vehicles. This can be explained by the very distinct nature of road passenger and road freight transport. In road freight transport, the high competition means that transport companies focus largely on reducing their costs. Given the significant share of fuel costs out of the total operating cost for commercial vehicles, fuel efficiency is an important purchase criterion. Nevertheless, transport companies were following a strict economic calculus when buying new trucks and in past periods of low oil prices they have not been ready to pay for "innovative technologies" as such. The situation is different in cars, where consumers' choice is influenced by a variety of factors. Cars are more exposed to a "differentiation and branding pressure", and innovative technologies can be one selling factor.

	R&D investment (€bn)		Sales (€bn)		Number of employees (million)	
	World	EU27	World	EU27	World	EU27
Automotive manufacturers	53.0	20.9	1,213	423	2.76	1.26
Automotive suppliers	19.6	9.5	437	156	2.33	0.98
Commercial vehicles and trucks	6.9	2.4	233	66	0.62	0.22
Aerospace &defence	15.6	7.5	379	129	1.74	0.55
'Transport' sector	95.1	40.3	2,262	774	7.50	3.00
All industries	431.0	130.0	13,897	5,712	45.10	21.00

Table 16.1. *R&D investments, sales and total number of employees related to the transport sector (2008) (Source: EU Scoreboard 2009, European Commission DG RTD, 2009)*

R&D investments in rail and maritime were more limited in 2008, comparing the absolute values with road and air. However, when setting the R&D investments in relation to the net sales of the sectors – i.e. estimating the R&D intensity – this heterogeneity becomes less pronounced. In 2008, R&D intensities in the road sector were around 5% (passenger cars: 5.3% and commercial vehicles: 3.5%; suppliers: 6%), while aviation (civil aeronautics) shows significantly higher (7.8%) and rail (3.9%) and waterborne (3.2%) slightly lower values.

Industrial R&D investments have been highly concentrated to a few players in 2008, with 15 companies accounting for 80% of the total transport-related R&D investments. EU-focused patent analysis carried out for hydrogen fuel cell technology, biofuels, hybrid and electric vehicles enablesus to identify leading countries and companies for such technologies. For example, patenting technologies relevant for hybrid and electric vehicles quadrupled for the period 2006–2009 compared to previous periods, with Germany and France taking the lead in EU (top of Figure 16.2) as well as German and French companies (2007 only, bottom of Figure 16.2).

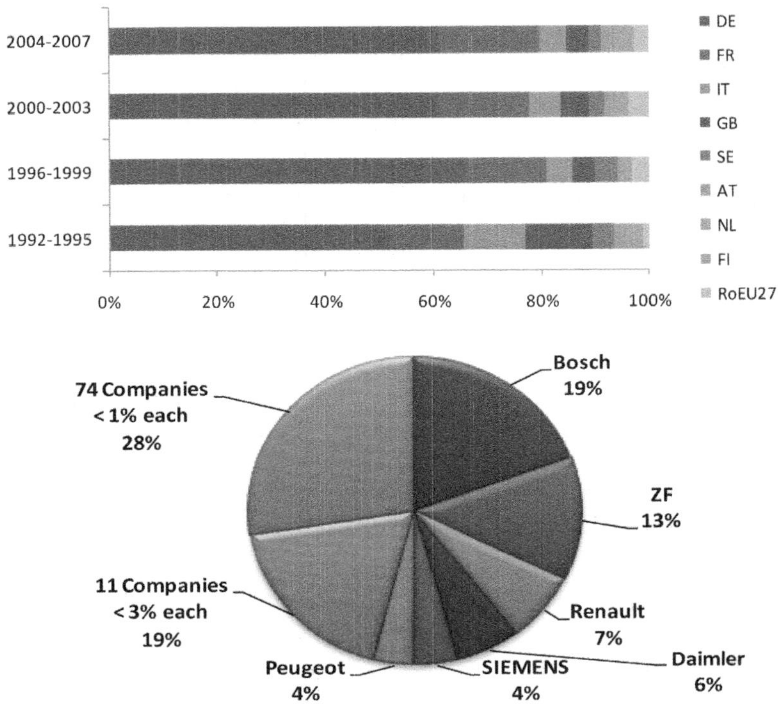

Figure 16.2. *Example of patent analysis for hybrid and electric vehicles ([LED 11] 2011 after Fraunhofer ISI). For a color version of the figure, see www.iste.co.uk/jacob/energy.zip*

The concentration of R&D investments, and to a lesser extent of patenting activity, can be explained by the market structure of the transport industry, which is mainly oligopolistic competition, and the fact that most of the technological development comes from inside the industry rather than being purchased (as is the case, e.g. in the energy sector). However, this picture changes for alternative fuels and new technologies other than conventional internal combustion engines. Here, specialized niche providers have entered the market as well as major industries from non-transport sectors such as electric utilities. Often, new coalitions between established car manufacturers and component suppliers and these newcomers emerged, leading to a relatively rapid sharing of the new knowledge and therefore accelerating innovation within the sector in a vertical way ("supplier path").

16.3. Scoping of GHG reduction measures and technical reduction potentials

The first activity to quantify GHG reduction potentials of the different modes and due to alternative fuels consisted of three steps to assess the GHG reduction potentials of individual measures by mode:

1) A so-called *common energy framework* was developed forecasting a reference scenario of energy demand by mode and by fuel type until 2050. Any GHG reduction potential of a single measure was assessed against this energy framework. The energy framework used inputs from European projects (iTREN-2030, ADAM) and tools (TREMOVE, Ex-TREMIS database). The objective of the framework was to obtain a transport energy demand scenario until 2050 that can be differentiated by modes, fuel types and regions as these categories were needed for the assessment of the different measures.

2) A scoping exercise generated *long lists of potential measures* and their reduction potentials. Measures included technologies (e.g. light-weight construction, solar panels on vehicle roofs, exhaust heat recuperation, sky-sails, etc.) and national or urban policies (e.g. fuel taxation, urban cordon charging, etc.).

3) Based on an initial assessment of the long lists of each mode, *short lists of promising measures* have been developed that seem effective and feasible to provide GHG reductions either until 2020 or until 2050. For these short lists, the technical potential for GHG reductions by mode was estimated in detail using the common energy framework. Potential measure impacts were calculated in both relative and absolute potentials with regard to quantities of the energy framework. Furthermore, information on practical, technical or political feasibility of a measure was
collected.

The measures were classified by four broad categories of how they are reducing GHG emissions (ASIF approach), which supports a decomposition analysis of major sources of GHG reductions:

– Activity reduction: means that the measure would reduce transport demand. Usually, these are valid for demand management measures (i.e. transport policy).

– Modal shift: means that a measure would affect the modal shift such that low carbon modes increase their modal share. Usually, this holds for demand management measures (i.e. transport policy and infrastructure policy).

– Energy intensity: means that a measure improves energy efficiency: usually valid for technical measures in vehicles (e.g. engine efficiency and rolling resistance).

– Carbon intensity of *fuels*: means that a measure reduces the carbon emissions per unit of fuel consumed. Usually this would mean to use alternative fuels, i.e. non-fossil or low carbon fossil (e.g. biofuels and CNG).

The elaborated shortlists contained 19 bundles developed from more than 60 measures concerning road technologies differentiated into car and truck measures, 26 measures related to urban and (national) road policies, 11 air measures, 11 rail measures and 10 shipping measures related to the four categories of the ASIF approach. Based on these short lists that largely neglect the interaction between different measures, the medium- and long-term *theoretical reduction potentials* by mode against the reference of the energy framework have been estimated. These represent maximum potentials and the realizable reductions are expected to be smaller, e.g. taking into account costs, barriers and interactions between measures. Such realizable potentials were later identified by the scenario assessment with the models (see section 16.5). The results of the analysis of single measures are summarized in Table 16.2. These figures constitute the highest potentials compared with the economic potentials and the potentials that were finally estimated by our scenario analysis.

Mode	Type of measures	2020	2050
Road	Technical cars*	-40 to -45%	-60 to -68%
	Technical trucks	-30 to -36%	-57 to -63%
	Urbanmeasures**	-43%	-70%
	National policies***	-40%	-70%
Rail	Technology non-urbantraffic	-10%	-42%
	Technologyurbantraffic	-8%	-55%
Air	Technology&policy	-15%	-41%
Shipping	Technology&policy	-5%	-20 to -25%
Biofuels	Technology****	-20%	n.a.

* Potentials are calculated using the reference energy mix for electricity. Potentials can be higher if electricity would be produced carbon free, as then upstream emissions of electric vehicles would become zero.

** Taking into account most relevant and compatible urban measures.

*** Assuming reasonable combinations of national policies.

**** Not considering the impacts of land use changes. The economically realizable potential for reductions by use of biofuels is significantly smaller than the theoretical technical potential. And it strongly depends on external factors like the price of fossil fuels.

Table 16.2. *Theoretical technical reduction potentials by mode based on aggregation of potentials of single measures, %-relative reduction to energy framework (Source: [AKK 10])*

16.4. Techno-economic assessment of cost of GHG reduction measures

The previous assessment step did not consider the cost of each measure. To consider the cost to assess GHG reduction potentials is relevant in particular for the technological measures as both have to compete with each other and vehicle brands as well as modes are in competition with other vehicle manufacturers and other modes, respectively. Looking at the time horizon 2020, 2030 and 2050, costs of technologies will not remain static. Depending on the innovation process and the production numbers, the cost will reduce over time due to learning-by-searching (e.g. effect of R&D investment), learning-by-doing (e.g. effect of accumulated production and sales) and/or economies of scale (i.e. effect of mass production at industrial scale).

The ideal approach to project cost trajectories for new technologies in dependency from R&D and sales would be implemented by applying learning curves. In particular, the two-factor learning curve seems suitable in which one factor would constitute an R&D-related parameter (e.g. R&D spending and patents) and the other factor a production or sales-related parameter. Only for very few technologies, sufficient data or studies were available to carry out such cost estimations. In much more cases, it was possible to derive a one-factor learning curve only with production or sales driving down the cost. In other cases, we could only identify cost assessment studies providing cost projections for one or a few technologies or had to derive plausible cost pathways from roadmaps. In particular for road transport technologies and biofuels, it was possible to implement learning curve models in the ASTRA model providing the vehicle fleet forecast and the POLES/BioPol model. Typical learning rates identified for young technologies lie in the range between 10 and 25%, while for mature technologies cost improvements occur at slower speeds between 3 and 6% (see Table 16.3). These orders of magnitude could also be confirmed for the analyzed transport technologies, with some of the alternative engine technologies being at the lower end of the range (e.g. hybrid vehicles and pure electric vehicles).

Level of innovation	learning-by-doing rate	learning-by-searching rate
Young technology	15-25%	10–15%
Mature technology	4% in average	3–6%

Table 16.3. *Comparison of learning rates of young and mature technologies (Source: [SCH 11])*

16.5. Stepwise scenario building and model-based assessment of scenarios and their GHG reduction pathways

The analyses undertaken in GHG-TransPoRD had to deal with complex interacting systems. Not only that, the different transport modes should not be analyzed separately as competition and modal-shift always needs to be considered. Also, the interactions and feedback of transport with the energy system, with the drivers of the economic system, the demographic dynamics and the global trade have to be taken into account. Hence, GHG-TransPoRD decided for two approaches to tackle the complexity: first, a set of four models was applied, partially integrated with each other as the two core models of the analyses, ASTRA and POLES, and partially sequential as the case for TREMOVE and MARS models that received inputs from the former two models.

– ASTRA (Assessment of Transport Strategies). A System Dynamics model applied for transport-economic integrated assessment of policy strategies of the EU27+2.

– POLES (Prospective Outlook for the Long term Energy System). A System Dynamics global sectoral simulation model for the development of global and European energy scenarios.

– TREMOVE. A policy assessment model to study the effects of different transport and environment policies on the emissions of the transport sector of EU27+4 running until 2030, only.

– MARS. An urban dynamic Land Use and Transport Integrated (LUTI) model (the version developed for Leeds has been used).

With this suite, the global, European, national and urban spatial levels were covered as well as the transport, energy and economic systems. The second approach to deal with complexity concerned to involve the relevant stakeholders and academics. The models also provide a tool for structured thinking and rethinking intermediate results of scenario analyses. Such results were presented to and discussed with stakeholders, whose feedback was then taken into account for modified scenario simulations, be it their propositions of further GHG reduction measures to be considered, their comments and support to the cost assessments, their proposal whose measures could be combined into a policy package and whose measure combinations might be contradictory as well as their criticism on our initial scenarios and the proposals to improve the policy packages of these scenarios.

The concept of the definition of scenarios for the model-based analysis was three-fold: (1) a stepwise approach was chosen commencing with low ambition scenarios and increasing the ambition of GHG reduction with each new round of scenarios, (2) the structure of scenarios should allow us to differentiate between the big drivers of impacts, i.e. technology, policy and behavioral changes, and (3) initial scenarios started with a low number of single measures integrated into their policy package and to generate further scenarios gradually further measures are added to the policy packages. Thus, three major levels of scenarios can be distinguished: (A) technology scenarios, (B) policy scenarios on top of (A) and ambitious regulatory scenarios on top of (B).

The technology scenarios either focusing on efficiency of conventional cars or on alternative technologies would deliver about 34 to 37% reductions of GHG emissions until 2050. Adding policies in the policy scenarios, in particular pricing policies to foster behavioral change, would roughly add another 10% reduction. But only if further ambitious regulations were added, i.e. the phase out of conventional fossil fuel cars around 2035 and a modal-shift of about 4% points away from road freight to rail and shipping, could the most ambitious of the analyzed scenarios called AMB_REG scenario (ambitious regulation scenario) deliver the 60% reductions of GHG emissions of transport by 2050. Concerning the energy system, all scenarios considered a very significant shift toward the use of renewables such that in 2050 electricity in EU27 would be generated by 80% from renewable energy sources.

The story of the AMB_REG scenario reads as follows for passenger transport. Demand in terms of pkm increases by 36% until 2050 compared to 2010 (see Figure 16.3). GHG emissions continuously decline from 2014 until 2050. Until about 2035, the decline of GHG emissions largely comes from reductions of intensity of energy (i.e. energy efficiency improvements), but it seems that around 2035 a plateau is achieved beyond which further efficiency improvements are hard to implement. Until that point is reached in 2035, the carbon intensity of fuels could be moderately reduced by about 15%. However, after 2035, the regulatory measures (in particular the ban of conventional fossil fuel cars) enable us to sharply reduce carbon intensity until 2050 and thus continuously reduce GHG emissions of passenger transport. This means on the passenger side, the GHG reductions result from combined improvement of energy intensity and carbon intensity. For freight transport, the main driver of reductions is the decrease in intensity of energy use, while reduction of carbon intensity plays a limited role.

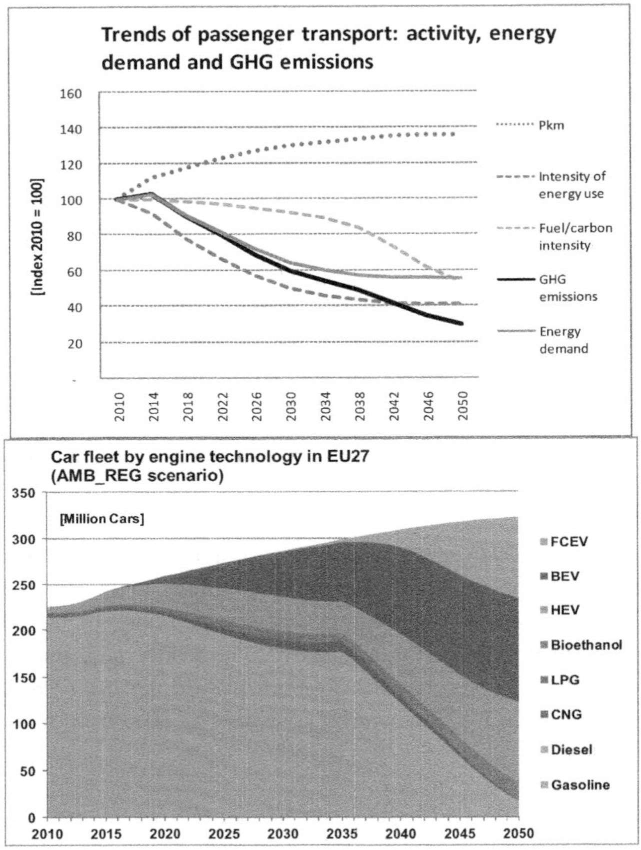

Figure 16.3. *Ambitious regulation scenario (AMB_REG): ASIF decomposition of passenger transport (top) and vehicle fleet structure (bottom, ASTRA model), FCEV: fuel cell electric vehicle, BEV: battery electric vehicle, HEV: hybrid electric vehicle, LPG: liquefied petroleum gas, CNG: compressed natural gas/biogas [FIO 12]. For a color version of the figure, see www.iste.co.uk/jacob/energy.zip*

Turning to discuss the fuel use in the AMB_REG scenario. In 2010, fossil fuels dominated, including only a minor share of blended biofuels. The strong influence of energy efficiency improvements sharply reduced the demand of fossil fuels until 2050. In parallel, the fraction of biofuels consumed increases, though a maximum amount of biofuel use of 50 mtoe, largely for air transport, is observed between 2030 and 2040. Afterward, biofuel demand declines to 40 mtoe. In 2050, about 40% of air energy demand would be supplied by biokerosene. Electricity demand of transport reveals the highest growth rates between 2020 and 2040, while hydrogen use starts to grow strongly toward 2040 and 2050. In 2030, biogas completely replaces natural gas, though due to limited uptake of gas vehicles in the fleet the demand side constrains an increased use of biogas.

Road transport, and in particular car transport, has to deliver the largest absolute reductions of energy demand and GHG emissions. With more than 90% of domestic transport GHG emissions accounting for road transport, this is obvious, as well. However, as road transport, and in particular cars and light-duty vehicles, disposes of the largest potentials to both reduce energy demand and to switch to low-carbon or carbon-free energy sources, these two findings of GHG-TransPoRD are consistent and fit together. Finally, it should be noted that the AMB_REG scenario achieving the 60% reduction target of the EU27 poses an abatement cost on transport users and corresponding a minor reduction of GDP but on the other hand it reveals a negative abatement cost for the society, or in other words an abatement benefit.

16.6. Synthesis on suitable GHG reduction strategy for transport

The Intergovernmental Panel on Climate Change (IPCC) and the International Energy Agency [IEA 11] both emphasize the requirement for a peak of GHG emissions during the current decade until 2020. This means early reductions of GHG emissions from European transport will be preferential compared with later reductions. In the next two decades, only the road mode will be able to contribute both significant and early GHG reductions. The other modes will mainly rely on operational measures to reduce their GHG emissions during that period due to slow fleet turnover. This can be identified by the tested maximum efficiency scenario (Max_E&M), which was designed by GHG-TransPoRD to implement the most effective efficiency technology in a fastest and market-driven way, which concerns technology choice. Until 2020, it generates close to 46% stronger reductions than a pure electric vehicle (EV) scenario, 60% stronger reductions than the hydrogen fuel-cell (HFC) scenario and 24% more reductions than the combined EV+HFC scenario. The driver of the substantial GHG emission reduction is massive introduction of efficiency technology of road vehicles to be either implemented due to the climate mitigation awareness of the automotive industry or by setting of strict CO_2 emission standards through European and global legislation. To ensure achieving the target, policymakers should choose the second option.

However, our analysis revealed that around 2035, conventional fossil fuel vehicles need to be banned from the market completely to achieve the 60% reductions until 2050. Only with such a policy, the alternative technologies like EVs, HFC and PHEV would diffuse into the market fast enough, and according to our analysis doubts must be raised if this would happen without such a ban. But this means, over the next 10–15 years, investments into efficiency technologies of conventional fossil fuel cars need to be substantial, despite these vehicles only being sold for another 20 years. Of course, some innovations like light-weight design and improved aerodynamics can be implemented into any car independent from its engine technology. Other technologies would constitute transitory techniques that bridge to the development of alternative fuel vehicles, while there would probably also be improvements of the combustion engines, which would become obsolete after conventional cars have been phased-out from the market. The challenge to identify those technologies that improve efficiency most effectively and could also be applied longer into the future has to be solved by the automotive industry, in particular after 2020.

The A(S)IF structure of the AMB_REG scenario demonstrates that looking at the full period until 2050 during the first two decades, the reduction of energy intensity (in other words, improvement of energy efficiency) constitutes the dominating source for reductions of GHG emissions, while between 2030 and 2050 the reduction of carbon intensity, in particular through electrification of transport and the parallel transition of electricity production to a renewable based system, will be the dominating source of GHG reductions. *In short, first quickly capture the energy efficiency potentials and then focus on the carbon-free energy potentials.*

However, this should not be understood in a way that alternative fuel vehicles should not be developed and brought to the market in the first two decades. But, it has to be considered that even in the automotive industry, being the sector with the biggest R&D budgets in Europe, these budgets are limited and investments need to be prioritized. Then, priority at least in this decade should be given to efficiency improvements, while from the climate mitigation point of view alternative fuel vehicles (in particular EVs and HFCs) could receive a lower priority. This should by no means lead to a halt of their development, but rather to a shift of their massive introduction to a few years later as this seems to be reasonable, at least under a constrained investment budget. Nevertheless, it must be ensured that these vehicles get onto their learning curve, e.g. by selling them only to early adopter markets, which in the case of EVs would be fleet operators in certain sectors, instead of intending to sell them to a mass market from the beginning of market diffusion. Such specific markets should also be considered when designing policies to foster alternative fuel vehicles.

Efficiency improvements of road transport in climate mitigation scenarios were much faster than the increase in energy prices, given the moderate increases estimated for the scenarios. Thus, in all scenarios that do not counterbalance efficiency gains by increasing other transport cost (e.g. fuel duties, road tolls and urban road charges), very strong rebound effects have been found leading to a strong modal-shift toward road transport and away from the more efficient rail mode and public transport. Such a rebound effect may consume a significant part of the GHG savings, such that counterbalancing measures need to be taken. The ambitious scenarios (AMB_TP and AMB_REG) both increase fuel duties and introduce urban charges such that they generated higher GHG emission reductions than the scenarios without such measures and a pure focus on efficiency and alternative technologies. Increasing road transport taxes and tolls brings about the co-benefit that government revenues from the transport sector are stabilized, while pure efficiency and alternative technology scenarios deteriorate the government revenues from the transport sector, which would in turn be required to fund infrastructure and operation within the transport sector, e.g. for railway infrastructure and public transport.

16.7. Conclusions

The most important conclusion to draw from the integrated analysis is that the 60% GHG reduction target for the transport sector could be achieved by 2050 compared to EU27 transport-related emissions of 1990. Of course, the target is ambitious such that most of the scenarios and policy packages tested by GHG-TransPoRD failed to deliver the required reductions. However, the scenario analysis concluded that scenarios combining:

– fast development of efficiency technology;

– alternative engine technologies able to build their energy supply on renewable electricity;

– ambitious policy-making to counterbalance rebound effects and maintain financial stability of government transport revenues;

– ambitious regulation phasing out fossil fuel cars around 2035 together with a moderate modal-shift from road toward more energy efficient modes;

– adaptation of the electricity system to become largely renewable based

will enable us to achieve these targets. Such a scenario was developed and tested in the AMB_REG scenario by GHG-TransPoRD. Sensitivity analyses confirm that with higher oil prices, even more stringent GHG reduction targets could be achieved. The largest reductions have to come from road transport, while in the medium term the slow fleet turnover of shipping and air mainly enables reductions through operational measures. For rail and waterway transport, it would be more relevant to be put in a position of being able to carry a larger share of demand than to focus only on technological or organizational improvements within the mode.

Building on the scenario analyses and the understanding of technological potentials, economic constraints and feedbacks between modes and measures, the GHG-TransPoRD project proposes GHG reduction targets for transport in EU27 as presented in Table 16.4. The targets are defined by mode as well as for the total transport sector. The table contains in the upper part reduction targets referring to a GHG emissions base calculated for the year 2010, as the measures implemented and tested in GHG-TransPoRD commence in 2011. The lowest row then presents proposed reduction targets for total EU27 transport in comparison with 1990, which is the base year usually applied in climate policy. It should be pointed out that Table 16.4 builds on absolute values of GHG emissions such that targets, e.g. for rail transport and road freight transport consider modal-shift from road to rail as it was observed in the scenarios.

		2020	2030	2050
Road	Passenger	-20% to -30%	-40% to -55%	-70% to -85%
	Freight	-10% to -20%	-30% to -45%	-40% to -60%
Air		0% to -5%	-10% to -20%	-40% to -55%
Ship		(+15% to 0%)	(+30% to 0%)	(+50% to -20%)
Rail		+10% to -10%	0% to -20%	-10% to -35%
Transport	(excl. ship)	-10% to -20%	-40% to -50%	-70% to -90%
Transport	vs. 1990	+10% to +5%	-20% to -30%	-60% to -70%

Table 16.4. *GHG reduction targets by mode for EU27 compared to emissions of 2010 (top) and 1990 (bottom). Proposal by GHG-TransPoRD (Source: [SCH 12])*

The transport strategy identified by GHG-TransPoRD provides both research and policy recommendations. R&D recommendations include to support R&D for biofuels, in particular for air transport, and R&D for cross-modal transport. The policy strategy suggests in early years a focus on road transport (i.e. efficiency of internal combustion engines and vehicle structures like aerodynamics and lightweight construction) and in medium-term time horizon on alternative energy technologies (i.e. electricity and hydrogen both produced from renewable electricity). Additionally, the transport strategy must include pricing measures and ambitious regulations (e.g. banning of selling conventional fossil fuel cars after 2035) and also changing behavior.

16.8. Acknowledgments

The project GHG-TransPoRD was co-funded by the European Commission 7th research framework programme. The project was coordinated by Dr. Wolfgang Schade at that time at Fraunhofer-ISI, Germany, and was undertaken in collaboration with four European partners, TRT from Italy, JRC-IPTS from Spain,

TML from Belgium and ITS Leeds from the UK. The project team included, in alphabetical order: Akkermans L., Fiorello D., Guehnemann A., Jopson A., Köhler, J., Krail M., Laird J., Leduc G., Martino A., Moizo A., Raganato P., Schade B., Schade W., Shepherd S., Sievers L., Tercero L., Tight M., Timms P., van Herle K., Weiss C., Wiesenthal T. All of them made valuable contributions to generate the results presented in this chapter. Any errors and omissions of course remain with the authors of the chapter.

16.9. Bibliography

[AKK 10] AKKERMANS L., VANHERLE K., MOIZO A. *et al.*, Ranking of measures to reduce GHG emissions of transport: reduction potentials and qualification of feasibility, Deliverable D2.1 of GHG-TransPoRD project co-funded by European Commission 7th RTD Programme. Transport & Mobility Leuven, Belgium, 2010.

[EEA 11] EEA – EUROPEAN ENVIRONMENT AGENCY, Laying the foundations for greener transport. TERM 2011: transport indicators tracking progress towards environmental targets in Europe, EEA Report 07/2011, Copenhagen, 2011.

[EUR 09] EUROPEAN COMMISSION DG RTD, The 2009 EU Industrial R&D Investment Scoreboard. JRC/DG RTD, available at: http://iri.jrc.ec.europa.eu/research/scoreboard_2009.htm, 2009.

[FIO 12] FIORELLO D., SCHADE W., AKKERMANS L. *et al.*, Results of the techno-economic analysis of the R&D and transport policy packages for the time horizons 2020 and 2050, Deliverable D4.1 of GHG-TransPoRD project co-funded by European Commission 7th RTD Programme. TRT Trasporti e Territorio SRL, Milan, Italy, 2012.

[IEA 11] IEA, INTERNATIONAL ENERGY AGENCY, World Energy Outlook 2011, Paris, 2011.

[KAH 08] KAHOULI-BRAHMI S., "Technological learning in energy-environment-economy modelling: a survey", *Energy Policy*, vol. 36, pp. 138–162, 2008.

[LEY 10] LEDUC G., KÖHLER J., WIESENTHAL T. *et al.*, Transport R&D Capacities in the EU, Deliverable D1.1 of GHG-TransPoRD project co-funded by European Commission 7th RTD Programme. Fraunhofer-ISI, Karlsruhe, Germany, 2010.

[OLI 12] OLIVER WYMAN, FAST 2025 – Future Automotive Industry Structure, Published by the German Association of the Automotive Industry (VDA), Berlin, 2012.

[SCH 11] SCHADE W., AKKERMANS L., FIORELLO D. *et al.*, Bottom-up quantifications of selected measures to reduce GHG emissions of transport for the time horizons 2020 and 2050: Cost assessment of GHG mitigation measures of transport, Deliverable D3.1 of GHG-TransPoRD project co-funded by European Commission 7th RTD Programme. Fraunhofer-ISI, Karlsruhe, Germany, 2011.

[SCH 12] SCHADE W., KRAIL M., Aligned R&D and transport policy to meet EU GHG reduction targets, Final Report. Deliverable D7.1 of GHG-TransPoRD project co-funded by European Commission 7th RTD Programme,Fraunhofer-ISI, Karlsruhe, Germany, 2012.

[WIE 12] WIESENTHAL T., LEDUC G., CAZZOLA P. *et al.*, Mapping innovation in the European transport sector, An assessment of R&D efforts and priorities, institutional capacities, drivers and barriers to innovation, JRC EUR 24771, Seville, Spain, 2012.

17

Why do CO_2 Emissions from Heavy Road Freight Transport Increase in Spite of Higher Fuel Prices?

The chapter analyzes why CO_2 emissions from heavy road freight transports increase in spite of higher fuel prices. Swedish time series data for 1990–2011 are analyzed with the help of indicators. The logistic efficiency and the energy efficiency improved, especially in the 1990s due to the allowance of heavier trucks. Since then, no major efficiency gains have been realized. Potentially cost-effective technologies exist but technical, institutional and financial barriers reduce the incentives for the transport firms to implement these. Split incentives caused by contract structures or ownership patterns can impede the employment of the technologies. If fuel savings are realized, rebound effects can appear that cancel out improved energy efficiency. Deeper knowledge of the impacts of policy measures is required in order to understand why CO_2 emissions increase despite increased fuel prices. A better understanding of the model simplifications in the Swedish national model Samgods is also needed, as well as an analysis of what is required to better mirror the contracts that exist.

17.1. Introduction

17.1.1. *Background*

Most of the CO_2 emissions within the transport sector are from passenger cars and heavy trucks. In spite of increased traffic, the CO_2 emissions from cars in Sweden decreased by around 105 between 1990 and 2011 [NAT 15]. This reduction can be explained by the use of more energy-efficient cars and an increased use of biofuels. This decrease was counteracted by the around 42% increase in the CO_2 emissions

---

Chapter written by Inge VIERTH.

from heavy trucks during the same period.[1] CO_2 emissions from heavy trucks decreased in 2012 and 2013 but are expected to increase again if no additional policies are implemented. Taxes on fuel based on their carbon content are used to control CO_2 emissions. Fuel taxes are, in contrast to administrative and informative measures, particularly useful when there is asymmetric information, e.g. regarding the available technology. This follows mainly from taxes and other economic instruments' ability to decentralize decision-making to the firms which are best equipped to make such decisions. The diesel price (including tax) increased nearly four times as much as GDP between 1990 and 2011 and was subject to larger fluctuations. The increase in the diesel price was partly due to the abolishment of the km tax for heavy trucks in the beginning of the 1990s. The observed development of higher fuel prices and higher CO_2 emissions from heavy trucks is somewhat puzzling. Our purpose is to investigate this and try to explain the historical developments in Sweden since 1990.

17.1.2. *Scope and structure of the chapter*

This chapter analyzes how CO_2 emissions from heavy road freight transport in Sweden have developed since 1990. Indicators are applied for transport demand, modal split and transport service prices (section 17.2), the logistic efficiency of road freight transport (section 17.3), their energy efficiency (section 17.4) and the CO_2 intensity of the fuel used (section 17.5). We discuss how different factors and circumstances influence the indicators. In section 17.6, we compare the in-Sweden observed price elasticities to modeled elasticities from the literature and the Swedish national freight model. In section 17.7, some conclusions are drawn.

17.2. Swedish freight transport market

17.2.1. *Freight transport intensity and modal split*

Freight transports have grown rapidly in both Sweden and the rest of Europe as well as the wider world. Increased transport volumes will *ceteris paribus* lead to increased demand for energy and more CO_2 emissions. Figure 17.1 shows that tonne-km increased faster than GDP in the 1990s, while it has been the other way round since 1999. This means the transport intensity has decreased and a decoupling process has begun. Similar developments are described for the UK [PAG 09] and Denmark [KVE 07]. Eng-Larsson *et al.* [ENG 12] explain the cyclic behavior of freight transport CO_2 emissions in Sweden 1990–2008 by applying the theory of economic growth cycles.[2]

1 CO_2 emissions from light trucks doubled between 1990 and 2011 [NAT 15].

2 Around 80% of the freight transport CO_2 emissions are from heavy trucks [NAT 15].

The amount of cargo transported in, to and from Sweden was quite stable over the period 1990–2011. The increase in tonne-km was caused by the increase in transport distance and the creation of centralized production and warehouse systems. McKinnon [MCK 10] suggests that the trend has been driven all over the world by (1) the wider sourcing of supplies and (2) the centralization of industrial production and warehouses (incl. increased foreign trade). He states that the expansion of the market area is now at an advanced state in mature countries but not in all parts of the world and that CO_2 taxes or emission-trading prices would have to be set at a high level to return to more decentralized systems. McKinnon emphasizes that it is necessary to carry out trade-off analyses related to CO_2 emissions in order to determine the net carbon impact of transport on one side and warehousing/production on the other side.

If we have a closer look at the different modes of transport, the tonne-km increase on the Swedish territory was, in most years, highest for heavy road freight transport. Rail transport has been quite stable during the 1990s and increased since the turn of the millennium. The development of sea transport to, from and between Swedish ports on Swedish territory was about the same as for the total tonne-km.

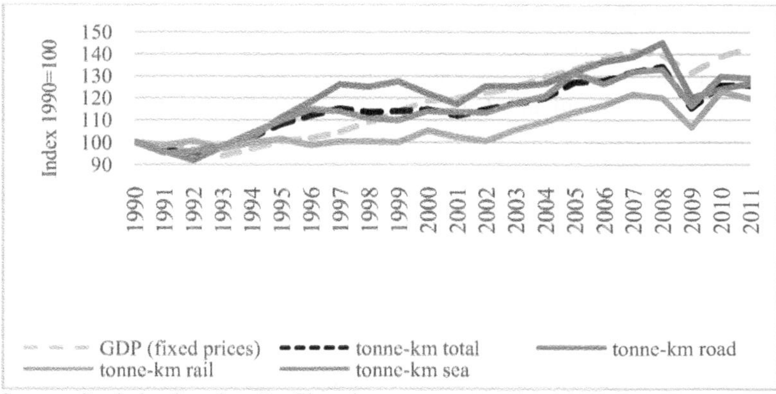

Source: Statistics Sweden, Trafikanalys

Figure 17.1. *Development of GDP (fixed prices) and tonne-km road, rail and sea and in total in Sweden 1990–2011. For a color version of the figure, see www.iste.co.uk/jacob/energy.zip*

The CO_2 intensity of the energy used, expressed in grams CO_2 per tonne-km, differs between the modes. Shifting from road to less CO_2-intense modes can reduce the CO_2 emissions per tonne-km. According to [EUR 11], 30% of the road freight over 300 km should shift to rail or waterborne transport by 2030, and more than

50% by 2050. In this context, we have to be aware that shippers take into account quality aspects (above the transports costs) when choosing transport solutions. Vierth and Mellin [VIE 08] find that there are large transfer potentials from long-distance road transport to combined road rail transport. However, it is noted that there are barriers in the form of insufficient rail infrastructure capacity and insufficient flexibility and co-operation along intermodal chains. Hjelle and Fridell [HJE 12] suggest that a modal shift to sea transport in many cases would demand increased use of less energy efficient roro-vessels and/or increased speeds and fuel consumption which means that modal shifts from road to sea or rail do not always lead to less CO_2 emissions and pollution.

17.2.2. *Transport service prices*

Statistics Sweden has published freight rates for road and sea transport since 2004 [SCB 12]. The prices are based on information from Swedish transport firms (road) and different stakeholders (sea). Figure 17.2 illustrates how sea transport prices fluctuate. The rates increased more than GDP between 2004 and 2008 and have dropped by about 30% points since then. The road transport prices, on the other hand, increased between 2004 and 2012. The growth rates for GDP and road freight transport prices were about the same until 2007; since then, road transport prices have grown faster than GDP. We have to be aware that different segments of the freight market use the road and the sea. Furthermore, the sea transport rates are to a high extent negotiated at a global level, and road transport at the national.

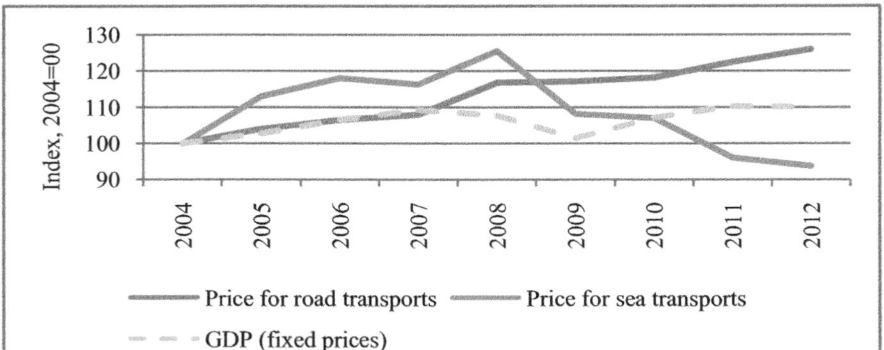

Source: Statistics Sweden

Figure 17.2. *Freight rates for road and sea transport 2004–2012, Index (2004=100). For a color version of the figure, see www.iste.co.uk/jacob/energy.zip*

17.3. Logistic efficiency of road freight transports

17.3.1. *Vehicle utilization*

The tonne-km/vehicle-km relation indicates how many vehicle-km are needed to perform a certain amount of tonne-km. If the vehicle utilization is increased, fuel consumption and CO_2 emissions per tonne-km decrease. The other modes also have a potential to increase vehicle utilization, but we focus on road transport. We also have to be aware of the fact that the commodity mix has changed over time toward more light and high-value goods.

Figure 17.3 shows that vehicle-km and tonne-km developed in the same way except for the periods 1993–1999 and 2008–2011. The allowance of heavier trucks in Sweden in 1990 (max. 56 tonnes instead of max. 51.4 tonnes total weight) and 1993 (max. 60 tonnes total weight) led to more efficient road transport due to a better use of the capacity per truck.[3]

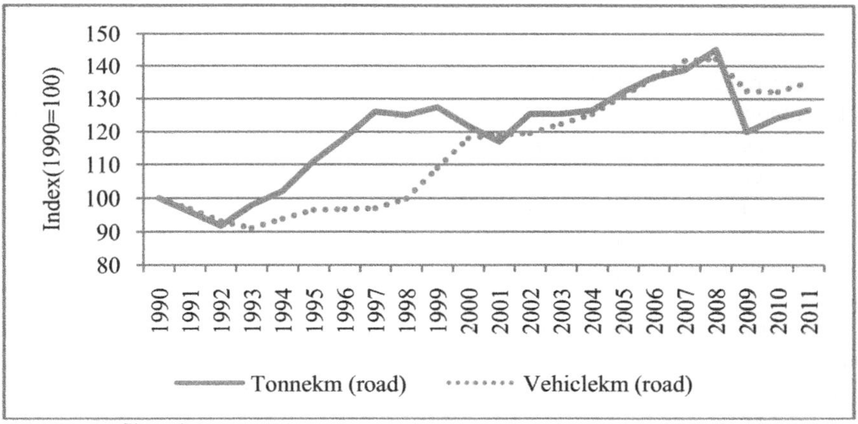

Source: Trafikanalys, HBEFA

Figure 17.3. *Road tonne-km and vehicle-km in Sweden 1990–2011*

Second-order effects that increase CO_2 emissions can appear in the form of shifts from rail to road. However, a review of the Swedish time series 1975–2005 for total transport shows that it is difficult to find evidence of road and rail taking volumes from each other [VIE 08]. However, there is more direct competition between the modes for international long-distance transport [VIE 12]. Beyond the use

3 Since 1 June 2015 Sweden allows trucks up to 64 tonnes total weight, see https://www.transportstyrelsen.se/sv/Nyhetsarkiv/andrade-bruttoviktstabeller-och-krav-for-64-ton-tunga-fordonskombinationer.

of heavier trucks, several measures can be applied to increase truck utilization. For example, McKinnon [MCK 10] lists improved back loading or use of more space-efficient handling systems and stresses the "inter-functional relationship" between freight transport and production, inventory management and that it is in many cases rational to prioritize other activities over improvements in transport efficiency.

In Figure 17.3, we observe that the load factor decreased during/after the recession in 2008. The drop in demand contributes to the lower load factor since 2008/2009. However, the share of domestic traffic without load decreased from 37% (2008) to 17% (2011). This indicates that the overall lower load factor cannot be explained by an increase in empty trips. The load factors, for the loaded trips and all trips, are tentatively higher for longer distances. The share of international transport, which are tentatively longer than domestic transport, has increased over time. The average distance of domestic road freight transport increased from 75 km (2000) to 103 km (2011).

17.3.2. *Consumed fuel per tonne-km*

Heavy road freight transport carried out in Sweden has become more efficient as the amount of fuel used has increased less than the tonne-km performed (see Figure 17.4). The efficiency improvement can be explained by e.g. the allowance of heavier trucks in the 1990s. There was a remarkable reduction in fuel consumption per vehicle-km during that period. The fuel savings were increased when focus shifted to the reduction of HC, NOx and PM emissions and the environmental classes (Euro classes) were introduced [KNE 14].[4]

The European Commission has developed a strategy to reduce the fuel consumption and CO_2 emissions from heavy vehicles [EUR 14]. To date, there are no standards for the fuel consumption of heavy vehicles in Europe as there are in Japan, the US, Canada and China. In contrast, there are fuel consumption standards for passenger cars and vans in Europe. The reduction of the fuel consumption per truck-km continued for several years after the 1990s because of the replacement of old trucks with a high fuel consumption. However, we have to be aware that our aggregated analysis does not include developments in different market segments. One phenomenon that should be studied in more detail is why the amount of consumed fuel grew faster than the tonne-km performed at the end of the 1990s. The overall drop in demand contributes to the higher fuel consumption per tonne-km since 2008.

4 According to [EUR 07], Euro class 6 (introduced in 2013) generates a 2–3% increase in fuel consumption compared to Euro 4.

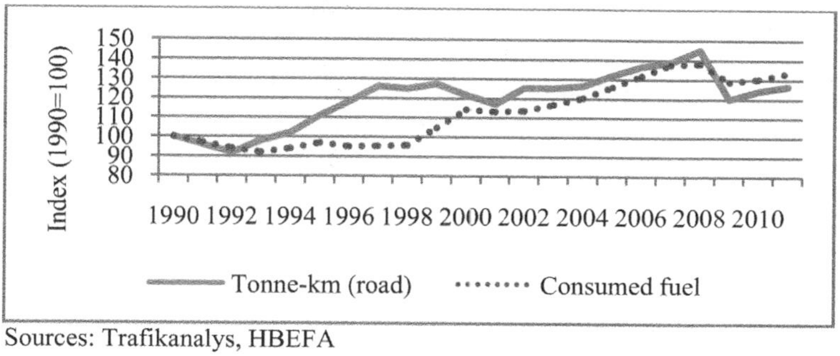

Figure 17.4. *Development of tonne-km and consumed fuel 1990–2011, Index (1990=100)*

Hovi and Andersen [HOV 10] studied Norwegian time series data for the period 1993–2008 and found that road freight transport has become more efficient due to better use of the capacity and the fact that longer transport has increased most.

17.4. Consumed fuel per vehicle-km

The development of vehicle-km and consumed fuel is highly correlated. Figure 17.5 shows that there was a decrease in the total vehicle-km and total fuel consumption during the 1990s and an increase since then. The energy efficiency (fuel consumption per vehicle-km) improved from the end of the 1990s as the total vehicle-km increased more than the total fuel consumption.

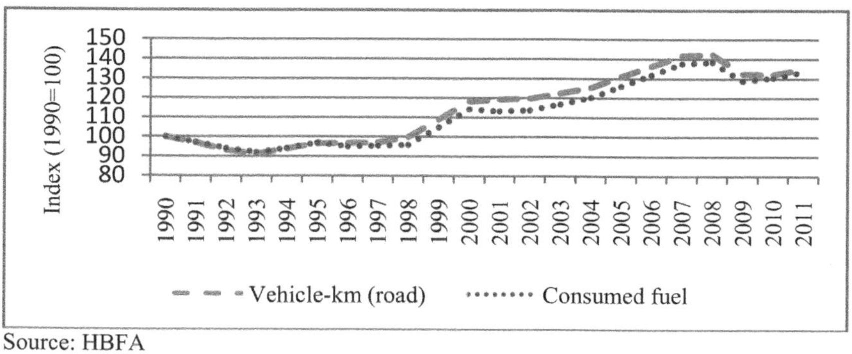

Figure 17.5. *Development of vehicle-km and consumed fuel 1990–2011, Index (1990=100)*

17.4.1. *Measurements to improve fuel efficiency*

The development of vehicle-km and consumed fuel shown in Figure 17.5 implies that factors other than the use of heavier trucks have counteracted the increase in fuel consumption per truck-km. These factors include the technical characteristics of the trucks, driving style, as well as design and maintenance of the roads. There have been only incremental efficiency gains from the refinement of truck technology since 1990.

The Ricardo study that was prepared for the UK Department for Transport [BAK 09] concludes that several near-term fuel technologies (improved engines and tyres, aerodynamic profiling, etc.) exist and estimates that the CO_2 emissions per truck can be reduced by at least 30%. Driver training, where transport firms monitor driving performances and give the staff incentives to drive fuel-efficiently, can reduce the fuel consumption per vehicle-km *ceteris paribus*. In Sweden, courses in “eco driving” have been mandatory for truck drivers since 2008. Liimatainen *et al.* [LII 14] state that “monitoring drivers’ performances is a complex matter” – in Finland, Sweden, Norway and Denmark typically done manually when filling the tank – and that performance data (but typically not tonne-km) are kept on computers. Shippers are becoming interested in the environmental performance of their hauliers.

Since 2006, Swedish truck and fuel providers, carriers, forwarders and shippers, researchers and public agencies co-operate within the KNEG network (Climate Neutral Freight Transportation) that has the goal to halve the climate impact of road freight between 2005 and 2020 by improving the efficiency of logistics, the efficiency of the trucks and fuel used and increasing the use of renewable fuels [KNE 09]. Mellin and Pyddoke [MEL 13] interviewed eight firms that are KNEG members and concluded that these firms apply several voluntary measures but cannot say what (1) the specific costs for these measures are or (2) whether these measures lead to significant reductions of CO_2 emissions.

17.4.2. *Barriers to the implementation of fuel-saving technologies*

A study performed by CE Delft on behalf of the ICTT (International Council on Clean Transportation) identifies barriers to the implementation of potentially cost-effective fuel efficiency improving technologies in the European freight sector [AAR 12]. The study finds that there are technological, institutional or financial barriers when it comes to the implementation of these technologies. Even though they are commercially available, there may be uncertainties related to certain technologies and market developments. Split incentives (principle agent problems) can impede the implementation of fuel-saving technologies as the firms that invest in these technologies have little incentive to do so. Split incentives can result from *contract structures*; i.e. when carriers can pass "their fuel costs" to shippers or logistics providers.

According to Greater Than [GRE 11], who performed questionnaires with 1,100 carriers and logistics providers in Europe, the following contract types are common in the road transport sector:

– *open book contracts*, where shippers and carriers agree on a *fixed operational margin*. These contracts remove much of the incentive for the carriers to improve the trucks' fuel efficiency (unless this is a specific criterion in the contract);

– contracts where carriers and shippers agree on a *fixed per unit price*. This price is protected by *fuel escalators* through which carriers can pass on increases in fuel costs to shippers;

– contracts where forwarders add a *commission* on top of the transport costs. This means the higher the transport costs, the higher the commission, which reduces the incentive to save fuel costs. Furthermore, logistics providers often work with *fuel surcharges* in order to pass on fuel cost increases to shippers; these surcharges also reduce the incentive to perform transport more energy-efficiently.

Greater Than [GRE 11] find that 92% of the 1,100 interviewed firms do not prioritize fuel savings and do not have fuel savings as a specific target during operations. However, when investing in new vehicles, fuel efficiency is one of the top priorities. Greater Than think that the low prioritization of fuel savings can be explained by the contract structures described above and by 'tradition', e.g. the transport sector's perception of fuel as a necessity that is hard to influence. Khan *et al.* [KHA 07] find that energy-related measures are generally given low priority when energy costs are only a small share of a firm's total costs.

Split incentives can also be caused by *ownership patterns* of trucks and trailers. Logistics providers do not usually own vehicles and therefore do not invest in fuel-saving technologies themselves. If they did invest, it would be the carriers that benefitted from the reduced fuel consumption.

Lack of information can also be an obstacle. Firms that are not aware of fuel-saving measures will obviously not invest in them. Aarnink *et al.* [AAR 12] mention that there is no commonly agreed standard for expressing the fuel efficiency of truck and trailer. According to [JOH 12], energy efficiency improvements claimed by the truck manufacturers are seldom confirmed in real traffic. This can probably be explained by the fact that different driving styles, e.g. using the same speed uphill as downhill, cancel out the energy efficiency improvements. A study by the Forestry Research Institute of Sweden reveals that the fuel consumption of timber trucks was the same in 2008 and 2013 [SKO 14]. According to the truck manufacturers, more in-depth analyses are needed to study this result.

Once an investment decision for a fuel-saving technology has been made, *additional transaction costs* such as costs for negotiation, implementing and monitoring contracts will result. Faber *et al.* [FAB 09] state that these aspects can be very costly and lead to *hidden transaction costs*. These transaction costs can act as a barrier, especially for smaller companies.

Financial barriers can limit the introduction of new solutions. According to McKinsey [MCK 09], fuel-saving technologies in the transport sector have relatively high upfront capital costs compared to other sectors. Goodyear Dunlop [DUN 12] found that 40% of the fleet managers in a sample of 400 that were surveyed lacked funds to replace older trucks. A payback period of 2–3 years is normally required for investments in fuel-saving technologies [AEA 11]. The Swedish road freight transport sector is characterized by a high share of small carriers. Around 40% are one-man firms and around 80% have fewer than five employees.

The level and development of fuel prices also influence the adoption of new fuel-saving technologies. The value of the technologies depends on the fuel price: the higher the fuel price, the shorter the payback period. Uncertainties about the fuel price can impede the adoption of fuel-saving technologies. In addition, the annual mileage of the trucks also influences the payback period. Dargay and Gately [DAR 97] conclude from an econometric study that "consumers do not necessarily respond in the same way to rising and falling (fuel) prices, nor in the same way to sudden and substantial price rises as to minor price fluctuations".

Another aspect that has to be kept in mind is that the road freight market is deregulated at the national and EU-level. The share of the fuel costs varies between carriers (and drivers) from different countries. Hovi and Hansen [HOV 11] find that staff costs are far more important than taxes and fees. Figure 17.6 illustrates how truck and fuel costs are quite similar in Norway, Sweden and Poland. However, staff costs are very different: their share of the total transport costs varies between 12% in Poland and around 40% in Norway.

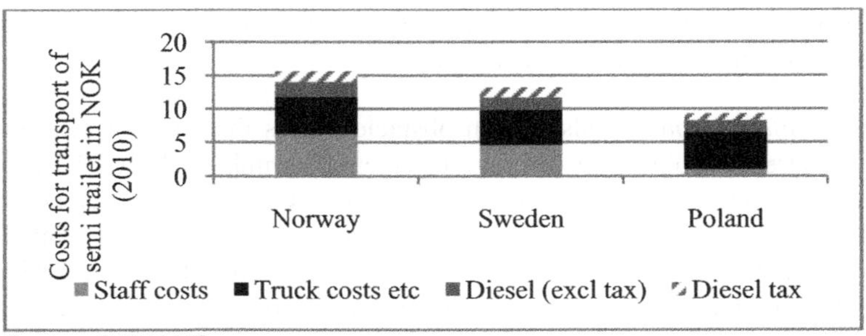

Source: [HOV 11]

Figure 17.6. *Transport costs for semi-trailer in NOK (2010) . For a color version of the figure, see www.iste.co.uk/jacob/energy.zip*

17.4.3. *Rebound effect and elasticities of fuel use in respect to fuel price*

Winebrake *et al.* [WIN 12] analyze the existing literature and find that there is little focus on the rebound effects, and that a framework for considering these studies in a policy context has not yet been established. The paper differentiates between direct rebound effects, indirect rebound effects and economy-wide rebound effects and focuses on the direct rebound effects. Winebrake *et al.* [WIN 12] further suggest that the "freight transport elasticity estimates are influenced by a number of factors of variability categorized by the specific nature of the shipping acticity, the macroeconomic influences involved, and the measurement tolls used to access the elasticities".

The paper of De Borger and Mulalic [DEB 12] is not included in Winebrake *et al.* [WIN 12]. It studies the determinants of fuel use in the Danish road transport industry using aggregate time series data over the period 1980–2007. An econometric model captures *the linkages between the demand for road freight services and the characteristics of the truck fleet* (that determines fuel efficiency and the demand for fuel) and allows them to identify and estimate the *rebound effect* of improvements in fuel efficiency. As fuel efficiency improves, the variable unit costs per tonne-km are reduced. This leads to lower prices for road transports (when a competitive market structure is assumed). In turn, the lower road transport prices stimulate the demand for road transport services, which raises fuel consumption. In other words, the reduction in fuel induced by higher fuel efficiency is partly offset and "the end total fuel use" responds less than proportionally to the exogenous changes in fuel efficiency. A rebound effect of approximately 10% in the short run to 17% in the long run is estimated. This means that a 1% improvement in fuel efficiency reduces fuel use by 0.90% (in the short run) to 0.83% (in the long run). The authors find that carriers react to higher fuel prices by (1) raising the average capacity of the trucks (which in turn increases the fuel consumption per truck so that the total fuel consumption increases) and (2) investing in more efficient trucks (so that the fuel consumption decreases). De Borger and Mulalic [DEB 12] estimate elasticities of the fuel use with respect to the fuel price of –0.13 (short run) to –0.22 (long run) and explain the relatively low elasticities with the rebound effect *and* the impact of the fuel prices on the size/age composition of the fleet.

17.5. CO_2-intensity of fuel used in the road freight sector

Finally, it is possible to reduce CO_2 emissions from road freight transport by using energy sources that contain no or low carbon, such as electricity, biofuel or biogas. Cullinane and Edwards [CUL 10] point out that the batteries which electric vehicles depend on are quite large and heavy and have a very limited range. KNEG

[KNE 12] discuss several obstacles for increasing the share of alternative non-fossil energies: access to raw material including distribution, production plants, and infrastructure for vehicles that can use the alternative fuel. In 2014, 97.5% of Swedish heavy trucks were diesel trucks and their average age was nearly 11 years [TRA 15]. The limited incentive to replace old trucks due to the quite low time-based infrastructure fee (Eurovignette) is one explanation for the high age of the fleet in Sweden [VIE 12].

17.6. Modeling policy impacts

17.6.1. *Observed elasticities*

Elasticities are usually derived from econometric models or network models and are defined using the *ceteris paribus* condition, i.e. are valid under the assumption that all other things do not change. Elasticities can also be calculated from observations of the impact of a change (e.g. a change in fuel price) where other things can change. Figure 17.7 comprises the development of the diesel price (incl. tax), the liters of diesel consumed, the price of road transport and road tonne-km for heavy trucks in Sweden 2004–2011. The fuel price fluctuated a lot and increased by 70%; the fuel consumption fluctuated and increased much less but largely followed the fuel price.

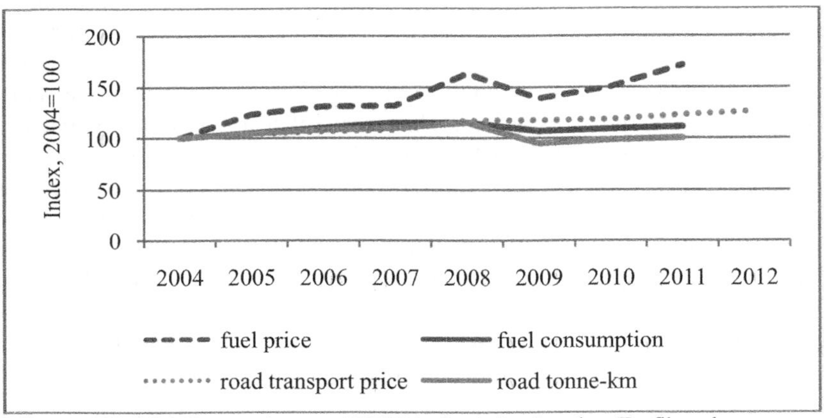

Source: Swedish Energy Agency, HBEFA, Statistics Sweden, Trafikanalys

Figure 17.7. *Development of fuel price (incl. tax), fuel consumption and price for road transports (in running prices) and road tonne-km. For a color version of the figure, see www.iste.co.uk/jacob/energy.zip*

Typically, demand for road transport decreases when fuel or transport prices for road transport increase and *vice versa* which gives negative price elasticities. The road transport price elasticity (demand elasticity) of tonne-km 2004–2008 is positive and around 1. We could consider that this is caused by omitted variables. In 2008, the amount of tonne-km dropped in the same way as the fuel price and increased again, but not as much as the diesel price. The fuel price elasticity of liters fuel consumed is also positive (other than in 2007) but less elastic than the road transport price (costs) elasticity of tonne-km. We are aware of the fact that we cannot draw firm conclusions from our back-of-the-envelope calculations for the relatively short and fluctuating period 2004–2011. But it can be shown that fuel consumption (and CO_2 emissions) and road tonne-km in general increased, in spite of higher fuel prices and road freight transport rates.

17.6.2. *Modeled elasticities*

To our knowledge, very little literature examines fuel price elasticities of heavy truck fuel demand and transport. The few existing studies include Winebrake *et al.* [WIN 15] for the US and Barla *et al.* [BAR 14] for Canada. When it comes to Europe, de Jong *et al.* [JON 10] compile long-term elasticities for long-distance road freight transport. They derived the following 'best guess' elasticities from the ranges in the literature (see Table 17.1). The *fuel price elasticity* with regard to fuel use includes changes in fuel efficiency, logistic efficiency and demand. The *vehicle-km price elasticities* consist of changes in mode, logistic efficiency and demand. The *tonne-km price elasticities* consist of changes in mode and demand. The best guess road transport price elasticity of tonne-km is –1.0 and the best guess fuel price elasticity of liters fuel used is –0.1. The elasticities have about the same absolute size as our roughly calculated elasticities in Figure 17.8, but the opposite sign.

Price change	Impact on		
	Fuel use	Vehicle km	Tonne km
Fuel price	-0.3	-0.2	-0.1
Vehicle km price		-0.9	-0.6
Tonne km price			-1.0

Source: De Jong *et al.* [JON 10]

Figure 17.8. *Price elasticities*

The Swedish national freight transport model Samgods assumes a fixed transport demand. The model system comprises a logistics module which allows taking into account shipment sizes, economies of scale and consolidation in the transport system, which makes the model more realistic than the former model that assumed constant shipment sizes. The model is deterministic and minimizes the annual logistics costs of hypothetical shippers of different sizes. For information about the model, see [JON 08]. With the help of the Samgods model, we calculated a 2.5% decrease in the road tonne-km due to a 10% increase in the distance-based road transport costs (an elasticity of –0.25). We also calculated shifts from smaller to larger trucks. The Samgods model does not distinguish between different time perspectives and does not take into account different types of actors (carriers, forwarders and shippers) and their contracts. Perfect competition is assumed, as is shippers taking into account e.g. the *changed* fuel taxes (that are faced by the carriers) when they choose transport solutions. It is assumed that major changes in the relative costs lead to revised transport solutions.

We see a need to analyze the implications of the simplifications in the model and consider what is needed to better mirror the circumstances we observe in reality. It is, for example, necessary to distinguish between the short-term perspective (with existing contracts and ownership patterns for vehicles) and the medium- and long-term perspectives (where contracts can be revised).

17.7. Conclusions

CO_2 emissions from heavy trucks in Sweden have increased between 1990 and 2011, even though they were reduced due to the economic recession. No major decreases are expected if no additional policies are implemented. CO_2 emissions from cars decreased by around 10% between 1990 and 2011. This is partly explained by the existence of fuel consumption standards for light vehicles but not for heavy vehicles within the European Union. Europe is after Japan, the United States, Canada and China, which have implemented fuel consumption standards for heavy vehicles.

Different potentially cost-effective technologies to further reduce CO_2 emissions from heavy road freight transport are available. However, technical, institutional and financial barriers may substantially reduce the incentives for the transport firms to improve the fuel efficiency of their trucks. Institutional barriers are related to the incentive structures that involve several stake holders. Split incentives can impede the employment of energy efficient technologies, as the firms that invest in the technologies have little incentive to do so. Split incentives can result from contract structures, where carriers can pass their fuel costs directly to shippers, ownership patterns for trucks and trailers, lack of information and transaction costs.

If fuel savings are realized, rebound effects can appear. Usually, fuel efficiency improvements imply lower unit costs per tonne-km, which in turn leads to lower road transport prices. The lower road transport rates can stimulate the demand for road transport, which raises CO_2 emissions. This calls for complementary price policies, e.g. increasing CO_2 taxes. Empirical analyses of the determinants of fuel consumption and the estimation of rebound effects are highly relevant for the evaluation of policy measures as higher fuel taxes, regulations for fuel efficiency and truck dimensions. More research is needed to understand the causes and scale of the rebound effects related to road freight transports. Policy makers and other stakeholders have to keep in mind that the road freight transport market differences in staff costs are generally more important than differences in taxes and fees.

CO_2 emissions from freight transport can be reduced by shifts from road to less CO_2-intense modes, but shippers minimize their total costs and value quality aspects above transport costs when choosing transport solutions. There are obstacles to increasing the share of non-fossil energies in the form of access to raw materials (including distribution), production plants, infrastructure for vehicles that can use the alternative fuels, etc.

A better understanding of the impacts of different policy measures to reduce CO_2 emissions is required in order to understand why CO_2 emissions for heavy trucks increase, despite increased fuel prices. More in-depth knowledge about the barriers to the adoption of fuel-saving technologies and the rebound effect that cancels out the effect of energy-efficiency measures is desirable. A better understanding of the implications of the lack of thresholds and other model simplifications in the Swedish national freight model Samgods is also needed, and an analysis of what is required to better mirror the contracts that exist.

17.8. Acknowledgments

The author would like to thank Håkan Johansson, Swedish Transport Administration and Björn Carlén, VTI, for fruitful comments on an earlier version of the chapter.

17.9. Bibliography

[AAR 12] AARNINK S., FABER J., DEN BOER E., Market barriers to increased efficiency in the European on-road freight sector, CE Delft report, Delft, 2012.

[BAK 09] BAKER H., CORNWALL R., KOEHLER E. *et al.*, Review of low carbon technologies for heavy goods vehicles, report prepared for the Department for Transport, RICARDO, 2009.

[BAR 14] BARLA P., GILBERT-GONTHIER M., TAGNE KUELAH J.-R., "The demand for road diesel in Canada", *Energy Economics*, vol. 43, pp. 316–322, 2014.

[CUL 10] CULLINANE S., EDWARDS J., "Benefits and costs of switching to alternative fuels", in MCKINNON A., CULLINANE S., BROWNE M. (eds), *Green Logistics: Improving the Environmental Sustaninability of Logistics*, Kogan, London, pp. 313–326, 2010.

[DAR 97] DARGAY J., GATELY D., "The demand for transportation fuels: imperfect price-reversibility?", *Transportation Research Part B: Methodological*, vol. 31, pp. 71–82, 1997.

[DEB 12] DE BORGER B., MULALIC I., "The determinants of fuel use in the trucking industry – volume, fleet characteristics and the rebound effect", *Transport Policy*, vol. 24, pp. 284–295, 2012.

[DEJ 08] DE JONG G., BEN-AKIVA M., BAAK J., Method report logistics model in the Swedish freight transport model system, Deliverable 6B for the Samgods group, Significance, 2008.

[DEJ 10] DE JONG G., SCHROTEN A., VAN ESSEN H. *et al.*, Price sensitivity of European road freight transport – towards a better understanding of existing results. A report for Transport & Environment, Significance & CE Delft, 2010.

[ENG 12] ENG-LARSSON F., LUNDQUIST K.-J., OLANDER L.-O. *et al.*, "Explaining the cyclic behavior of freight transport CO2-emissions in Sweden over time", *Transport Policy*, vol. 23, pp. 79–87, 2012.

[EUR 07] EUROPEAN COMMISSION, COM 851, Annex to the proposal for the regulation of the European Parliament and of the Council on the approximation of the laws of the Member States with respect to emissions from on-road heavy duty vehicles and on access to vehicle repair information, Commission Staff Working Document, 2007.

[EUR 11] EUROPEAN COMMISSION, Roadmap to a single European transport area – towards a competitive and resource efficient transport system, white paper, 2011.

[EUR 14] EUROPEAN COMMISSION, Strategy for reducing heavy-duty vehicles' fuel consumption and CO_2 emissions, Communication com(2014) 285 final, 2014.

[FAB 09] FABER J., MARKOWSKA A., NELISSEN D. *et al.* Technical support for European action for reducing greenhouse emissions from international maritime transport, CE Delft, 2009.

[GOO 12] GOODYEAR DUNLOP, Driving fleet fuel efficiency – the road to 2020, Goodyear Dunlop Tires Europe B.V., Belgium, 2012.

[GRE 11] GREATER THAN, Analysis of the European road freight market: business models and driving forces influencing its carbon footprint, Greater Than AB, Stockholm, 2011.

[HAN 09] HANSEN F., Åkerinäringen, hastigheten och den ekonomiska beslutssituationen, Report 2009:11, Dalarna University, 2009.

[HJE 12] HJELLE H., FRIDELL E., "When is short sea shipping environmentally competitive?", in OSSTHUIZEN J. (ed.), *Environmental Health – Emerging Issues and Practice*, Chapter 1, INTECH Open Access Publisher, pp. 2–20, 2012.

[HOV 10] HOVI I.B., ANDERSEN J., Utveckling i Transportnyttelse og miljø for godsbiler, Transportøkonomiskt Institutt TØI, TØI report 1063/2010, Oslo, 2010.

[HOV 11] HOVI I.B., HANSEN W., Rammbetingelser i transport-og logistikkmarkedet, TOI report 1150/2011, Oslo, 2011.

[IEA 12] IEA, *Technology Roadmap, Fuel Economy of Road Vehicles*, International Energy Agency, Paris, 2012.

[JOH 12] JOHANSSON H., Minskade utsläpp av koldioxid från vägtrafiken, Trafikverket, Borlänge, 2012.

[KHA 07] KHAN J., HARMELINK M., HARMSEN R. *et al.*, From theory based policy evaluation to SMART policy design (summary report of the AID-EE project), Ecofys, Utrecht, 2007.

[KNE 09] KNEG, On the road to climate neutral freight transportation, cooperation for a better climate, KNEG, 2009.

[KNE 12] KNEG, Hinder och drivkrafter för minskad klimatpåverkan från godstransporter, KNEG, Trafikverket, Calmers, 2012.

[KNE 14] KNEG, Ställ krav för klimatsmartare transporter, KNEG Resultatrapport 2014, KNEG, available at: http://kneg.org/wp-content/uploads/KNEG-Resultatrapport-2014_150dpi-7-MB.pdf, 2014.

[KVE 07] KVEIBORG O., FOSGERAU M., "Decomposing the decoupling of Danish road freight traffic growth and economic growth", *Transport Policy*, vol. 14, pp. 39–48, 2007.

[LII 14] LIIMATAINEN H., NYKÄNEN L., ARVIDSSON N. *et al.*, "Energy efficiency of road freight hauliers –a nordic comparison", *Energy Policy*, vol. 67, pp. 378–387, 2014.

[LIM 13] LIMATAINEN H., POLLÄNEN M., "The impact of sectoral economic development on the energy efficiency and CO2 emissions of road freight transport", *Energy Policy*, vol. 27, pp. 150–157, 2013.

[LUM 11] LUMAN R., Wegtransport uitgedaagd tot hogere vrachtrpis: vrachtvulumes herstelln, strategische keuzes noodzakelijk: Sectrowisie Wegtransport, ING, Belgium, 2011.

[MCK 09] MCKINSEY & COMPANY, Pathways to a low carbon economy, CEPS, Brussels, 2009.

[MCK 10] MCKINNON A., "Green logistics: the carbon agenda", *Electronic Scientific Journal of Logistics*, vol. 6, pp. 1–9, 2010.

[MEL 13] MELLIN A., PYDDOKE R., Frivillia klimatåtgärder på den svenska godstransportmarknaden, En intervjustudie, VTI report 778, 2013.

[NAT 15] NATURVÅRDSVERKET, Utsläpp av växthusgaser från inrikes transporter 1990-2013. Hämtat från, available at: http://www.naturvardsverket.se/Sa-mar-miljon/Statistik-A-O/Vaxthusgaser-utslapp-fran-inrikes-transporter/#, 2015.

[PAG 09] PAGNOLUCCI, BONILLA D., "UK freight demand: elasticities and decoupling", *Journal of Transport Economics and Policy*, vol. 43, pp. 317–344, 2009.

[SCB 12] SCB, Hämtat från Producentprisindex för tjänster (TPI), available at: http://www.scb.se/Pages/SSD/SSD_TablePresentation_340486.aspx?rxid=ad2a2b0b-4ca8-4f8a-b11e-f422fc60a6ca, 2012.

[SKO 14] SKOGFORSK, Bränsleförbrukning hos rundvirkesfordon 2008 och 2013, Skogforsk (Arbetsrapport nr. 820-2014), 2014.

[TRA 15] TRAFIKANALYS, Lastbilars klimateffektivitet och utsläpp, Trafikanalys, report 2015:12, available at: http://trafa.se/Projekt/Egna-projekt/Nulagesanalyser/Lastbilars-klimateffektivitet-och-utslapp/, 2015.

[VIE 08a] VIERTH I., MELLIN A., Svensk godsstudie baserad på nationell och internationell litteratur, VTI report 629, 2008.

[VIE 08b] VIERTH I., BERELL H., MCDANIEL J. *et al.*, The effects of long and heavy trucks on the transport system, Report on a government assignment, VTI report 605A, 2008.

[VIE 09] VIERTH I., LORD N., MCDANIEL J., Representation av det svenska godstransport- och logistiksystemet, VTI note N17A, 2009.

[VIE 12a] VIERTH I., KARLSSON R., Effekter av längre lastbilar och godståg i en internationell korridor, VTI report 764, 2012.

[VIE 12b] VIERTH I., SCHLEUSSNER H., Impacts of different environmentally differentiated truck charges on mileage, fleet composition and emissions in Germany and Sweden, CTS working paper No. 2012:22, Stockholm, 2012.

[WIN 12] WINEBRAKE J.J., GREEN E.H., COMER B. *et al.*, "Estimating the direct rebound effect for on-road freight transportation", *Energy Policy*, vol. 48, pp. 252–259, 2012.

[WIN 15] WINEBRAKE J.J., GREEN E.H., COMER B. *et al.*, "Fuel price elasticities in the U.S. combination trucking sector", *Transportation Research Part D*, vol. 78, pp. 166–177, 2015.

18

A Study on Super Credits and their Impact on Fleet-Average Real-World CO_2 Emissions

In this chapter, the authors investigate the impact of applying super credits to CO_2 emissions with respect to the new fleet and the total passenger car fleet. The study assumes different electric vehicle penetration scenarios in the time period up to 2050 in the EU-27 region. The modeling is realized using custom-designed software and the evaluation is carried out for type – approval CO_2 emissions, and extended to include real-world CO_2 emissions. Results show that super credit policies can have a decisive effect on fleet-average emissions depending on the e-mobility penetration share in the total fleet, the eligible vehicle development pattern, the new registrations growth rate, the technology distribution and any limitations applied according to each scenario. This correlation is also straightforward in terms of total fleet emissions, especially when real-world emissions are considered.

18.1. Introduction

According to the the European Commission white paper on Transport [EUR 11a], which is a roadmap setting the targets for the evolution of transport in the European Community until 2050, efficient, sustainable, and border-free transport in the European Community is envisioned. In this context, related key policies in the area are addressing carbon dioxide (CO_2) emissions and energy efficiency in order to provide the fundamental mechanisms to decouple transport growth from fossil fuel consumption. For instance, the 20-20-20 is a collection of robust policies aiming to achieve 20% lower CO_2 emissions in 2020 compared to 1990. For light duty

Chapter written by Petros KATSIS, Thomas PAPAGEORGIOU and Leonidas NTZIACHRISTOS.

vehicles, two main policies in the framework of 20-20-20 include Regulation (EC) No 443/2009 and Regulation (EC) 510/2011, and Directive 2009/30/EC which provide targets for the efficiency improvement of passenger cars, light commercial vehicles and the penetration of biofuels, respectively.

According to Regulation 443/2009 [EUR 09], two key targets are set for the mean CO_2 emission rate of new passenger cars sold in Europe. First, a gradual decrease in average emission levels to 120 g/km to be fully achieved by 2015, and then a further step to reach 95 g/km by 2020. The corresponding targets for light commercial vehicles set a target of 175 g/km by 2017 and then 147 g/km in 2020. Directive 2009/30/EC allows higher blends of biofuels in fossil fuels, allowing up to 10% bioethanol in conventional gasoline and up to 7% biodiesel in conventional diesel.

These regulations for M1 and N1 vehicles do not establish any targets beyond 2020 and both contain specific clauses related to their future review. It is obvious that further lowering CO_2 emission limits will be required to support the longer-term visions of the community, which currently extend to the 2050 horizon. The up-to-date vision for 2050 is to reduce road transport CO_2 emissions by 60% with respect to 1990 levels [COM 12]. Therefore, the next targets for emission limits for M1 and N1 vehicles will soon have to be decided. Indicative targets as low as 70 g/km have already appeared in relevant policy documentation.

Passenger cars are currently responsible for almost 12% of Europe's carbon dioxide (CO_2) pollution. In view of the aforementioned goals and in order to provide manufacturers with additional incentives to suppress emissions, there is a proposal to grant super credits for low emitting vehicles. Each such ultra low-emitting vehicles (ULEV) will be counted as multiple cars toward achieving the imposed CO_2 targets.

Super credits were first introduced in the last round of fuel efficiency regulations in 2008, to entice car companies into manufacturing vehicles with emissions below 50 g/km. Each low-emitting car would be counted as 3.5 vehicles in 2012 and 2013, 2.5 in 2014, 1.5 vehicles in 2015 and then one vehicle from 2016 onwards. This approach could help manufacturers to further reduce the average emissions of their new car fleet, yet all manufacturers would be required to achieve the same level of reduction. Super credits with a multiplier of 1.3 would apply in 2020–2023 for vehicles emitting less than 35 g/km for a limited number of cars per manufacturer. Recent impact assessments [HIL 13] have shown that the application of super credits could increase total well-to-wheel (WTW) greenhouse gas (GHG) emissions from around 1% in 2020 to as much as 14.5% in 2030.

Within this framework, electricity stands out as an appealing candidate for ultra low-emitting vehicles. The term "electromobility" (or e-mobility) often refers to the concept of using electric powertrain technologies and infrastructures to enable the electric propulsion of vehicles and fleets. Electromobility efforts are motivated by the need to address corporate fuel efficiency and emission requirements, as well as market demands for lower operational costs. E-mobility is mainly intended for passenger cars and light commercial vehicles as efficiency drops with mass increase. Hybrid electric vehicles (HEVs), plug-in hybrid electric vehicles (PHEVs), battery electric vehicles (BEVs), EVs with range extender (EREVs) and fuel-cell EVs (FCEVs) are some representative electromobility technologies which can, up to a different extent, minimize or even eliminate oil dependency and tailpipe CO_2 emissions, provided that the production supply chain does not emit too much CO_2. PHEVs, EREVs, FCEVs and certain EVs are generally considered vehicles that fall into the category of ultra low emitters, at least at a type-approval level.

In this work, the authors present a study on the impact of a variety of super credit scenarios on type-approval and real-world CO_2 emissions with respect to the new fleet and the total passenger car fleet. The study assumes different electric vehicle penetration scenarios upon which the super credit scenarios are based. The case study investigates the time period up to 2050, while super credit application typically involves the 2016–2023 period. The impact of super credits is simulated using custom software, and the evaluation is carried out bearing in mind the effect not only on new cars in the market, but also on the full passenger car fleet. Emission calculation is initially performed using type-approval CO_2 emissions. However, the chapter extends the traditional super credits impact assessment by examining the consequences in terms of real-world CO_2 emissions, which are generally higher than type-approval ones [MEL 11, NTZ 12].

This chapter is organized as follows: section 18.2 describes the methodology used to simulate the impact on CO_2 type-approval (TA) emissions due to the introduction of super credits. Section 18.3 analyzes the model for the simulation of the previous scenarios regarding real-world emissions. Section 18.4 discusses the results of the study and section 18.5 wraps up the conclusions of this work.

18.2. Methodology

Super credits are awarded with the aim to enhance e-mobility penetration; each low-emitting vehicle (some type of electric vehicle in this simulation) is awarded a multiplier, i.e. it counts as more than one vehicle when the fleet-average CO_2 emissions are calculated. As a result, this process would create additional "phantom" vehicles. Thus, total emissions would be allocated to a larger fleet and average emissions (gCO_2/km) would be reduced. This would provide more flexibility for non-ULEVs, allowing for a decrease in emission limits. Apart from this main concept, several other options could be considered; super credits could be either spent in the year acquired, or accumulated and distributed later on, in order to allow more time for conventional vehicles to lower their emissions.

In this work, the super credits period of application will be from 2016–2023. With the latter approach, a super credits "banking" scheme, each electric vehicle in the time span 2016–2019 will count as one vehicle for CO_2 averaging purposes (no supercredits) and the remaining credits (multiplier-1) will be saved to be distributed in the 2020–2023 period. Extra credits can also be awarded after 2019.

This study focuses on the calculation of ICE-average CO_2 emissions (g/km per vehicle) after introducing super credits (SC) under the assumption of different electric vehicle stock developments. The ULEVs under consideration will comprise BEVs, PHEV and EREVs, while the remaining vehicles will be deemed internal combustion engine (ICE) passenger cars. The banked electric vehicles are calculated by applying the respective SC multiplier – 1 upon the EV new registrations for each year. This number of vehicles is saved for later use. The accumulated banked electric vehicles can then be distributed as super credits.

The following equation is used to describe the required ICE-average CO_2 per year (in g/km) without using super credits based on new registrations:

$$CO_{2(ICE-average)}\Big|_i = \frac{N_{ALL} \cdot CO_{2(Fleet-average)} - \sum N_{EV} \cdot CO_{2(EV)}}{N_{ICE}} \qquad [18.1]$$

where i is the year of interest, N_{EV} denotes the number of new electric vehicles, i.e. all vehicles eligible for super credits, N_{ALL} denotes the entire new fleet and N_{ICE} denotes the ICE new fleet, $CO_{2\ (EV)}$ denotes the emissions for each EV and $CO_{2\ (ICE\text{-}average)}$, $CO_{2\ (Fleet\text{-}average)}$ refer to the average emissions from ICE and all vehicles, respectively, while all CO_2 levels are type-approval values. The total electric vehicles are modified to include the super credits, i.e. the total vehicles are increased for the purpose of this calculation:

$$N_{ALL} = N_{ALL(OLD)} + (a_{SC}-1)\ N_{EV}, \qquad [18.2]$$

where a_{SC} is the super credits multiplier for the year of interest, in order to reflect the addition of accumulated electric vehicles. The previous modified equation when compared to the original (no super credits) equation yields the CO_2 gain in g/km for ICE vehicles acquired in a year. The same notion can be applied to calculate the fleet-average CO_2 gain. When this gain is expressed with regard to the combined new fleet (ICE and electric vehicles), the actual average CO_2 due to accumulated super credits is given by:

$$CO_{2(average)} = \frac{\left(N_{ICE} + N_{EV} + N_{EV,SC}\right) \cdot CO_{2(Fleet-average)} - \sum N_{EV,SC} \cdot CO_{2(EV)}}{N_{ICE} + N_{EV}} \quad [18.3]$$

It is apparent from the previous equations that ICE vehicles are assumed to exploit the entire margin offered by the application of super credits. Average CO_2 targets for new vehicles in 2016–2019 are assumed to follow an interpolation from 130 g/km in 2015 to 95 g/km in 2020 according to the current mandated policy. After 2020, there are no safe estimations about future regulations and thus, in this projection the 95 g/km limit remains for the entire considered time span. According to the initial assumptions, the fleet-average CO_2 emissions for 2010–2050 can be seen in Table 18.1. An interpolation was used to extract the CO_2 emissions prior to 2015 by setting the 2010 emissions equal to 140.3 g/km, which is the EU-27 type-approval value for passenger cars in 2010, in accordance with the CO_2 monitoring database [EEA 11].

Year	*CO_2 (g/km)*
2010	140.30
2011	138.24
2012	136.18
2013	134.12
2014	132.06
2015	130.00
2016	123.00
2017	116.00
2018	109.00
2019	102.00
2020	95.00

Table 18.1. *Assumed fleet-average CO_2 levels*

This assumption, in conjunction with the e-mobility penetration projections compiled previously, is expected to result in a relatively more relaxed ICE-average CO_2 emission limit after 2025, when electric vehicles will be a large part of the new registrations fleet, thus driving ICE vehicle emission actual limits higher.

18.2.1. *E-mobility penetration*

The distribution of electric vehicle new registrations according to electric vehicle type, size and secondary conventional fuel was based on [TNO 11]. This source also includes information regarding the CO_2 emissions in g/km for all the types of electric vehicles used for the type-approval simulation part.

Regarding the number of EVs as a part of the total fleet, an overview is provided in Figure 18.1. The base case was compiled on the assumption of the baseline scenario in the software tool used (more information will be provided in section 18.3). Two e-mobility penetration scenarios will be used, serving as the basis for the study of the super credit scenarios. The e-mobility schemes will be denoted as EV base and EV scen 1 hereinafter, as shown in Figure 18.1.

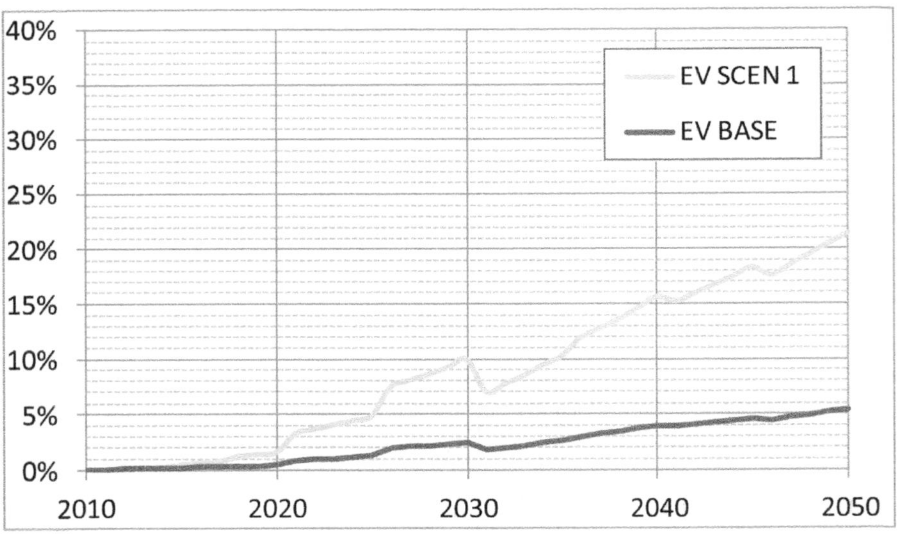

Figure 18.1. *E-mobility penetration in terms of new registrations percentage. For a color version of the figure, see www.iste.co.uk/jacob/energy.zip*

18.2.2. *Super credits*

The e-mobility penetration scenarios are combined with the super credit schemes by using the model described in the methodology, and they can be seen in Table 18.2. Two super credit schemes have been implemented, scenario A and scenario B, while the base scenario contains no super credits.

Year of application	**2016**	**2017**	**2018**	**2019**	**2020**	**2021–2023**
Base	1	1	1	1	1	1
Scenario A	3.5	3.0	2.5	2.0	1.5	1
Scenario B	3	2.5	2	2	1.5	1.3

Table 18.2. *Development of super credits for very low-emitting vehicles*

A multiplier of 1 implies that no super credits are awarded. The banking scheme, i.e. the allocation method of the extra EVs due to super credits, can be found in Table 18.3.

Year of application	**2020**	**2021**	**2022**	**2023**
Base	N/A	N/A	N/A	N/A
Scenario A allocation	40%	30%	20%	10%
Scenario B allocation	80%	10%	10%	0%

Table 18.3. *CO_2 banking gain distribution (%)*

As a result, a total of three SC scenarios and two EV penetration scenarios will be examined in conjunction.

18.3. Simulation

18.3.1. *Software*

The methodology analyzed in the previous sections was integrated into a custom software application [EMI 13], a scenario-evaluation tool capable of assessing different approaches which aim to mitigate CO_2 emissions from road transport. The model is based on the concept that road transport emission projections can be delivered by utilizing a different technology/fuel mix, which also considers technology efficiency development as input. The underlying database is calibrated on the output of forecasting energy-economic models and includes emission factors based on real-world driving conditions.

The software has a modular base, consisting of a series of virtually independent and sequential modules (Figure 18.2), each of them being responsible for the calculation of a certain metric (e.g. stock, mileage distribution, energy consumption, etc.). It uses a bottom-up approach in forming total activity, rather than a top-down approach that would come from a demand module. Traditionally, new registrations are calculated as the difference of the size stock minus the surviving stock from the previous year.

The model employs a series of modules to estimate the vehicle stock and activity and then extracts emissions. The first four modules estimate the vehicle stock (with respect to year, age and class). Then by applying the mileage and speed distributions, the activity can be accurately estimated.

A new scenario starts by formulating the vehicle stock projection, which can be manually altered based on desired assumptions on market growth, stock growth or vehicle utilization patterns. The average mileage combined with the previously estimated stock yields the total vehicle-kilometres. The classification of vehicles in different categories then provides the average projected mileage per technology and energy source.

The calculation of activity per vehicle type in conjunction with fuel consumption and emission factors can then lead to a bottom up energy consumption and emission estimation. The model contains emission and consumption factors for a variety of modern vehicle technologies, based on extensive powertrain software modeling and simulations of real-world driving cycles [AND 04], as well as actual vehicle data. Real-life vehicle characteristics were used for the development of the vehicle models, and the advertised performance and fuel consumption specifications served as feedback to calibrate the results. A correlation diagram has been obtained for each vehicle category (and fuel type for multi-fuel technologies), by simulation and testing procedures as well as standard Tier 3 methodology [EME 13, IPC 06]. These energy consumption factors are speed and trip-length dependent in the case of advanced powertrains. The model includes a vehicle efficiency modifier to reflect technology evolution over time.

The utilized baseline scenario is consistent with the PRIMES 2012 reference model. Since the simulation includes two alternative EV penetration scenarios, both of them have been utilized.

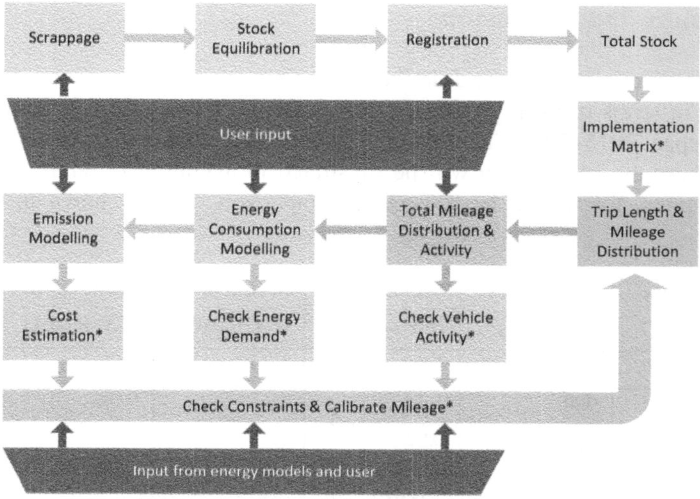

Figure 18.2. *Block diagram of the modeling process*

18.3.2. *Super credits estimation*

Using the previous e-mobility penetration assumptions and the super credit schemes, the calculated total "phantom" vehicles which are added to the fleet are provided in Table 18.4.

Year	Scenario A (EV base)	Scenario A (EV scen 1)	Scenario B (EV base)	Scenario B (EV scen 1)
2016	342576	114192	91354	274061
2017	330601	110200	82650	247951
2018	392646	98161	65441	261764
2019	307807	76952	76952	307807
2020	179838	44959	44959	179838
2021	0	0	55958	223832
2022	0	0	63384	253537
2023	0	0	71279	285116
Total	1553468	444465	551977	2033905

Table 18.4. *Extra vehicles due to super credits*

18.4. Results and discussion

The scenario combinations described in the previous sections were realized and compared in terms of average CO_2 emissions per year and total CO_2 emissions for the time span 2015–2050. The estimations are delivered for both type-approval values, where the comparison will be illustrated in average CO_2 and in real-world CO_2 emissions, for which the respective comparison will be shown in total CO_2 difference.

Year	Scenario A (EV base)	Scenario A (EV scen 1)	Scenario B (EV base)	Scenario B (EV scen 1)
2020	0.54	1.90	1.33	4.96
2021	0.40	1.43	0.17	0.62
2022	0.26	0.95	0.16	0.62
2023	0.13	0.47	0.00	0.00
Total	1.33	4.74	1.66	6.21

Table 18.5. *Average annual CO_2 gain in gCO_2/km for ICE (non-ULEV) new vehicles*

18.4.1. *Type-approval results*

The average CO_2 gain for each year is illustrated in Table 18.5 for all the super credits – EV penetration combinations. This is obtained by using expression [18.1], i.e. this gain constitutes the extra margin that ICE vehicles can exploit due to the super credits application instead of the standard requirement for each different EV stock structure.

Year	Scenario A (EV base)	Scenario A (EV scen 1)	Scenario B (EV base)	Scenario B (EV scen 1)
2020	0.53	1.86	1.33	4.88
2021	0.40	1.38	0.16	0.60
2022	0.26	0.91	0.16	0.60
2023	0.13	0.45	0.00	0.00
Total	1.32	4.61	1.65	6.09

Table 18.6. *Average annual CO_2 gain in gCO_2/km for all new vehicles*

Likewise, the effect on the full fleet as mandated by expression [18.3] is provided in Table 18.6. The gain only differs slightly between the two SC scenarios for the same EV penetration scheme; the relative gain change increases sharply as the e-mobility penetration rises, which is expected due to the overwhelming number of ICE vehicle new registrations compared to electric vehicle new registrations in the early years.

Figures 18.3(a) and (b) demonstrate the effect of applying SC scenario 1 on the average type-approval CO_2 depending on the number of EVs in the market. From Figure 18.3(a), it is clear that if a rather optimistic EV penetration curve is actually followed, the extra CO_2 gain for ICE vehicles can rapidly rise after 2025. In Figure 18.3(b), the effect on the total fleet is depicted.

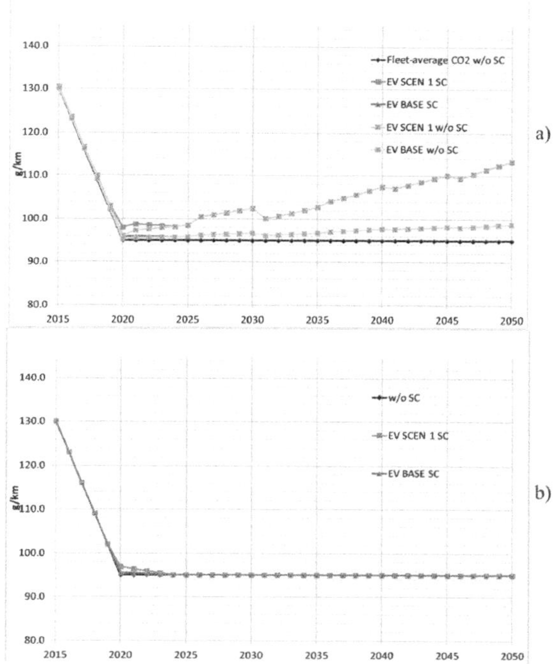

Figure 18.3. *a) Average TA CO_2 for ICE vehicles and b) average TA CO_2 for all vehicles for SC scenario A. For a color version of the figure, see www.iste.co.uk/jacob/energy.zip*

From the second SC scenario application (B), it is quite evident that application of the rapid EV penetration also has a limited long-term effect; in the slow EV penetration case, however, the 2020 limit is given a significant allowance, as shown in Figures 18.4(a) and (b).

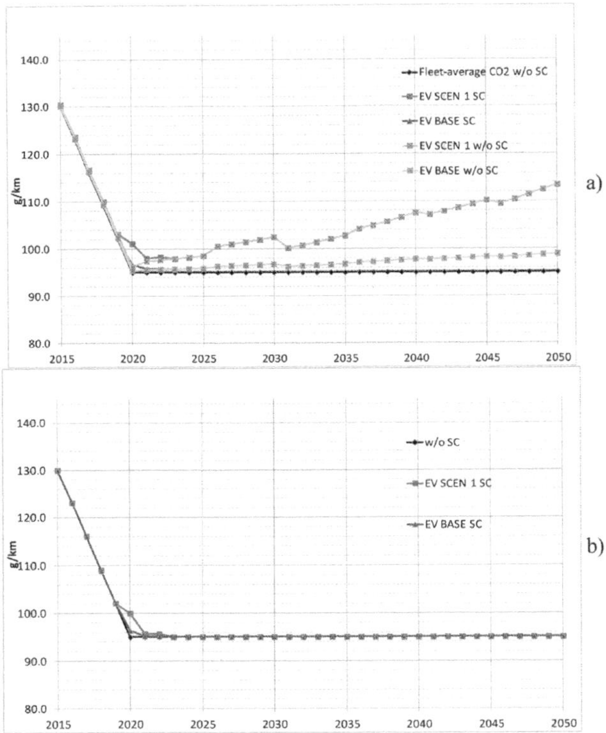

Figure 18.4. *a) Average TA CO_2 for ICE vehicles and b) average TA CO_2 for all vehicles for SC scenario B. For a color version of the figure, see www.iste.co.uk/jacob/energy.zip*

18.4.2. *Real-world results*

In this section, the real-world emission factors were used to evaluate the total CO_2 emissions on both the new fleet and the entire passenger car fleet. For this calculation, the EV development curve is defined by EV scenario 1, i.e. rapid e-mobility penetration. Therefore, the results involve three cases: base case, scenario A and scenario B regarding super credits; they are illustrated in Figures 18.5(a) and (b). The first is a comparison of new registration emissions, while the second regards the total passenger car fleet.

The additional CO_2 real-world tailpipe emissions for the period 2015–2030 in the total fleet reach 15.12 Mt for scenario A and 20.85 for scenario B; if the time span is extended to 2050, the additional emissions are 24.07 and 31.59 Mt, respectively.

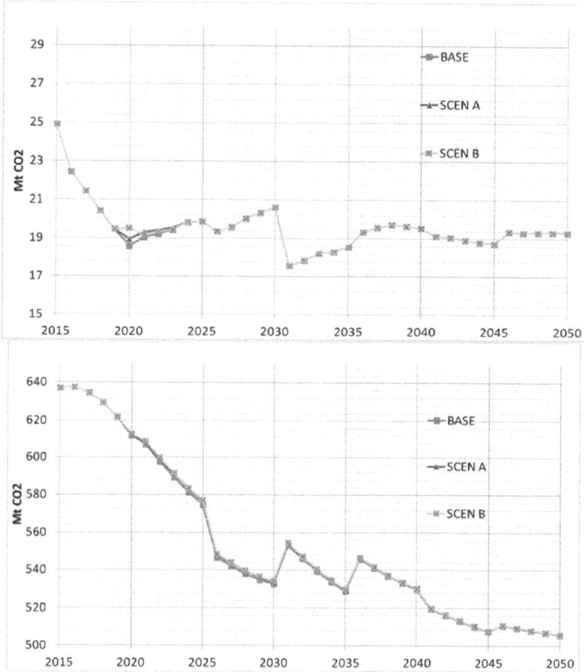

Figure 18.5. *a) Total CO_2 emissions from new vehicles and b) total CO_2 emissions from all vehicles. For a color version of the figure, see www.iste.co.uk/jacob/energy.zip*

As a result, the increase in percentage CO_2 emissions for the period 2015–2030 in the total fleet reaches 0.13% for scenario A and 0.16% for scenario B, while in 2015–2050 range, the percentages are 0.16 and 0.22%, respectively.

18.5. Conclusion

In this chapter, the impact of super credit scenarios on CO_2 emissions with respect to the new fleet and the total passenger car fleet was studied, on the assumption of different electric vehicle penetration scenarios in the 2010–2050 period for the EU-27 region. Results show that the effect of super credit scenarios on average emissions depends heavily on the EV penetration and less on the actual super credit scheme. When real-world emissions are taken into account, this is also apparent, but the actual influence of super credits on total CO_2 emissions is less pronounced.

Although it is unclear how the practical application of such policies will affect market growth and/or provide the impetus for the penetration of ULEVs, further research could include Well-to-Wheels or even lifetime emissions in order to investigate the full effect of SC policies.

18.6. Bibliography

[AND 04] ANDRE M., "The ARTEMIS European driving cycles for measuring car pollutant emissions", *Science of the Total Environment*, vols. 334–335, pp. 73–84, 2004.

[EME 13] EMEP/CORINAIR, Atmospheric Emission Inventory Guidebook, 3rd edition, European Environment Agency, Copenhagen, available at http://www.eea.europa.eu//publications/emep-eea-guidebook-2013, 2013.

[EMI 13] EMISIA, SIBYL 2.0, Thessaloniki, Greece, available at http://www.emisia.com/sibyl/SIBYL.htmlEnergy-Economy-Environment Modelling Laboratory, 2013.

[EUR 09] EUROPEAN PARLIAMENT, Setting emission performance standards for new passenger cars as part of the Community's integrated approach to reduce CO_2 emissions from light-duty vehicles, Regulation EC No. 443/2009, 2009.

[EUR 11a] EUROPEAN COMMISSION, Roadmap to a single European transport area – towards a competitive and resource efficient transport system, White paper, Brussels, 2011.

[EUR 11b] EUROPEAN ENVIRONMENT AGENCY, Monitoring of CO_2 emissions from passenger cars – Regulation 443/2009, available at: http://www.eea.europa.eu/data-and-maps/data/co2-cars-emission, 2011.

[HIL 13] HILL N., NEWMAN D., KIRCHE F., Low emission car measures Under the EU's CO_2 regulations for passenger cars, report for Greenpeace and Transport & Environment, 2013.

[INT 13] INTERGOVERNMENTAL PANEL ON CLIMATE CHANGE, IPCC Guidelines for National Greenhouse Gas Inventories, Energy, vol. 2, Hayama, Japan, available at http://www.ipcc-nggip.iges.or.jp/public/2006gl/vol2.html, 2013.

[MEL 11] MELLIOS G., HAUSBERGER S., KELLER M. *et al.*, Parameterisation of fuel consumption and CO2 emissions of passenger cars and light commercial vehicles for modelling purposes, *Joint Research Centre report,* 2011.

[NTZ 12] NTZIACHRISTOS L., MELLIOS G., SAMARAS Z., "What is the real-world CO_2 reduction benefit of the 95 g/km passenger car average emission target to be reached by 2020"? *Procedia – Social and Behavioral Sciences*, vol. 48, pp. 2048–2057, 2012.

[PRI 13] PRIMES model, available at: http://www.e3mlab.ntua.gr/ e3mlab/index.php?option=com_content&view=category&id=35&Itemid=80&lang=en, 2013.

[TNO 11] TNO, Support for the revision of Regulation (EC) No 443/2009 on CO_2 emissions from cars, available at http://ec.europa.eu/clima/policies/transport/vehicles/cars/docs/study_car_2011_en.pdf, 2011.

19

A Study on Co-modality and Eco-driving Mobility

The last decade has brought on a revolution in the private automotive sector, whereby increasingly location-based information has become available. All these big data sets open up a huge potential for advanced data mining and enrichment activities. Based on mobility data arising from a large private vehicle fleet (300 users, 60 car types, 20 sec GPS traces, 9 months, over 110,000 recorded trips), we assessed the modal shift potential by coupling these data with available public transportation data in the region around Brussels, Belgium. Our results indicate that less than 8% of all trips have a suitable public transport alternative. We furthermore concluded that the potential CO_2 reduction in city centers is very limited, and by itself not enough to reach either of the presented CO_2 reduction targets. In contrast, there is a high potential on motorways as the majority of kilometres were driven there. Public transportation in itself has great difficulty in reaching the emission reduction targets. Finally, there are future business opportunities for the automotive sector in order to remain competitive by having a relevant and realistic outlook on the future mobility market.

19.1. Introduction

The last decade has brought on a revolution in the private automotive sector, whereby increasingly location-based information has become available. Both private companies and governmental bodies are using GPS "track n trace" vehicles in order to optimize their own fleets, or even develop and increase the performance of traffic management systems. As all these big data sets are growing, it opens up a huge potential for advanced data mining and enrichment activities. Our study was in particular concerned with a huge dataset made available by Toyota Motor Europe (TME). By using and developing the necessary computational tools, we analyzed these data in such a way that an understanding of the underlying structures emerged

Chapter written by Sven MAERIVOET, Lars AKKERMANS, Kristof CARLIER and Muriel DESAEGER.

(i.e. extraction of trip legs and trip chains). In the following sections, we first detail how we collected and processed the relevant mobility data. We then explain how these data were coupled with a custom-developed model with the available public transportation data (bus, metro and train) in the region around Brussels, Belgium. The coupling allowed us to assess the modal shift potential, by calculating how many trips could be replaced with the available public transportation, both from an optimistic and realistic point of view. Finally, we present a quick-scan of promising technologies and services within the different modal shift scenarios that we previously created. We then identify the most important consequences for personal mobility, linking them to modal and/or technological choices that are necessary to obtain the target proposed by the European Commission. All details of this study are reported in [AKK 12].

19.2. Analyzing the GPS-based mobility data

In the next sections, we first describe the general structure of the available GPS-based mobility data, after which we explain how the mobility patterns were extracted by means of trip legs and trip chains. We finally show how we calculated the CO_2 emissions per trip.

19.2.1. *General information*

All available GPS-based data arised from an internal TME testproject that followed 295 unique users (with their consent, guaranteeing their anonymity by means of proper privacy and security measures), covering a fleet of some 60 different vehicle types. The latter top 10 included Toyota Prius, Avensis, Yaro, Aygo, Corolla Verso, Rav4, Auris and IQ with engine sizes ranging from 1,400 over 1,800 to 2,000 and 2,200 cc. Most of these were diesel cars, although some of them were petrols and hybrids. The data spanned a time period of approximately 1 year between September 2010 and July 2011, which in total made up some 324 days. The database was converted into MySQL (Ora), whereby each entry corresponded to a 20 s GPS measurement of one vehicle. After elimination of duplicate and malformed records, the database contained 10,213,297 valid 20 s GPS records in Belgium. All the records were *a priori* map-matched by TME onto four NAVTEQ-based layers. These latter corresponded to different road types, from motorways to local roads and city centers. Considering the locations of all trips, they all departed (and arrived) in the same regions around Brussels. The top departure location is Brussels as TME's office location clearly stands out. Only few departures and arrivals originated and ended in the Walloon region, with the majority of the departures arising from Antwerp, Brussels and Flemish-Brabant.

Looking at the spatio-temporal structure of the data, we can see how various congestion patterns arise on the road network. There is a clear distinction between morning and evening rush hours, and the night and daytime off-peak periods. By plotting a sequence of all recorded GPS positions, coloured according to their speed, and time-dependent, we obtained a qualitative view on the congestion throughout the day. Figure 19.1 gives the examples of these phenomena by means of successive hourly time lapses; all GPS records were taken from all Thursdays in the dataset, and plotted on top of each other. The dots are coloured according to the median speed, with red dots slower than 30 km/h, orange to yellow between 30 and 60 km/h, green between 60 and 90 km/h, light blue between and 110 km/h, and dark blue faster than 110 km/h.

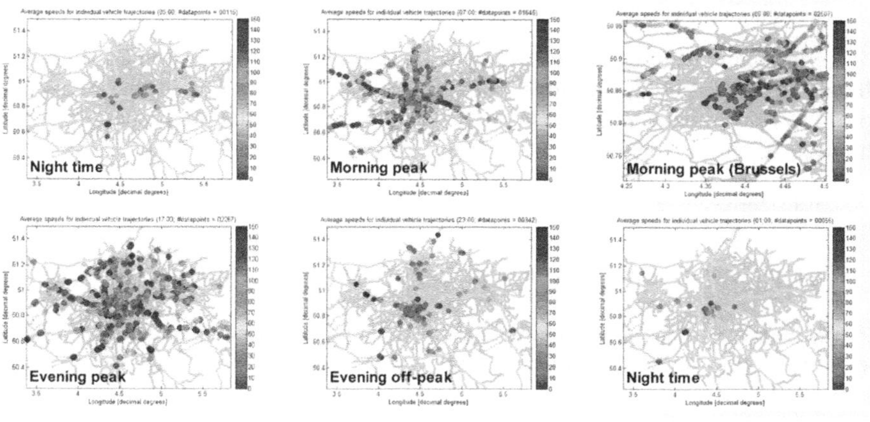

Figure 19.1. *Examples of spatio-temporal congestion patterns during the off-peak periods and morning and evening rush hours. Blue colours denote high average speeds, green colours denote medium speeds and red colours indicate slower speeds (which on motorways are most likely attributed to congestion). For a color version of the figure, see www.iste.co.uk/jacob/energy.zip*

We also analyzed the spatio-temporal spread of the different vehicles according to their fuel type, split among the morning peak (defined from 06:00 to 09:00) and the evening peak (defined from 16:00 to 19:00). The results are presented in Table 19.1.

	Diesel			Petrol			Hybrid		
Road type	On-peak	Off-peak	Total	On-peak	Off-peak	Total	On-peak	Off-peak	Total
RT 1	27%	24%	51%	29%	23%	52%	22%	27%	49%
RT 2	7%	9%	16%	6%	7%	14%	7%	10%	17%
RT 3	11%	12%	23%	12%	12%	24%	11%	13%	24%
RT 4	5%	5%	10%	6%	5%	11%	5%	5%	10%

Table 19.1. *Total distances driven over all 20 s GPS records (procentual)*

From Table 19.1, we see how diesel cars typically drove 51% of all their distances on motorways (road type 1), and 10% of their distances within city centers (road type 4). Comparing these numbers between fuel types, we observed that the usage patterns of hybrids were typically the same as diesel or petrol cars. As such, we concluded that hybrids as a technology did not cause a change in behaviour.

19.2.2. *Extraction of mobility patterns as trip legs and trip chains*

In order to determine the mobility patterns contained within the dataset, we extracted trip legs and trip chains. These were based on the following assumptions:

– trip chains and legs were derived for each user in the database;

– trip chains were complete and began and ended at a user's home location. All (longitude and latitude) locations that lay within a 100 m radius of the user's home location were considered as the home location (because matching home locations on the street level was too complex);

– trip chains started after 5 h and ended before 5 h the day after; they spanned at most 1 complete day;

– a trip chain consisted of one or more trip legs.

In order to derive each user's home location, we assumed this location corresponded to the user's point of first departure during the morning commute on working days. Taking for each user the location having the highest number of such departures gave us his most probable home location. An example of the extraction of home locations near Brussels can be seen in Figure 19.2. The black dots denote all the locations where a user first departed during the morning, whereas the blue dots denote the calculated users' home locations. The red dots show for each user the average locations of all departures within each city. Note the scatter of the black dots in some regions, implying that users did not always park their vehicles at the same spot or street, but rather stayed in the vicinity.

Once all home locations were known, TME processed all GPS records in order to extract trip legs and trip chains. Most trip chains contained two trip legs, with lower probabilities occurring for increasing numbers of trip legs per trip chain. The rationale behind this was that most users travelled directly to and from their work. Based on the distribution of the number of trip legs per trip chain, we estimated that some 80% of all trip chains were composed of four trip legs or less. Furthermore, irrespective of the number of trip legs in a trip chain, all trip chains typically consisted of two long trip legs (mostly from home to work and vice versa), and were interspersed with some smaller trip legs that covered a significantly lesser distance. As the number of trip legs per trip chain increased, we estimated that the distances of the short legs did not change as there were more or less legs in a trip chain. We also believed that the amount of short trip legs remained the same, because people tended to shop either in the neighbourhood of their home location or their work location. A possible explanation was the fact that in most cases, people knew these home and work neighbourhoods quite well, and were therefore more keen to have activities there, as opposed to "somewhere in the middle" of their trip chain.

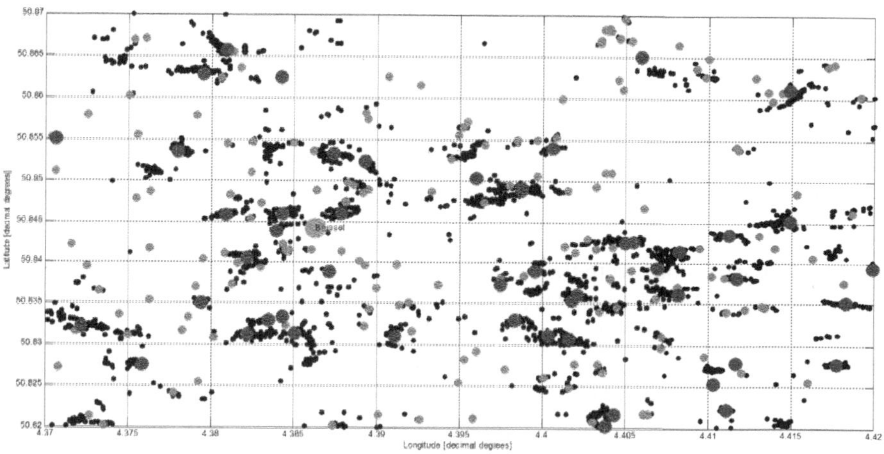

Figure 19.2. *Determination of users' most probable home locations (blue dots), based on their departures in the early morning. For a color version of the figure, see www.iste.co.uk/jacob/energy.zip*

Most of the trip chains covered distances of less than 100 km. Their distribution had a very long, flat and low tail, implying that only few users made very long trips. The mean lay around 44.3 km with a standard deviation of 55.7 km; the median lay around 25.4 km.

The distribution of total time spent between departing from and arriving again to a users' home location contained two noticeable peaks, with one for very small times (i.e. less than 2 h) and the other slightly over 10 h. The former peak was indicative of very short trips, whereas the latter one was indicative of typical home → work → home trips (with optional destinations in between). It also implied that users probably spent some 8 h at work, with some time travelling to and from, including optional intermediate stopping points in between. When we excluded the time a vehicle is stopped, and only considered the time users spent in their cars driving around, the mean lay around 1.0 h with a standard deviation of 1.0 h; the median lay around 0.8 h. The majority of users were spending around a maximum of 1–2 hours in their cars. There was a very long, flat and low tail testifying that only a few users spent long times driving around in their cars, with some of them driving for a very long time (i.e. over 6 h).

Considering the distributions of the trip speeds leads us to a very distinct peak around 34.9 km/h on average with a standard deviation of 19.7 km/h; the maximum at the tail corresponded to some 135.3 km/h. The median lay around 31.2 km/h, corresponding with what we found in literature according to [MAE 06].

19.2.3. *Calculating CO_2 emissions per trip*

Based on all trip legs in the database, we calculated the CO_2 emissions and costs of driving:

– the CO_2 emissions were derived from the travelled distance for each vehicle, its median speed, the road type travelled on, the vehicle type and the estimated fuel consumption. The latter was expressed through the emission factors per road type and vehicle speed range, which were adjusted to match the EU27 fleet; the CO_2 emissions were then expressed in g/km. TME provided all the speed-dependent CO_2 emission factors for the various vehicles in the database. It was based on TME's default available fuel consumption and CO_2 emission cycle data;

– the costs were in turn derived from the travelled distance for each vehicle, the vehicle type and the cost per kilometre (which TME provided as a total cost of ownership (TCO) per kilometre); they were expressed in euros/km.

The CO_2 emissions were dependent on the speeds at which vehicles drove. We split the CO_2 calculations out based on different road types, to see on which roads there was more or less CO_2 being emitted.

Almost half of all the distances were driven on motorways, contributing to exactly half of all the emissions, implying that there is a huge potential to be gained here. From our calculations, we could deduce the average EU27-adjusted CO_2 emission of a vehicle in the TME fleet, which corresponded to some 164 g/km. As was expected, the total amount of CO_2 emitted increased as the distance per trip leg increased. For shorter distances, there were on the one hand very high CO_2 emissions per kilometre, and on the other hand a large spread which decreased with longer trip leg distances. This is because on a long leg, there was a higher probability that you would drive on more different types of road, and hence encounter more congestion.

19.3. Assessing modal shift potential

In the next couple of sections, we show how we assessed the modal shift potential of the TME fleet, based on the users' mobility patterns extract in section 19.2. We first explain what public transport data we collected, after which we elaborate on the development of a public transport model. The latter was then applied to the mobility patterns from section 19.2, yielding the modal shift potential.

19.3.1. *Collecting public transport data*

We collected public transport data from the four main public transport companies in Belgium, i.e. De Lijn (Flanders bus and tram), MIVB/STIB (Brussels bus and tram), TEC (Walloon bus and tram) and NMBS/SNCB (Belgium train). The requested data were "static" (i.e. the planned schedules, not the schedules as they were actually driven which would include delays and cancelled trips). From each of the companies, a schedule for a regular week was requested. A week was considered "regular" if it was not during the holidays, if there was no high day during the week and if there were no special services for non-recurring events (such as concerts, tourist trains to the coast, etc.). For our analyses, we used a working day, a Saturday and a Sunday. The data were composed of a table with for all the stops including at least the coordinates of its location and its name, as well as information on which service passed by at which stop and at what day of the week and time instant. The locations of all public transport stops in Belgium are shown in Figure 19.3.

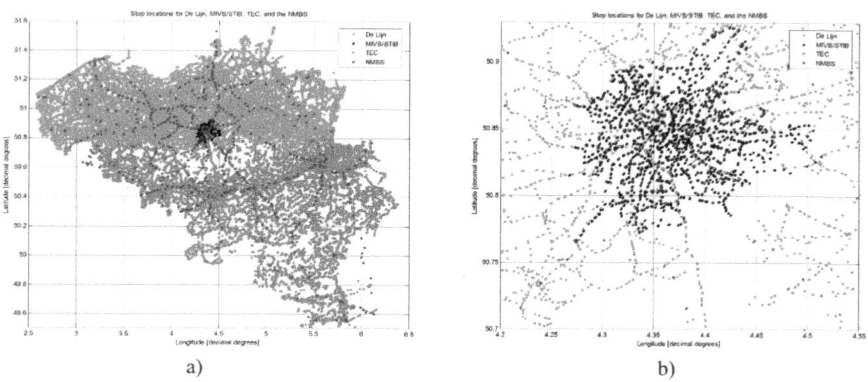

a) b)

Figure 19.3. *a) Geographic overview of all public transport stops in Belgium, coloured according to the public transport operator. The blue dots are Belgian trains, which are fewer in number than the total number of busses and trams. b) A closeup of Brussels containing a high density of stops which are operated by the MIVB/STIB. For a color version of the figure, see www.iste.co.uk/jacob/energy.zip*

19.3.2. *Development of the public transport model*

The general goal of the public transport (PT) model was to find the best matches between transport modes (taking time and location restrictions into account), given various scenarios and constraints. The results from section 19.2 were linked to the public transport data by means of the PT model. In order to investigate for which car trips in TME's dataset a viable public transport alternative was available, we calculated the shortest public transport route for each trip. After such a public transport route was found, it was compared with the original car trip to determine whether or not it was a viable alternative.

The public transport router first generated a "hyper-network". Based on the data in the schedules of the public transport providers, a spatio-temporal network was created, which consisted of:

– travel links: these were the direct translation of the bus/tram/train services: a link between two stops with a departure and arrival time;

– boarding links: virtual links representing the boarding distance, the boarding travel time (the distance and time to travel from the origin of the trip to the first public transport stop on the shortest route) and a part of the waiting time of the trip (the time the traveller spent waiting at the first stop);

– egress links: similar to boarding links, but from the last public transport stop to the final destination;

– transfer links: virtual links that represent the transfer from one public transport service to the next. This can be a transfer between two different lines at the same stop, or at nearby stops. The transfer link represents the waiting time, and – when it is a transfer between different stops – the distance between the stops and the required walking time.

When generating these links, some constraints were immediately applied in order to rule out unrealistic public transport routes and limit the number of boarding, egress and transfer links:

– boarding and egress links were generated only to public transport stops within a certain radius around the origin/destination. If less than three stops were found within this region, the three nearest stops were used;

– transfer links were generated between stops within a given radius and between services that guaranteed a minimum transfer time (3 min plus the time required to walk from the one stop to the other, if any) and did not exceed a maximum transfer time.

The model also took walking and cycling into account in order to access the public transport alternatives. Our model thus simulated a habit in user choice. The final hypernetwork for 1 day contained about 40 million links. Finally, after the generation of the spatio-temporal hypernetwork, the A* shortest route algorithm was used to find an appropriate public transport route within this network [HAR 68].

In order to determine the shortest route, not every part of the public transport trip was valued in the same way. Surveys investigating travel behaviour reveal that people's perception of travel time depends on the purpose of their trip and the nature of their actual "travelling" activity (e.g. driving a car vs. waiting at a public transport stop. Our PT model reflected this perception within its travel times.

19.3.3. *Modal shift analysis*

In order to assess the modal shift potential, we envisaged three different scenarios for which the PT model was ran:

Scenario 1: maximized public transport:

Here, we used the PT model to identify possible public transport alternatives as a replacement for car trips, without very strict practical restraints (access and egress time, number of transfers, etc.);

– Scenario 2: emission capping:

Here, we focused solely on the capping of emissions as a result of the (theoretical or practical) implementation of emission targets;

– Scenario 3: time restrictions (acceptable time losses):

Like in scenario 1, we also made use of the PT model but this time used a more stringent set of practical restraints to mimic more realistic modal choice behaviour.

The constraints within each scenario represented actual policy choices that are currently being considered or already actively implemented by administrations and policy makers. We compiled extensive background policy information and parameterization to allow for a correct and realistic quantification of the different policies as well as the results of the different modal shift analyses.

For scenarios 1 and 3, we estimated the modal shift potential within the context of an optimally used, maximally extended public transport system (i.e. it is assumed that there is ample capacity to accommodate all the trips). The concept of a modal shift analysis is depicted in Figure 19.4, exhibiting a trip chain that consisted of five trip legs. The top picture shows the result for a single user at the trip leg level, whereas the bottom picture shows a different result for the same user but this time at the trip chain level. At the latter level, more stringent constraints apply, e.g. it should be possible to make the various trip legs in succession, within reasonable time frames. From the bottom picture, we see how the user could now have only made two trip legs by public transport (i.e. the second and the last, coloured green). For the first trip leg, no viable public transport option was found (coloured red). For the third and fourth trip legs, the user might have selected a public transport option, if he would have relaxed some restrictions (coloured yellow).

Note that when calculating the CO_2 emissions for public transport, we obtained the emission factors from TREMOVE (Transport & Mobility Leuven, 2011): for buses and trams, we used an average of 21.75 g/passenger km, and for trains we used an average of 3.31 g/passenger km. In order to compare the total cost of ownership, we obtained values from a previous MIRA external costs study [DEL 10]: for busses and trams, the cost was 26.51%/passenger km, for trains the cost was 6.36%/passenger km (= 6 cent + 6% VAT).

Running the PT model for scenario 1 (maximized public transport) gave the following comparison with the base case from section 19.2.3 in which no trip legs were substituted by public transport:

– a total distance of 1,976,375 km (119%, higher than the base case);

– a total CO_2 emission of 186,343,203 g (68% of the base case);

– a total cost of 461,581 euro (nearly equal to the base case).

When running the PT model for scenario 3 (time restrictions), we obtained the following comparison with the base case in which no trip legs were substituted by public transport:

– a total distance of 1,780,047 km (107%, higher than the base case);

– a total CO_2 emission of 228,941,205 g (84% of the base case);

– a total cost of 453,481 euro (98% of the base case).

A more detailed analysis of the amount of trip chains replaced by public transport yielded following modal shift potential in Table 19.2, incorporating results from the CO_2, travel time and cost impact calculations.

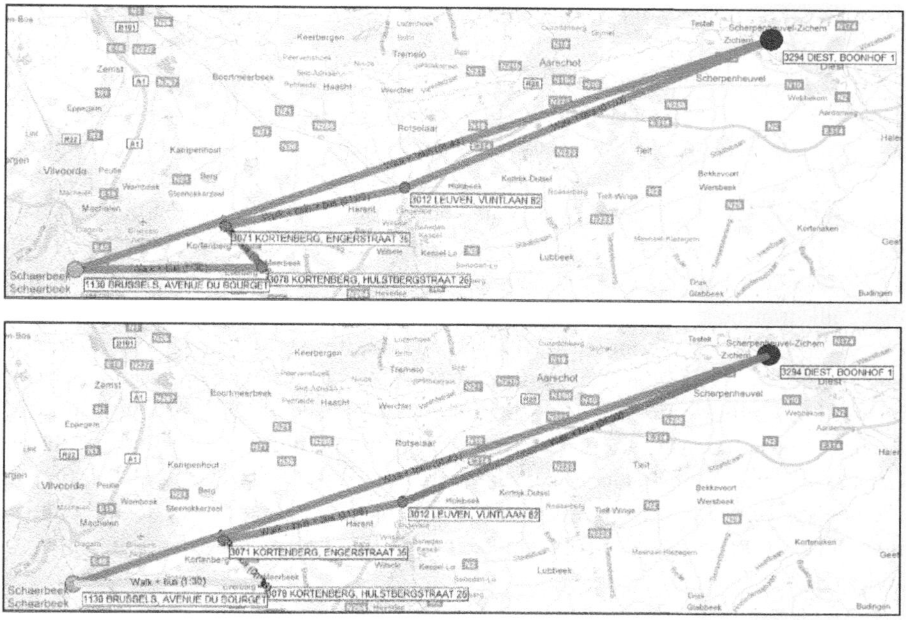

Figure 19.4. *Illustrative depiction of the modal shift potential for a single user. Top: the potential to find a suitable alternative for the car trip for each trip leg separately is high (all green lines). Bottom: the potential to find a suitable alternative for the car trip for the entire trip chain is limited (e.g. the red line denotes a trip leg that cannot be replaced with a public transport alternative, given a set of constraints). For a color version of the figure, see www.iste.co.uk/jacob/energy.zip*

Public transport	Modal shift	CO_2	Travel time	Cost impact
Scenario 1	19%	–9%	+52%	none
Scenario 3	8%	–3%	+54%	none

Table 19.2. *Calculated modal shift potential*

It is clear that the potential drops to 8% when looking at realistic conditions, i.e. scenario 3 in combination with only trip chains. This leads us to believe that there is a huge benefit to be gained here, as now only one out of 13 daily car trips can be replaced by a combination of PT alternatives. Interestingly, the CO_2 emissions only drop by 3%, which also leaves a large benefit to be gained here.

One of the often suggested policy options for triggering a modal shift in passenger transport is the capping of emissions related to private transport. In practice, however, this intention could be translated in various ways. At the moment, a plethora of emission targets are being discussed, addressed, formulated, implemented, etc., by different policy makers. In order to maintain a strong link with the EU background, the following targets were considered and discussed with TME: EU White Paper targets and EU "20-20-20" targets [EUR 11]. First, the EU stated that "all transport related emissions need to be reduced with 60% by 2050, *compared to 1990*". Within reasonable boundaries, it was proposed to extrapolate "a 2020 target of 30% reduction, compared to 1990". In a next step, this target was rescaled to represent first the Belgian situation and, second, the TME fleet in order to make a valid estimation of the effort that would be needed within a fleet that is comparable in usage, but not in technology status. Whether we used the first or second rescaling for further estimations was not of importance due to the relative high level of representation of vehicle use of the TME fleet when compared to Belgian's fleet usage.

We then looked at four different strategies for scenario 2 (emission capping), which might commonly be considered by administrations of different levels to have a direct or indirect effect on CO_2 emissions as a result of passenger car transport:

1) Impose restrictions on locations for the use of conventional internal combustion engines.

2) Introduction of more energy and/or carbon efficient technologies on vehicle or fleet level.

3) Impose restrictions on time periods for the use of conventional internal combustion engines.

4) Impose restrictions on behaviour linked to increased CO_2 emissions (speed).

One of the main results here is the CO_2 reduction potential for technologies, as shown in Figure 19.5.

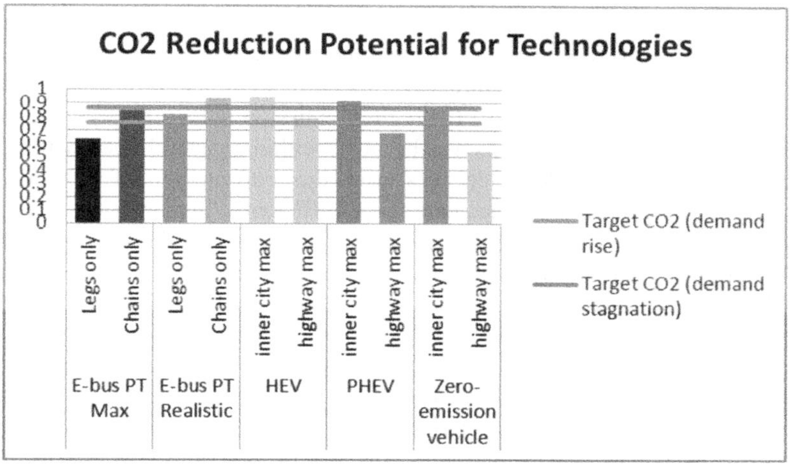

Figure 19.5. *Overview of the CO_2 reduction potential for various engine technologies. For a color version of the figure, see www.iste.co.uk/jacob/energy.zip*

19.4. Vehicle choice, technology and automotive service

Throughout this study, a number of conditions were constructed to allow for a very broad overview of the costs and results of the implementation of a wide set of public transport policies as well as the introduction of a set of existing technologies. For our quick-scan, we looked at different conditions for public transport, going from least over moderate to maximized uptake. In addition, we considered the use of an electric bus as a complete replacement for the traditional bus system. Likewise, we introduced four car passenger vehicle technologies, i.e. hybrid electric vehicles (HEVs), plug-in hybrid electric vehicles (PHEVs), battery electric vehicles (EVs) and hydrogen fuel cell vehicles (HFCVs). For each of the vehicle technology applications, costs were estimated based on TCO values. For HEV and PHEV, these costs were provided by TME and validated within the calculation model. For EV and HFCV, two TCO values were used (a low and a high estimation), resulting in separate subconditions for these vehicle technologies.

Our quick-scan indicated that the increased implementation of vehicle technologies (HEV, PHEV, EV and HFCV) allows for reaching the most stringent emission reduction target under specific conditions (motorway usage). Only HEV, PHEV and EV manage to do this within a reasonable cost increase margin (10%) compared to current costs. However, in the case of, for example, EVs, the high cost of development should not be underestimated, as this is currently blocking the ramping of EVs on the market.

Most noteworthy within the current analyses were the circumstances wherein a particular technology was applied and had an important effect on the outcome in terms of emissions saved. Measures that focused on reducing emissions in city centers (in general) did not receive the highest emission reduction target, nor would they have reached the lowest. This was because, within a general fleet that was used in the same way as the TME fleet was used, the total amount of CO_2 emissions within a built-up area did not account for a sufficiently high portion of the total emissions. Although emissions per vehicle kilometre were relative high, the total vehicle kilometres ran within a city center were relative low. The opposite was the case when specific attention was given to CO_2 emissions levels on the primary road network (motorways). Although individual vehicles tended to perform at their best on higher speed roads, at least in terms of emissions per vehicle kilometre, it was exactly on these roads that the highest number of kilometres were ran. Because of this, a relative high percentage of the total CO_2 emissions took place on these primary roads. Under these conditions, all four technologies reached the lowest emission reduction target (HEV, PHEV, EV and HFCV). Three technologies reached the highest emission reduction target: PHEV, EV and HFCV.

Finally, the introduction of automotive services needed to take care of quite a lot of the responsibilities that now fell to the vehicle owners: maintenance, cleaning, insurance, (initial) fueling, repairs, parking, etc. These elements fell well beyond the scope of the current study, but may have warranted further analyses of driving and mobility behaviour.

19.5. Conclusions and recommendations

A comprehensive analysis of the kind presented in this study, i.e. calculating the mobility behaviour based on detailed GPS data and then correlating this with the available public transport schedules, has not been done before. In this sense, our study highlights some of the shortcomings of and opportunities for the current state of public transport in Belgium. Our results indicate that less than 8% of all trips have a suitable public transport alternative, which opens up a lot of room for improvement given that the Belgian government is thinking about charging road users. We furthermore concluded that the potential CO_2 reduction in city centers is very limited, and by itself not enough to reach either of the presented CO_2 reduction targets. In contrast, there is a high potential on motorways as the majority of kilometres were driven there. Public transportation in itself has great difficulty in reaching the emission reduction targets. Finally, there are future business opportunities for the automotive sector in order to remain competitive by having a relevant and realistic outlook on the future mobility market. One of them is the increased implementation of modern vehicle technologies, by which we can reach the most stringent emission reduction targets within a reasonable cost increase margin (10%) compared to current costs. Nevertheless, in the case of EVs, the high cost of development should not be underestimated, as this is currently blocking the ramping of EVs on the market.

19.6. Bibliography

[AKK 12] AKKERMANS L., CARLIER K., MAERIVOET S., Co-modality Study Toyota, Confidential Report for Toyota Motor Europe, April, 2012.

[DEL 10] DELHAYE E., DE CEUSTER G., MAERIVOET S., Internalisering van externe kosten van transport in Vlaanderen, Study for the Vlaamse Milieumaatschappij, MIRA, MIRA/2010/10, Transport & Mobility Leuven, 2010.

[EUR 11] EUROPEAN COMMISSION, European 2020 Targets – as set by Member States in their National Reform Programmes in April, available at: http://ec.europa.eu/europe2020/pdf/targets_en.pdf, April 2011.

[HAR 68] HART P.E., NILSSON N.J., RAPHAEL B., "A formal basis for the heuristic determination of minimum cost paths", *IEEE Transactions on Systems Science and Cybernetics*, vol. 4, no. 2, pp. 100–107, 1968.

[MAR 06] MAERIVOET S., Modeling traffic on motorways: state-of-the-art, numerical data analysis and dynamic traffic assignment, PhD Thesis, Katholieke Universiteit Leuven, Belgium, 2006.

[ORA 06] ORACLE, MySQL 5.5.15 – The world's most popular open source database, available at: http://www.mysql.com/, 2006.

[TRA 11] TRANSPORT & MOBILITY LEUVEN, TREMOVE run version 3.5.2, 2011.

20

Harmonizing Carbon Footprint Calculation for Freight Transport Chains

The European Commission has set as a target a reduction of 60% in transport greenhouse gas emissions by 2050 [EC 11]. This includes freight transport emissions, which present a particular challenge due to the forecast increase in goods transport linked to future economic growth, the current trend of lengthening supply chains and the global nature of these supply chains. A wide range of methodologies and tools are available for calculation of the carbon footprint of freight transport, based on varying emissions data, routing and allocation assumptions and conflicts between actual and assumed vehicle loading performance. In spite of recent developments such as the publication of a European standard EN 16258 "Methodology for calculation and declaration of energy consumption and GHG emissions of transport services (freight and passengers)", there is a clear demand from industry, for harmonization of calculation methods applicable along the supply chain, across various modes, and at a global level. The COFRET project is working to support this process using real-life case study supply chains.

20.1. Introduction

20.1.1. *Scope and objectives for harmonizing the carbon footprint calculation of freight transport*

With transport and logistics currently contributing to an estimated 20–25% of overall global CO_2 emissions, and the potential for increasing supply chain lengths and complexity and further internationalization of production processes, governments are interested in improved efficiency of supply chains. On the regional level, the European Union (EU) has stated that European countries are expected to

Chapter written by Alan LEWIS, Verena EHRLER, Heidi AUVINEN, Hedi MAURER, Igor DAVYDENKO, Antje BURMEISTER, Saskia SEIDEL, Andreas LISCHKE and Jan KIEL.

reduce their annual greenhouse gas (GHG) emission by at least 20% by 2020 and by 60–80% by 2050, compared to 1990 emissions level [FIS 11].

There is also a clear incentive for businesses, whether actively involved in the transport of goods or as a generator of freight transport as a shipper or receiver of goods, to reduce CO_2 emissions because of the direct link between the emissions and the consumption of more and more expensive fuel or other energy resources; hence, there is a high-level alignment between the objectives of authorities and business community to improve freight transport efficiency and cut emissions, fuel use and costs.

In order for targets and actions designed to improve freight transport efficiency to be meaningful, there needs to be a clear baseline and an associated method for calculating comparable emission values, and a range of mechanisms available to those involved in shipping and transporting goods by which emissions might be reduced. A realization of potential gains in efficiency of supply chains requires the identification of best practice and a comparison of different approaches. Such a comparison is only possible if comparable analysis methods are applied. With so many transport operations and supply chains being at the global level, it is important that the methods applied for the calculation of emissions be comparable not only on a national, but also on an international basis.

This is fundamental for those involved in implementing choices at a practical level as they need to be able to apply their decisions in a consistent manner across what are often global supply chains. At the moment, transport companies have at their disposal a range of possible modes by which to transport goods. Different countries may have their own legislation in respect to GHG accountancy; variations in technology and working practices are common. A range of initiatives to calculate the emissions from freight transport have been developed: sometimes originating on a modal basis, at other times by region; some companies have felt the need to create their own systems, further adding to the number of methods and tools in use. Having such a wide range of calculation methods and tools available makes direct comparison of the results, even for the simplest supply chains, impossible. The multiplicity of tools and methodologies results in insecurities about the "correct" approach and impacts upon the (perceived) accuracy of the calculations, especially in the case of complex multi-modal supply chains [AUV 11, DER 12]. As a result, there is often no confidence or clarity in the results obtained.

All these factors are at the heart of the need for harmonization. Internationally applicable guidelines, supporting tools, data quality standards and methods for the calculation of transport-related emissions, therefore, need to be developed and implemented now. This requires several aspects to be taken into account, which are the focus of the research and future approach to the problem described in the remainder of this chapter:

– global acceptance by all relevant parties (primarily authorities and the business community, as well as those setting procedural and academic standards in this area);

– a pragmatic approach building on the best parts of the status quo, drawing together the main stakeholders on a common basis so that no one feels unduly disadvantaged, so as to avoid the risk of ongoing fragmentation;

– a practical and staged approach to the application of a standard methodology, recognizing differing capabilities between global regions and individual businesses and reflecting the conflicting balance between simplicity, accuracy and flexibility, while also ensuring sufficient transparency in the method applied to avoid future conflict and ongoing suspicion about the calculation results.

20.1.2. *Objectives*

With this in mind, the COFRET project, which was cofinanced by the European Commission, had as its main objective the task of providing industry, shippers and logistics providers with the opportunity to remove the current uncertainty over calculating the carbon footprint of freight transport by:

– reviewing existing methodologies for the calculation of the carbon footprint and greenhouse gas emissions of freight transport and logistics in the context of supply chains and evaluation of their compatibility with the European standard EN 16258 "Methodology for calculation and declaration of energy consumption and GHG emissions of transport services (freight and passengers)" [CEN 12];

– identifying gaps and ambiguities in the calculation guidelines within the EN 16258 standard relating to coverage of freight transport and logistics and prioritizing suggestions for possible approaches to achieve comparability for the calculation of emissions along supply chains, with a special focus on the identified gaps and ambiguities, supported by the means of real-life case studies;

– facilitating the next steps needed in order to achieve global harmonization of calculation principles and comparable reporting as part of a process to support global alignment of standardization.

20.1.3. *The current situation*

Currently, there is no single globally recognized and accepted standard for the calculation of the carbon footprint of a freight transport supply chain. There are numerous initiatives that could contribute to such a standard: the International

Standardization Organization (ISO) established the environmental management standard ISO 14064-1 in an initial effort to address the problem of incomparability. Another standard is the Green House Gas (GHG) Protocol (I-III), announced by the World Resources Institute (WRI) and the World Business Council for Sustainable Development (WBCSD), providing guidelines for the calculation of greenhouse gas emissions at a company level. These methods enable companies to calculate their corporate carbon footprints on a general level and include the topics of transport and supply chains as part of their general recommendations.

Another standard in preparation by the ISO at the moment is ISO 14067 [ISO 11], which focuses on the carbon footprint of products, which includes transportation elements at various stages within the production lifecycle (e.g. raw material, component and finished product transportation).

Besides these schemes and approaches, the European Committee for Standardization (CEN) applied itself to the topic of carbon footprint of transport: in 2012, CEN published the European norm EN 16258, which is, most probably, the closest thing to a supply chain emission calculation standard at the moment, although it carries no statutory weight. Various industry-led initiatives, such as Green Freight Europe, already work on an alignment of their calculation methodologies to the broad rules set out in EN 16258.

An example of country-level initiative is the French requirement for CO_2 reporting for transport services (both passenger and freight) starting, finishing or passing through France, active from October 2013 (Ministère de l'Écologie, du Développement durable, des Transports et du Logement, 2011). Other governments, such as the UK, are taking a different approach, requiring a small number of the biggest companies to calculate the CO_2 emissions from their full business operations, including transport.

Moving away from legislative and standardization approaches, one of the most established participation schemes aimed at collecting data from freight transport companies, calculating an indication of the carbon intensity of the transport with a focus on driving change through comparative reporting and certification, is the Smartway programme, organized by the US EPA. Initiatives modeled on Smartway are now growing in Europe and Asia under the name "Green Freight", although with a more industry-led focus due to the acknowledgment by industry of the commercial and environmental benefits of such an approach.

Further organizations which provide emission calculation tools and services on an international scale, often based on a broad membership of commercial organizations with mode-specific origins, applications or specializations, are the Clean Cargo Working Group, ECO TransIT and IATA, to mention just a few.

Despite all the aforementioned efforts and the identified need for harmonization, various gaps within the coverage of the supply chain, accuracy or consistency exist when it comes to the calculation of its GHG emissions using these approaches: ISO (draft) 14067 and the GHG Protocol do not focus on transport directly. As a result, their regulations about the accounting of transport are not sufficiently specific to transport application to ensure direct comparability between compliant calculations [SCH 12]. In contrast, EN 16258 focuses specifically on transport with a logistics component; even so, its vehicle operation system (VOS) approach, while being a positive first step, needs to be specified further in order to ensure consistency and comparability. Furthermore, EN 16258 lacks coverage of terminals and warehouses, which form an important part of the supply chain. Hence, companies are left without guidance on how to calculate the handling and storage of goods in the process of transport and logistics; such guidance would be particularly important for temperature controlled transportation. Although it is uncertain what the exact contribution of terminals and warehouses is within a typical supply chain, it could prove an important variable when making comparisons between different options and should not be ignored until it has been considered on a directly comparable basis. Finally, the issue of default data sources in case no measured fuel consumption is available is not fully standardized, leaving space for interpretation and incomparable calculations as a consequence. This topic is of particular relevance in the case of subcontracting, which is to be found more and more often within commercial supply transport chains, leaving the primary transport provider and their customer without access to first-hand information for its emission calculations (e.g. if you cannot measure it, you cannot manage it [KAP 96]).

20.2. The basis for standardization

In spite of the shortcomings identified in the previous section, the European EN 16258 standard provides the most comprehensive framework to date for a global emission calculation standard applicable to the carbon footprinting of freight transport. Stakeholders have quickly become aware of EN 16258 and are keen to understand its requirements and the implications for the way that they currently approach their carbon footprint calculations.

20.2.1. *The technical approach*

In searching for a harmonized solution, it will not be possible to accommodate all the subtleties that come from the wide range of factors that influence reporting requirements or energy consumption such as local legislation, mode-specific working practices, company specific data handling, varying climate conditions throughout a year, etc. At the same time, the general approach needs to be practical and pragmatic in order to ensure a high user acceptance rate. A high level of user acceptance for any emission calculation method that is developed to become a global standard is crucial as there is no organization to enforce the use of it – effectively, it will rely on industry acceptance and use to become a self-policing norm, depending on the accreditation processes that accompany it. As a result, any further development of a GHG emission calculation method needs to bridge the fragile trade-off between simplicity, accuracy and flexibility, and all within an envelope of transparency of approach, as shown in Figure 20.1.

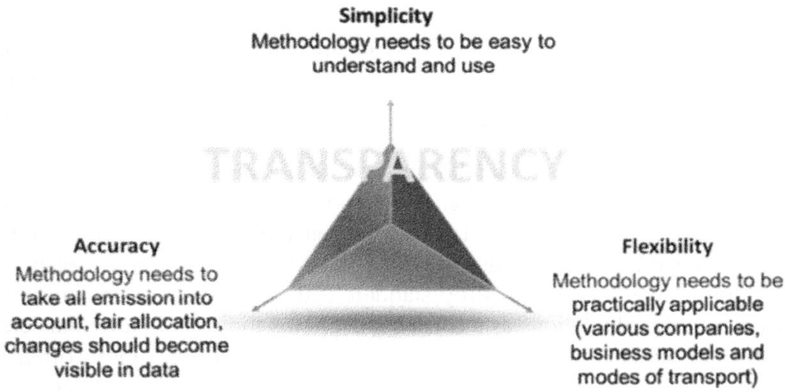

Figure 20.1. *User trade-offs*

When calculating GHG emissions, it is important that the entire supply chain, from an item's point of production to its place of consumption or further processing, is covered. This item, or a unit of freight, can be a product, a package, a pallet, bulk product (measured in tonnes) or a container. This approach to GHG emission calculation allows computation of emissions in door-to-door settings, thus allowing shippers a choice in the selection of their transport solutions. Furthermore, product-level door-to-door emission computations also allow benchmarking of supply chains and labeling of the products in respect to GHG emissions coming out from transport and logistics operations on the way from production to consumption.

The COFRET project has introduced the notion of supply chain element (SCE) as a modular and independent operation that brings the goods close to their final destination. Figure 20.2 presents an example of a transport chain composed of SCEs. It is worth noting that not only transport operations are considered as SCEs, but terminal and warehousing operations are also treated as standalone SCEs. The resulting GHG emissions at the product level are, therefore, a sum of the emissions resulting from the SCEs that constitute the transport chain.

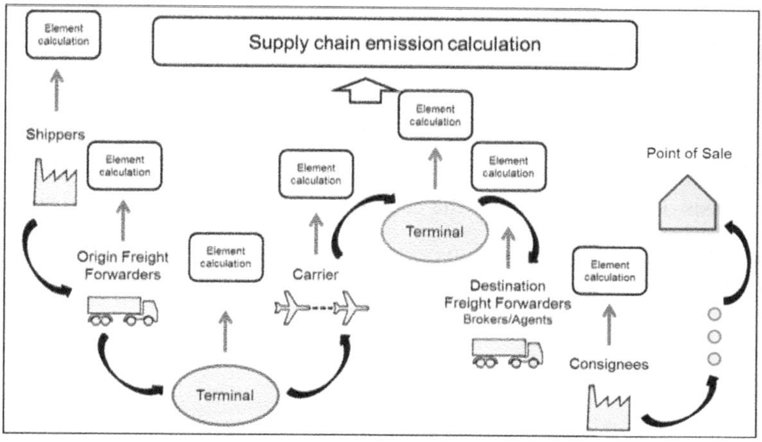

Figure 20.2. *Example of a supply chain split into supply chain elements [HER 13]*

20.2.2. *Gaps and ambiguities*

The division of any supply chain into a number of sequential SCEs greatly simplifies the effort necessary to compute chain-level emissions. Any supply chain can be composed of a limited number of SCEs, such that the building blocks (SCEs) can be used and reused in any arbitrary situation. It also simplifies conformity to the GHG calculation standard, EN 16258, as it applies to the individual transport-related SCEs. Product-level emissions should be computed for each constituent, transport-related SCE.

However, GHG emission calculation at the product level presents an intrinsic challenge, which is related to data availability and different perspectives along the transport chain. Product-level emission calculation requires information on GHG emission from all constituent logistics chain SCEs. Furthermore, product-level emissions are related to the shipper or consignees, who send or receive the products, while a number of transport service providers are executing transport activities. Thus, in many cases, it is impossible for the shipper or consignee to determine total product-level emissions accurately without information from a range of different sources at the SCE level.

The EN 16258 standard does not provide the necessary level of guidance for this, as it prescribes how GHG emissions should be computed within an individual transport-related SCE, rather than joining them to form a supply chain. Three major issues that have been identified (among others) which remain to be addressed are:

– the definition of a directly comparable basis for vehicle operations for the transport elements (the vehicle operation system, or VOS) for which GHG emissions can be calculated;

– the basis of fair allocation of the calculated GHG emissions to the cargo;

– determination of the emissions relevant to each product shipment within the overall cargo.

20.2.3. *The vehicle operation system*

The VOS is defined as a consistent set of vehicle operations relevant to the transport leg (i.e. the individual supply chain element) being calculated. The VOS is defined taking into account factors which affect the scale and composition of the VOS such as number and type of vehicles to be included, the extent of the route covered and the time when the activity occurs. In all cases, the VOS will include the empty trips related to the chosen vehicle operations. For the highest level of accuracy, the GHG emissions are calculated from the primary fuel or energy input, where its information exists; in cases where this information is not available, then assumptions will need to be made and noted in order to assess the potential inaccuracy that is introduced.

Because, according to EN 16258, the VOS may be selected according to criteria chosen by the user, the actual VOS chosen could be based on:

– the whole activity of the fleet of a transport operator over 1 year, including all legs performed by this fleet, irrespective of location;

– all the return journeys between two specific locations per quarter;

– a single leg in a collection and/or delivery round trip.

This large degree of freedom means that in the most extreme cases, the basis for emission calculations could be a single vehicle trip or an annual operation of the whole vehicle fleet and leads to the potential for a fundamental lack of comparability between different sets of calculations.

The definition of the VOS aggregation level has its own advantages and drawbacks, from an operational management perspective as well as an impact on the resulting computations, for instance:

– if the form of VOS is based on individual trips, then the emissions allocated to a cargo unit would vary day-by-day, as the vehicle used, the degree of loading and the routes followed may vary. From a commercial perspective, such a detailed level contains the risk of revealing sensitive information to customers and/or competitors (e.g. the opportunity to infer fuel usage for the cargo and compute related costs);

– in contrast, a highly aggregated VOS that includes many vehicles (including different vehicle types) covers very long periods, or includes routes with differing characteristics, would not help GHG optimization, which is at the core of the efforts that the standard is intended to facilitate. This high level of aggregation would only indicate how efficient the service provider is in respect to overall network tonne-kilometer-related emissions, which may well be a useful output, for benchmarking, for example, but a detailed analysis of the GHG emissions of specific shipments would not be possible using this approach.

20.2.4. *Allocation principles*

Once the GHG VOS-level emissions are calculated, it is necessary to allocate those emissions to a particular cargo load. The default recommendation in EN 16258 is that the transport activity should be quantified by multiplying the mass of the cargo by the real distance travelled; the corresponding allocation unit is tonne-kilometre. Collection and distribution trips are designated as an exception in the standard in that the distance travelled can be either the great circle distance or the shortest feasible distance. However, the EN 16258 standard allows the use of other parameters and units, particularly if it is more relevant to the vehicle's capacity limit (so mass may be replaced by another unit in cases where this can be justified, such as volume in cases where bulky, low-density product causes a vehicle to reach capacity, or lane meter for the specific case of ferry traffic, etc.). Although these options appear logical, there is no clear guidance for when and where to use each alternative.

The prescription to use actual distances except in the case of distribution trips would increase the GHG allocation to shipments that are using a service provider's fixed network. An extreme example of this might be where the service provider diverts from the direct route to pick up or offload other cargo. Consignments making the full journey would travel extra kilometers just to fit into the provider's network and therefore be associated with additional allocated emissions for a routing which is not necessary for the consignment's own route but which is purely imposed by the specific provider's network. The use of actual fuel use in deriving the GHG calculation should in itself be sufficient to take account of the effect of this diversion.

20.2.5. *Determination of shipment-level emissions within the full transport supply chain*

Calculation of the product or shipment-level emissions depends on that product's share of the total tonne-kilometers within the VOS. When considering a full supply chain, the individual values should be computed for each link and each node of the chain (the SCEs) and then need to be summed together. In general, the data for an SCE are only available to the service provider for that SCE. In many cases, this complexity means that it would be infeasible for a shipper to calculate accurately the GHG emissions for all SCEs along the chain. Only the service provider can compute shipment-related emissions with this approach, especially as customers and shippers have no insight into the transport providers' VOS systems, as in the particular case of subcontracting. This places emphasis on the way in which data are transferred within the supply chain between service providers and their customers. It appears that current arrangements are often not adequate, but also that for this situation to be fixed a clear description of how chain-level emissions should be calculated and communicated needs to be provided.

20.2.6. *Addressing the gaps and ambiguities*

Possible ways to address the above-mentioned and other ambiguities have been presented by Davydenko *et al.* [DAV 14]. Ultimately, the choices made, and the extent of their impact on the calculations, will need to be acceptable and workable to those purchasing and supplying freight transport services. This is a critical part of the ongoing work being carried out in partnership with the COFRET Advisory Board and with the case study partners in real-life situations.

20.3. Working with real-life supply chains

The COFRET case studies have been chosen to cover a broad selection of shippers and carriers across various modes and different locations. The sectors covered by the case studies include food and non-food distribution, logistics services, forestry and paper industry, container shipping, liquid bulk and dry bulk (cement). The case studies will demonstrate the use of EN 16258 when expanded to cover transports as well as nodes (terminals, warehouses, etc.) along the supply chain. Furthermore, the case studies will pinpoint the critical parts of the carbon footprint calculation process, where gaps and ambiguities lead to incomparability. The case studies were designed to show viable solutions to solve these issues by testing, for example, suggestions that narrow down alternative calculation approaches.

A standard approach has been defined using the following seven stages:

– defining specific areas of interest per test case based on the EN16258;

– defining a standard procedure for the test cases (incl. reporting guidelines);

– contacting the industry partners;

– calculating the test cases by applying an approach corresponding to EN 16258;

– sensitivity analysis regarding alternative emissions calculation methods;

– testing impact of possibilities to improve EN 16258;

– reporting.

Points that are worthy of particular note include:

– the EN16258 leaves a great degree of freedom on how to define VOS in practice. Together with the COFRET Advisory Board, a first step has been taken to classify vehicle operating systems, thus making the application of the EN16258 more practical. In the case studies, particular care will be taken to work with the case study partners to define the VOS according to their existing procedures (if they already conduct this type of calculation) or to explain the options and allow them to make the decision based on their own preference – as is allowed under EN 16258. In this way, we will be able to gauge the inclination of different stakeholders to choose an approach based on specific journeys, regular line hauls or a network approach;

– the standard procedure used within the test cases follows the approach laid out in EN 16258 and has been embedded in a spreadsheet approach that will ensure conformity. Table 20.1 presents the steps of this process;

– to identify and overcome the uncertainty due to the variety of possible allocation calculations, the COFRET case studies took the following simplified approach: the first choice of unit for quantity in all case studies is mass. Thus, transport activity will be expressed as tonne kilometers (tkm). To allow further analysis, each case study may perform a parallel, additional calculation with another choice of measure (e.g. volume instead of mass). However, the parallel calculation needs then to expand over the entire supply chain in a consistent manner;

– the emission calculations were supported by a calculation module based on the Map and Guide application [PTV 13] that provides the user with the opportunity to vary the basis by which the emissions are calculated. This allowed the difference in results to be identified between three commonly used options, namely:

- according to the "Handbook on Emission Factors for Road Transport (HBEFA 3.1)", a default value-based approach,

- according to the EN 16258 approach,

- according to the French Décret n° 2011-1336;

– it is interesting to note that the content of the presented emission set and structure varies according to the calculation approach used, which in itself may influence the choice of preferred approach by the end users of the case studies.

Step Number	Description
Step 1	Identification of all SCE
Step 2	Data collection on vehicle tour or equipment operations
Step 3	Data collection on fuel or energy consumption
Step 4	Converting consumed energy into CO_2e
Step 5	Allocation
Step 6	Presentation of result

Table 20.1. *Explanation of the step-wise approach in the standard procedure*

20.4. Outlook: requirements toward a future structure for the harmonized approach

The need for comparability of calculation of emissions along supply chains is acknowledged by the transport industry, its related organizations as well as by governmental bodies and researchers. Furthermore, the various guidelines and standardization efforts such as EN 16258, ISO 14067, the GHG Protocol, Green Freight Europe etc., reflect the desire for establishing a reliable standard in this context. The approaches developed so far still contain ambiguities and possible conflicts and there is no organizational body that could put a globally applicable regulation in place. The COFRET project, with its independent character and its industry-research partnership, is an ideal platform for contributing to a further development of the existing approaches toward a global standard.

Building on the detailed reviews of user needs and existing methods, tools and data for calculation of the carbon footprint for logistics chains, and a consultation exercise conducted in the early stages of COFRET with a selection of organizations responsible for shipping and transporting goods, one of the main achievements of COFRET in its first 18 months was the establishment of a very active Advisory Board that comprised a mix of companies responsible for shipping, receiving and transporting goods, the developers of several existing GHG calculation tools and the convenor of CEN Technical Committee responsible for EN 16258.

This group played an active role in developing and applying carbon footprint methodologies and tools to freight transport and has been very supportive in providing a strategic direction to COFRET's work that meets industry needs.

COFRET was able to reach out globally to those who are active in similar topic areas in other parts of the world, particularly North America and Asia. In this way, COFRET was able to establish the nucleus of a global network, where interested parties could discuss the appropriateness and compatibility of the main methodologies applied across freight modes and global regions. This situation provided a basis opportunity to create a broader framework to achieve a set of global harmonized methodologies for carbon footprinting (with the potential for expansion to other environmental impacts, especially air pollution) across freight modes.

Implementation of this framework will require establishment of a long-term initiative and co-operation framework to continue the work conducted to date and to build on it, integrating it into future activities.

For this to be successful, several structural and practical issues will need to be confirmed, and steps are currently underway to gain consensus as to the best way to achieve the following:

– involvement of representation of the key international partners;

– transparency and co-operation to develop a common approach aligning all existing stakeholders, modal representatives and global regions on an equal basis;

– direct involvement of recognized international standardization bodies;

– an appropriate decision-making process that balances industry opinions with the needs of legislators;

– a technical work programme that is backed up by real-world testing leading to future methodology revisions;

– identification and use of international channels for dissemination and exploitation so that the global potential is maximized;

– adequate funding.

While the basis of this framework will be to focus on those organizations and networks that already provide the most advanced and best-accepted emission calculation tools and services on an international scale, such as the Clean Cargo Working Group, ECO TransIT and the airfreight carbon initiative, for the successful global emission calculation standard the continued strong research-industry partnership is essential, as the works conducted by the COFRET project show.

20.5. Conclusions

A single globally recognized and accepted standard is required to support the European Commission's ambitious target of reducing GHG emissions from transport by 60% by 2050 compared to 1990 levels. This process can be supported and sensibly monitored only if a unified method is in place which ensures the consistency and comparability of GHG emissions on a shipment, not on a national but a truly global basis. COFRET is contributing to a further global harmonization of carbon footprint calculation for freight transport chains. Basing its work on the currently most comprehensive framework for global emission calculation, the European Standard EN 16258, COFRET offers a practical and staged approach to the application of the standard methodology. COFRET provides further guidance on elements of the EN 16258 which receive little or no coverage, such as the treatment of terminals and warehouse and the definition of vehicle operation systems. The present and future efforts undertaken by COFRET to create a structure for a harmonized approach in the calculation of the carbon footprint for freight transport chains build on a close partnership with established initiatives and services on an international level.

20.5. Acknowledgments

The research is based on the work of the COFRET project consortium and its Advisory Board. The COFRET project (see: www.cofret-project.eu) was cofinanced by the European Commission Directorate General for Research and Innovation as part of the 7th Framework Program. The authors express their gratitude for the support received by these partners.

20.6. Bibliography

[AUV 11] AUVINEN H., MÄKELÄ K., LISCHKE A. *et al.*, COFRET Deliverable 2.1 Existing methods and tools for calculation of carbon footprint of transport and logistics, 2011.

[DAV 14] DAVYDENKO I., EHRLER V., DE REE D. *et al.*, "Towards a global CO_2 calculation standard for supply chains: suggestions for methodological improvements", *93rd Annual Meeting Transportation Research Board (TRB)*, Washington, United States of America, pp. 12–16, January 2014.

[DER 12] DE REE D., TON J., DAVYDENKO I. *et al.*, COFRET Deliverable 3.1 Assessment and typology of existing CO_2 calculation tools and ethodologies, 2012.

[EC 11] EUROPEAN COMMISSION, Roadmap to a Single European Transport Area – Towards a competitive and resource efficient transport system, White Paper, Brussels, 2011.

[EHR 13] EHRLER V., DAVYDENKO I., DE REE D. *et al.*, "Standardized emission calculations along supply chains as a basis for smart global transport solutions in a 'reality of less' – Is an approach within reach", *Transportation research record: Journal of the Transportation Research Board (TRR), 92nd Annual Meeting Transportation Research Board (TRB)*, Washington, United States of America, pp. 13–17, January 2013.

[EN 12] EUROPEAN COMMITTEE FOR STANDARDIZATION, EN 16258: Methodology for calculation and declaration of energy consumption and GHG emissions of transport services (freight and passengers), Brussels, 2012.

[FIS 11] FISCHER-KOWALSKI M., SWILLING M., VON WEIZSÄCKER E.U. *et al.*, Decoupling natural resource use and environmental impacts from economic growth, A Report of the Working Group on Decoupling to the International Resource Panel, United Nations Environment Programme (UNEP), 2011.

[KAP 96] KAPLAN R.S., NORTON D.P., *The Balanced Scorecard: Translating Strategy into Action*, Harvard Business School Press, Boston, MA, 1996.

[ISO 11] ISO, ISO/CD 14067.3: Carbon footprint of products. Requirements and guidelines for quantification and communication, Working Draft, Geneva, 2011.

[MIN 11] MINISTRY OF ECOLOGY, SUSTAINABLE DEVELOPMENT, TRANSPORT AND HOUSING, Information on the quantity of carbon dioxide emitted during transport, Decree no. 2011–1336, 2011.

[PTV 13] PTV, Map & Guide, available at: http://www.mapandguide.com/en/, 2013.

[SCH 12] SCHMIED M., WÜTHRICH P., "Der neue standard EN 16258", *Internationales verkehrswesen*, vol. 64, no. 2, pp. 12–14, 2012.

[STR 79] STRUNK W., JR. WHITE E.B., *The Elements of Style*, 3rd ed., Macmillan, 1979.

[VAN 00] VAN DER GEER, J., HANRAADS J.A.J., LUPTON R.A., "The art of writing a scientific article", *Journal of Scientific Communications*, vol. 163, pp. 51–59, 2000.

21

Carbon Intensity of French Shippers

Greenhouse gases are an important issue, especially in freight. In order to reduce them, it is necessary to quantify them and to identify the activities that are the most emissive. To estimate CO_2 emitted by a French shipper for the transportation of its shipments, we use the French shipper survey. In this survey, we only have three shipments per shipper and carbon intensity per shipment is highly variable. So the CO_2 per shipper was estimated indirectly, using two different approaches: first, we set up a typology of the shippers, in 10 types of 'homogenous' establishments, and estimated in each type an average level of emissions per shipper. In a second approach, carbon intensity per tonne shipped has been modeled, using only the shipper characteristics as explaining variables. Then for each shipper, multiplying this average CO_2 emission per tonne by its yearly tonnage shipped gave the amount of emissions for its shipments during that year. The two approaches gave reasonably similar results. On average, a French shipper emits a little more than 620 tonnes per year for the transport of its shipments, and this quantity varies from 78 tonnes for the 'lower' type of emitter up to 4,700 tonnes for the 'upper' type.

21.1. Introduction: context and objective

Over one quarter of greenhouse gas (GHG) emissions in France come from the transport sector and this share is growing. Furthermore, within the transport sector, the share of emissions from freight activity is growing and there is no sign of saturation [EME 10]. Therefore, deep changes are necessary to reach the target of dividing by four the anthropologic GHG emissions, either through carbon efficiency improvements or by reduction of transport demand. The link between transport emission and climate change has been analyzed in Schwanen *et al.* [SCH 11]; here we are focusing on freight carbon intensity.

Chapter written by Christophe RIZET, Cecilia CRUZ, Matthieu DE LAPPARENT and Martine VROMANT.

The carbon efficiency of freight transport and also the determinants of freight transport demand are not well understood. We claim that an in-depth knowledge of the factors that influence firms in their logistical choices as well as the factors that influence carbon efficiency would make freight GHG mitigation much easier. This lack of knowledge is due to several factors, including the theoretical complexity of the problem, the insufficiency of resources that have been made available for freight compared with passenger transport, and the inadequacy of the existing data. The French shipper survey developed in 2004 is an attempt to fill this gap in data. The survey methodology is discussed in Rizet *et al.* [RIZ 04], and [GUI 10] give an overview of the results.

Our objective in this chapter is, using this survey, to estimate the level of CO_2 emissions and carbon efficiency of freight transport per shipper. After presenting the survey (section 21.2) and developing the way we estimate transport energy consumption and CO_2 emission per shipment (section 21.3), we present the two methods used to estimate the yearly quantity of freight transport CO_2 emissions per shipper. First, we set up a typology of shippers, classified 10 types of 'homogenous' shippers and estimated in each type an average level of emissions per shipper (section 21.4). In a second approach, we set up a model explaining the CO_2 emitted per tonne shipped, using only the shipper characteristics as explaining variables. Multiplying this average CO_2 emission per tonne of the shipper by the yearly tonnage gives us the amount of emissions of this shipper for its shipments during that year (section 21.5). Then, we compare the shipper emissions estimated through these two approaches and consider the variability of carbon efficiency among shippers (section 21.6).

21.2. Data: the 2004 French shipper survey

The two major characteristics of the French shipper survey are the description of the shippers' organizational features influencing the transport choices, and the tracking of a selection of shipments from their departure from the plant up to their arrival to the consignee. In this survey, information is collected at three levels:

– Shipper establishment: after a few questions about the volume and structure of the company's ingoing and outgoing transport flows and its own fleet of vehicles, a face-to-face interview is administered with the logistics manager of the company. The questions concern the economic characteristics of the firm: production, distribution, storage practices, relationships with its customers and suppliers, and the management and communications systems it uses. A "transport" section is added to this description of the firm's industrial and logistical organization. It deals with the firm's relationships with carriers, terms of access to the various types of infrastructure and how responsibility for transport is shared between the firm and its partners.

– Shipment level: at the end of the establishment questionnaire, the last 20 shipments are listed, three of which are randomly selected and then traced until they reach their final consignee. The consignment questionnaires are filled in either with the manager mentioned above or with the manager in charge of dispatching; they deal with the economic relationship between the shipper and the consignee and the terms of business between the two. This relationship between the shipper and his clients is described not only through the physical and economic characteristics of the shipment but also with information about the quantity of goods sent to this client per year (yearly tonnage to this consignee), the yearly number of shipments to this consignee, the split of responsibilities with regard to transport organization, and the contractual allocation of transport costs and associated services. The first information required in order to reconstruct transport chains is also collected at this level, with the identification of the consignee.

– Transport operator and journey link: for this part of the survey, questionnaires are administered by telephone. Questions deal with the economic characteristics of the operator, including the operator to whom the shipment was handed and the characteristics of the transport leg: mode and vehicle type, load, etc. The next operator is in turn questioned up to the final consignee and the description of the whole transport chain. The transport chain is split into as many legs as there are changes to another vehicle.

The transport chain is thus reconstructed throughout Western Europe and includes an interview with the consignee. For a shipment which travels beyond the limit of Western Europe, the transport chain is surveyed only until the first transfer point after the frontier of Western Europe. Furthermore, the 2004 shipper survey has been adapted to enable the quantification of energy consumed and CO_2 emitted, and to relate them to the determinants of freight transport demand [RIZ 04]. Primary and final energy consumption have been computed per leg, as the product of distance per vehicle energy consumption (for trucks: liters per km), divided by the load and multiplied by the shipment weight; empty backhauling is estimated using the results of the French national freight vehicles survey. Energy consumption is converted into grams of oil equivalent (goe) and shipment energy consumption is the sum of energy per leg. Both Tank to Wheel (TW) and Well to Wheel (WW) CO_2 emissions are then computed as the product of energy by an emission factor. Dividing CO_2 by the tonne-kilometer of the shipment gives the carbon efficiency of the shipment, in kg

CO_2/tkm. The method used to compute CO_2 per shipment is described in Rizet *et al.* [RIZ 13]. In this chapter, we present the methods used to estimate the yearly quantity of freight transport CO_2 emissions per shipper. This was done with two approaches: first, we set up a typology of shippers, in 10 types of 'homogenous' shippers and estimated the average level of CO_2 emissions in each type; in the second approach, we modeled the carbon intensity per tonne shipped and multiplied this by the yearly tonnage shipped by a shipper.

21.3. Estimating shipper yearly emissions from a typology of shippers

In the survey, the sample of shipments is obtained by a first sampling among shippers (about 3,000 surveyed shippers) and a second sampling of shipments among the surveyed shippers. Only three shipments are surveyed per shipper on average, which does not allow direct estimation of the volume of emissions of this shipper. So, we grouped a significant number of shippers having 'profiles' as close as possible, to obtain a number of shipments which allow us to estimate the volume of emissions of this type of shippers.

21.3.1. *Establishing a typology of shippers*

Our objective thus is to establish a typology of shippers using the variables which characterize the volume of emissions of these shippers. After different attempts, we finally used a typology realized by 'analysis of the multiple correspondences' (AMC) and a 'non-hierarchical classification', in 10 types [CLA 98]. We used the following variables:

– the number of employees;

– the annual number of shipments;

– the distribution of all the shipments according to their weight;

– the distribution of all the shipments according to their destination;

– the distribution of all tons shipped according to the main transport mode.

AMC allows realizing a typology based on resemblance. Two shippers will be closer if they share a large number of modalities (variables). The observation also concerns the association between modalities, establishes using the SAS CORRESP procedure. The output of this procedure is a set of new variables which correspond to the coordinates of the individuals (shippers) and the variables in a multidimensional space. This space is defined by more or less axes according to the number of variables or the quality of the analysis. The quality of the explanation of the cloud of points will be a function on behalf of slowness of this cloud, which is explained by axes. For the classification we used six axes, which bring the part of slowness explained to 51%. The typology is then realized by the method of classification of dynamic clouds applied to the coordinates of the shippers on each of these axes.

The next step is to group shippers by types, according to their positioning on the axes. Two shippers which are close according to these axes will be grouped in the same type. The dynamic clouds method is then applied to the coordinates of shippers on each of these axes.

Shippers are grouped in a limited number of types in such a way that shippers of the same group are alike from the point of view of variables used for the analysis. We chose to limit the number of types to 10. Starting from 10 cores and by successive iterations, the procedure supplies a set of types by grouping the closest establishments. The SAS CLUSTER and FASTCLUS procedures were used to define the types from the coordinates arising from the AMC. We thus obtained 10 types of shippers with more than 200 shippers in each type, plus a group of 120 shippers with missing variables, classified as 'without type'.

21.3.2. *Estimating the quantity of freight transport CO_2 emissions per shipper of each type*

For each type of shipper so obtained, we calculate then (see Table 21.1):

– the total quantity of CO_2 emissions of freight transport as the weighted sum of emissions of the shipments of this type;

– the average emissions of a shipper of this type, by dividing the total emissions of the type by the weighted population of shippers.

Levels of emissions are very variable among types. The biggest emitters are in type 2 with an average of 4,764 tonnes of CO_2 per year for the transportation of their shipments; the lowest emitters are in type 1, with an average of 79 tonnes per establishment and per year. To appreciate the dispersal of these values of average emissions per shipper according to the type, we calculated for each type, the ratio between the average emissions per shipper of this type to the emissions of the least emitting type of shipper (i.e. type 1): according to this ratio (last column in Table 21.1), a shipper of type 2 emits on average 56 times more than a shipper of type 1. The various types do not all differ significantly by their average level of emissions: types 5 and 9 on the one hand and types 7 and 10 on the other hand seem to have close levels of yearly emissions.

Type number	Sample size (number of observed shippers)	Population (weighted number of shippers)	Total emissions of the type (thousands tCO_2WW)	Average emissions/ shipper (tCO_2WW)	Ratio
0 (no type*)	120	1,869	1,970	1,054	13.3
1	287	13,731	1,089	79	1
2	265	2,228	10,614	4,764	60.3
3	227	4,999	1,261	252	3.2
4	399	7,260	4,632	638	8.1
5	300	4,626	1,860	402	5.1
6	324	3,340	5,189	1,554	19.7
7	246	3,810	4,612	1,210	15.3
8	236	15,127	1,306	86	1.1
9	261	4,897	2,421	494	6.2
10	214	7,335	8,031	1,095	13.8
TOTAL	2879	69,222	42,985	621	

*:Shippers with missing data

Table 21.1. *The 10 types of shippers and their CO_2 emissions*

21.3.3. *Precision of the estimated quantity of CO_2 per shipper*

To calculate the precision of this estimator, we used the 'bootstrap' method: in each type, we randomly sampled important groups of shipper establishments, and on each of these subsamples we calculated a new value of CO_2 by establishment with the previous method. The distribution of these indicators from the various subsamples of the same type allows us to calculate the precision of the estimation of the CO_2 per shipper of this type. The SAS SURVEYSELECT procedure allows making this poll of subsamples and so facilitates this method of 'bootstrap'. Figure 21.1 shows the distribution of the average CO_2 by establishment so estimated in the various subsamples of type 1.

The average values and the standard deviation of these estimators resulting from the bootstrap enable us to calculate the limits of the 95% confidence interval. These limits are shown for each type in Figure 21.2.

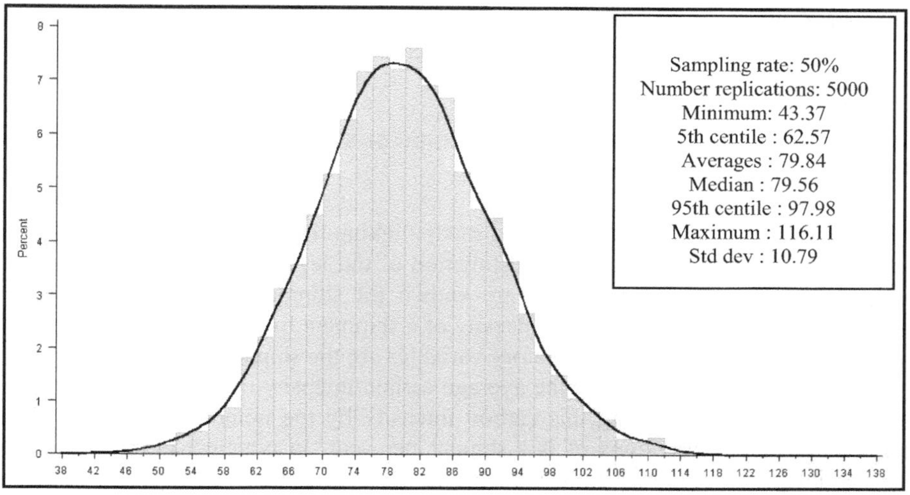

Figure 21.1. *Distribution of the estimators of CO_2 per shipper of type 1, in tCO_2/shipper/year*

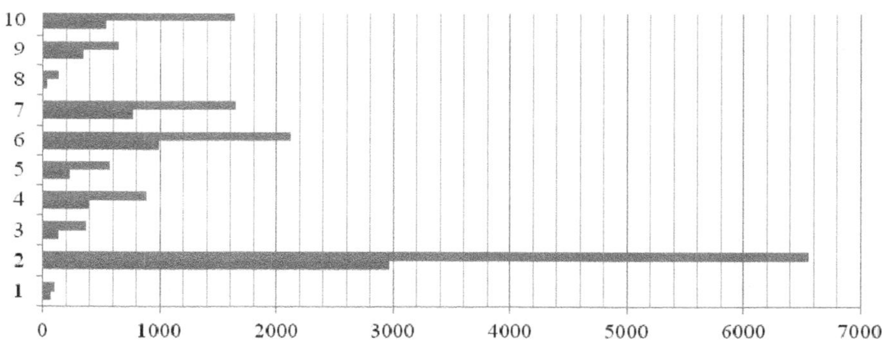

Figure 21.2. *Precision of the CO_2 estimator per shipper according to its type in tCO_2/shipper (the blue bars correspond to the minimum of the 95% confidence interval and the red ones to the maximum). For a color version of the figure, see www.iste.co.uk/jacob/energy.zip*

Figure 21.2 shows that only type 2, with an average of 4,770 tonnes of CO_2 per shipper and per year, differs from other types for the volume of emissions. Types 1 and 8, which have a low level of emissions, are very close to one another, with an average of, respectively, 80 and 87 tonnes and confidence intervals which widely recover. Confidence intervals of the other types also recover: types 3, 4, 5 and 9 on the one hand and types 6, 7 and 10 on the other hand.

21.4. Estimating yearly shipper emissions with a model of carbon intensity per tonne shipped

We want to develop and calibrate a model linking the CO_2 emitted to the shipper characteristics. If we disregard the estimation of the levels of emissions by type of shippers, transport CO_2 emissions are known for shipments. So, various models were tested to explain the carbon intensity of a shipment by the characteristics of the shipper. As far as the only explanatory variables are the shipper characteristics, such a model enables us to estimate the average carbon intensity of a shipment from this shipper and, by multiplying this carbon intensity by the number of shipments, to estimate the total emissions of this shipper per year. In practice, we modeled the carbon intensity per tonne shipped rather than per shipment.

21.4.1. *A model of carbon intensity per tonne shipped*

Two quantities are known for all shippers and could allow estimation of their freight transport emission: the yearly number of shipments and yearly tonnage shipped. After various attempts, we decided to model the carbon intensity per tonne, rather than the intensity per shipment, because the estimate is more accurate. Various forms of models were tested, in particular combinations of logarithms and levels.

The potential explanatory variables are the characteristics of the shipper from the survey, in order to be able to estimate an average level of carbon intensity per tonne for each shipper, independently of the characteristics of the shipment. In these 300 available variables, about 15 were found explanatory without counting too many missing. The class variables have been transformed into dummies. The final model is:

$$\text{Log}\ (CO_2/\text{tonne}) = a1 * x1 + a2 * x2 + a3 * x3 + ... + a15 * x15$$

where x1, x2, x3... are the shipper characteristics.

This is based on 7,573 observed shipments with all the explanatory variables; the coefficient of correlation is 0.40 and 14 variables are explanatory at the level of 5%. These 15 variables are the characteristics:

– of the distribution of the shipper shipments according to their weight (a proxy for the shipment weight);

– of the distribution of the shipper shipments according to their geographical destination (a proxy for distance);

– of how much this shipper is using the various transport modes (a proxy for the mode);

– of the shipper 'business sector' (a proxy for the type of goods shipped).

Estimated carbon intensity per tonne varies from about 10 kg to some 3 tonnes, with the majority below 1 tonne.

21.4.2. *Precision of the estimated carbon intensity per tonne*

The 95% confidence interval of carbon intensity was calculated with SAS, using the option *lclm* = lower limit and *uclm* = upper limit. In Figure 21.3, every establishment is represented by three points which have for abscissa the estimated CO_2/tonne of the shipper and, for ordered points, the same value of carbon intensity per tonne shipped by this shipper, as well as the lower and higher limits of the 95% confidence interval of this estimated value. To allow better legibility, Figure 21.3 focuses on carbon intensities lower than 1 tonne.

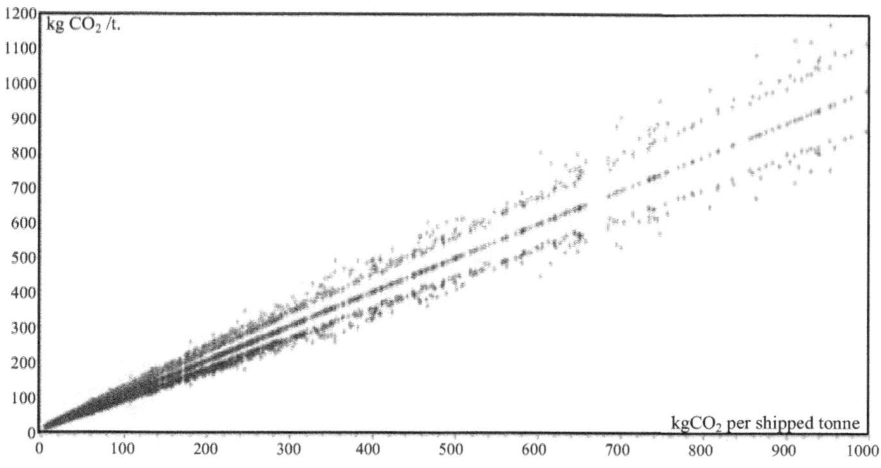

Figure 21.3. *Carbon intensity per tonne shipped and the 95% confidence intervals*

The limits of the confidence interval are about 20% above and below the estimated value. This confidence interval varies from one establishment to another, but this variability itself is limited: thus, there is an important difference between establishments on the left of the figure, which emit less than 100 kg by transported tonne and the establishments on the right, which emit more than 900 kg by tonne.

21.4.3. *Carbon intensity per shipper*

The CO_2 emitted by a shipper per year is calculated as the product of the CO_2 per tonne, estimated by shipper according to the method described above, by the yearly tonnage shipped by this establishment. The yearly tonnage of the shipper is known from the survey, so the limits and the confidence interval for the CO_2 per shipper are calculated as the product of the limits of the confidence interval of the CO_2 by tonne multiplied by the yearly tonnage of the shipper.

21.4.4. *Precision of the estimated carbon intensity per shipper*

According to this approach, the CO_2 emitted by a shipper in 1 year for the transport of its shipments varies from approximately 1 tonne (on the left of Figure 21.4) up to more than 10,000 tonnes (on the right). The confidence intervals are important in absolute value, but considering the dispersal of these estimated values of CO_2 per shipper, it is possible to define groups of establishments clearly differentiated from the point of view of emissions, with little overlapping of confidence intervals.

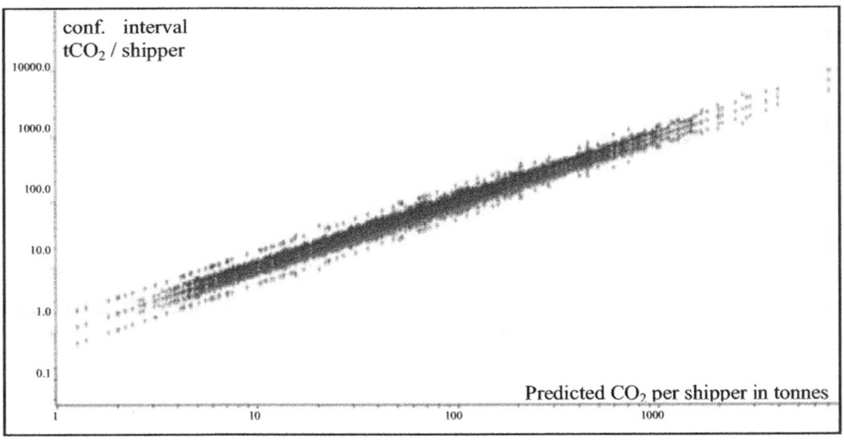

Figure 21.4. *Estimated tonnes of CO_2 per shipper and the 95% confidence intervals*

21.5. Comparison of carbon intensity per shipper estimated by the two methods

We now have two values of CO_2 emitted by shipper for the transport of yearly shipments; one established by the typology and the other one by the model. Figure 21.5 compares these indicators by type of shippers, for those shippers for which a level of CO_2 has been estimated in both ways. The biggest emitters (shippers of type 2), with 4,700 tonnes per year, emit 60 times more than those of type 1 (78 tonnes). A CO_2 mitigation policy aimed at shippers should clearly focus on these 'big emitters'. On average, the CO_2 emitted by a shipper is a little more than 620 tonnes a year.

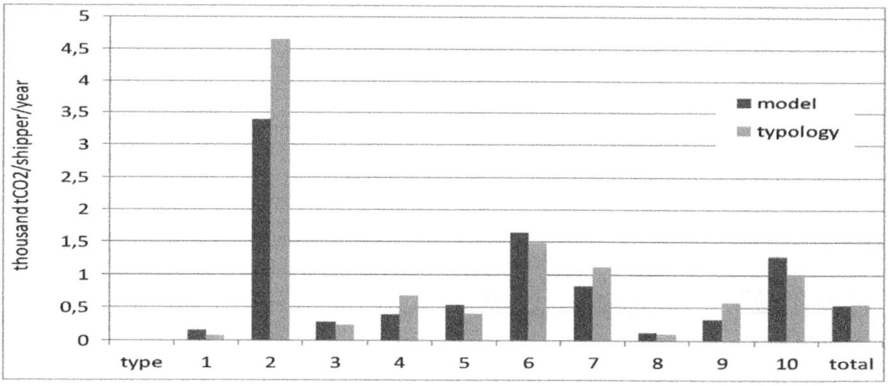

Figure 21.5. *A comparison of shipper carbon intensity estimated by two methods*

21.6. Carbon intensity, carbon efficiency and the impact of a carbon tax on shipper competitiveness

Quantifying CO_2 emitted by shippers is important for assessing the potential impact of a carbon tax policy. If we consider that such a carbon tax would be distributed on all the tonnes sold by this company, then the main indicator for the impact on shipper competitiveness should be the shipper CO_2 per tonne, i.e. its carbon efficiency. If we now consider that the amount of carbon taxes will be used to reduce the cost of labor (health, unemployment and pension), in order to improve the competitiveness of the national workers, then the benefit of such a policy for a company would be proportional to its number of employees: the more employees, the more cost reduction.

Our 10 types of shippers not only have different carbon intensity per shipper and per employee but they also correspond to different carbon efficiencies per tonne (from 12 up to 208 kg CO_2/tonne) and per tkm. Carbon efficiency per tkm is less variable (from 0.046 up to 0.678 kg CO_2/tkm). By employee, the average level of emission is 16.5 t CO_2 per year and varies from 4.3 t for type 10 up to 27.2 t for type 9; that is, a 1 to 6 ratio.

Type	N Obs	Thousands tCO_2	tCO_2 emitted		$kgCO_2$ emitted	
			per shipper	per employee	per shipped tonne	per tkm
0	120	1,970	1,054	13.0	28.9	0.078
1	287	1,089	79	4.3	11.8	0.110
2	265	10,614	4,764	12.3	89.6	0.111
3	227	1,261	252	4.4	61.0	0.102
4	399	4,632	638	9.6	52.3	0.124
5	300	1,860	402	6.9	18.5	0.082
6	324	5,189	1,554	7.9	44.8	0.046
7	246	4,612	1,210	16.5	24.0	0.136
8	236	1,306	86	4.7	59.5	0.151
9	261	2,421	494	6.1	38.6	0.089
10	214	8,031	1,095	27.2	208.4	0.678
Total	2,879	42,985	621	10.2	46.7	0.108

Table 21.2. *Carbon intensity and carbon efficiency per type of shipper*

Indeed, these four indicators give rather different views of the burden of a freight transport carbon tax among the different types of shippers:

– a transport carbon tax might be supported by the shippers according to their level of emissions. With a tax of € 100 per tCO_2, for example, the tax per shipper would range from € 7,900 for shippers of type 1 up to € 476,400 for shippers of type 2;

– these shippers have very different levels of sales or shipments; related to their tonnage shipped a tax of € 100 per tCO_2 would then range from € 1.18 sold tonnes for shippers of type 2 up to € 20.8 sold tonnes for shippers of type 10;

– if the product of this carbon tax is used to reduce the cost of labor, those benefited would then be the big employers. The number of employees is then an indicator of the cost reduction for the shipper of such a policy;

– the ratio tCO_2 emitted per employee is then a good indicator of the real cost for a given shipper or type of shipper. Multiplied per the ratio [tax (in €/tCO_2)/labor cost reduction (in €/employee), it gives an idea of the impact of such a policy for this type of shipper. If the product of these two ratios equals one, the benefit is as important as the cost. If this product is higher than one, then the cost is higher than the benefit; and if this product is smaller than one, then the reduction of cost is more important than the tax paid. We don not know the 'tax per tonne/reduction of cost per employee' ratio but we can make the hypothesis that this ratio will be identical for all shippers. Then, the comparison of the tCO_2/employee ratio is an indicator of the split of the burden among the different shippers. The winners of such a policy would then be the shippers of types 1, 3 and 8; the shippers of type 10 would be the main losers;

– carbon efficiency gives another view of the comparison of emissions between very heterogeneous shippers. Type 10 for example has a very low carbon efficiency per tonne and per tkm.

21.7. Conclusion

CO_2 emissions per shipper have been estimated with two different approaches, using the French shipper survey data: first, we set up a typology of the shippers, in 10 types of 'homogenous' establishments and estimated for each type an average level of emissions per shipper. In a second approach, carbon intensity per tonne shipped has been modeled, using only the shipper characteristics as explaining variables. Then, for each shipper, multiplying this average CO_2 emission per tonne by its yearly tonnage shipped gave the amount of emissions for its shipments during that year. The two approaches gave reasonably similar results.

This in-depth knowledge of the shipper CO_2 emissions can be used to compare the carbon efficiency of different types of shippers and the impact of a carbon tax applied to freight transport on shipper's competitiveness. Four indicators of carbon performance have been quantified that give very different results for our 10 types of shippers:

– carbon intensity per shipper and per year varies from 78 tCO_2, for the 'lower emitter type' up to 4,700 tonnes for the 'upper' type. The average is 620 tCO_2 per shipper and per year and a shipper of the 'upper' type emits 60 times more CO_2 than the 'lower' ones;

– the 10 types of shippers also have very different carbon intensities per tonne: from 12 up to 208 $kgCO_2$/tonne (a 1 to 17 ratio) with an average of 47 $kgCO_2$/tone;

– per tkm, carbon intensity varies according to the shipper types, from 46 up to 678 gCO_2/tkm (a 1 to 15 ratio), with an average carbon intensity of 108 gCO_2/tkm;

– finally, carbon intensity per employee could be a significant indicator. If the carbon tax receipt would be used to reduce the indirect cost of labor, shippers would pay taxes proportional to their emissions and would benefit from the reduction of labor costs according to their number of employees. Then shippers of type 1, with a ratio of 4.3 tCO_2 per employee per year, would be the winners while companies of group 10, with a ratio of 27.2 tons per employee, would be the main losers of such a policy.

21.8. Acknowledgments

The results presented here are part of an Ifsttar research program funded by the French Ministry of transport (DGITM); we have used the SAS software for data analysis and figures.

21.9. Bibliography

[CLA 98] CLAUSEN S.E., *Applied correspondence analysis*, Sage, 1998.

[EME 10] EMEP/EEA, Emission inventory guidebook European Environment Agency, Technical report no. 9, 2009, updated June 2010.

[GUI 10] GUILBAULT M., GOUVERNAL E., "Transport and Logistics Demand: new input from large shippers surveys in France", *Transportation Research Record: Journal of the Transportation Research Board*, no. 2168, pp. 71–77, 2010.

[RIZ 04] RIZET C., ARMOOGUM J., MARCHAL P., "Energy consumption estimation with a shipper & transport chain survey", *Proceedings of the International Conference on Transport Survey Quality and Innovation*, Costa Rica, 2004.

[RIZ 13] RIZET C., CRUZ C., LAPPARENT M., "Impact of logistical choices on freight transport carbon efficiency", *13th World Conference on Transport Research*, Rio de Janeiro, July 2013.

[SCH 11] SCHWANEN T., BANISTER D., ANABLE J., "Scientific research about climate change mitigation in transport: a critical review", *Transportation Research Part A*, no. 45, pp. 993–1006, 2011.

PART 5

Air Pollution

Introduction to Part 5

Air pollution remains a major public health concern in urban areas, and transports are now clearly singled out due to their significant contribution, suggesting a shift to clean modes and low-emissions vehicles and fuels. However, despite progressively tighter exhaust emission standards, despite significant improvements in the vehicles' technologies, in fuels and in traffic and mobility management, and despite the implementation of city-wide measures, a large number of urban areas in Europe are still facing air pollution exceeding PM10, PM2.5 and NO_2 limit values.

Real-world implementation of measures such as the Low Emission Zone in London did not show any significant improvement of air quality due to high background concentrations, while real-world NOx or NO_2 emissions did not decrease as expected by the emission standards.

A large range of measures is envisaged to decrease the transport contribution to air pollution (technologies and fuel improvement, soft modes promotion, traffic management, transport organization as well as various incentives), but their assessment remains highly difficult due to large uncertainties and limits of the mobility, traffic, emission and air quality models, lack of local data, insufficient knowledge on certain emission phenomenon and pollutants, and finally due to the dilution of benefits from transport mitigation measures, because other sectors and regional contributions are also significant.

The chapters of this part address two main topics: recent progress on depollution, through fuel composition and aftertreatment technologies, and new results regarding emission factors, input data, and tools for estimating pollutant emissions from transports.

Introduction written by Michel ANDRÉ.

Depollution techniques

The first chapter studies and demonstrates the positive effect of increased fatty acid methyl esters (FAME) content in diesel fuel blends on diesel engine pollutant emission, particularly the induced decrease of particulate matters, the decrease of the diesel particulate filters (DPF) regeneration frequency, and the reduction of the fuel consumption associated with soot loading and regeneration phases. FAME rates at 30% and 50% could increase the distance between regeneration by 40 and 90% (PM emission being decreased in the same proportions), while the decrease of fuel consumption would be around 1–2% (experiment on a single cylinder engine and NEDC test cycle).

The second chapter studies and proposes a configuration of exhaust aftertreatment technologies to remove NOx emissions and minimize CO_2. Apart from the diesel oxidation catalyst and particulate filter (upstream), the configuration consists of a NOx storage catalyst (NSC) and a bypass, a passive selective catalytic reduction (SCR) and an integrated fuel reformer (producing H_2 and CO from exhaust gas and diesel + air). The system operates discontinuously in 2 modes: 1- NOx absorption on the NSC, while NOx slip is continuously reduced in SCR by NH_3, 2- regeneration of the NSC, the stored nitrate and nitrite being reduced by reactions with HC, CO and H_2 (from the reformer), while the produced NH3 will be stored by the SCR. In all the configuration demonstrated a higher NOx efficiency with low fuel consumption penalty.

The last chapter proposes NOx/SOx depollution from marine diesel engines, through a power controlled microwave reactor. An experimental unit is developed, tested in combination with a marine diesel engine, and demonstrates efficient removal (90%) of NOx and SOx, even if these early results were obtained over a short duration (30 s).

Pollutant emissions estimation

Several emissions phenomenon and pollutants are insufficiently known to take into account in air pollution estimation. That is the case for non-exhaust road transport particulates emissions, addressed by the 4th chapter of this part, and more specifically emission factors linked to tyre wear, brake wear, road wear and resuspension of particles. Pollutants considered included PM2.5, PM10, particle number, elemental carbon, organic matter, 7 metals and 28 PAHs. A complete database of non-exhaust emission factors was thus elaborated from literature data and in complement of the COPERT methodology for exhaust emissions, thus constituting a important step towards taking into account the total emissions from road transports in air pollution assessment.

Local estimations of road transport emissions and their spatial distribution are a prerequisite for assessing local air pollution, traffic contribution, and potential health impacts while considering human exposure. With that aim, development and adaptation of the COPERT emissions calculation methodology has been envisaged for calculation at a street level. This work is described in the next chapter. Currently, the COPERT Micro or Street Level was demonstrated for its ability to compile city-scale and micro-inventories, and also to assess local measures such as Low Emission Zones.

Finally, input data are key elements of pollutant emission estimation, and particularly for local scale assessment. The last chapter of this part examines specifically the vehicle fleet composition in that process. Different estimations of the French fleet showed significant differences and thus demonstrated the need of explicit approaches and assumptions. Due to their high sensitivity to the vehicle fleet, emissions estimations could thus be questionable. A systematic review and update of the statistics is conducted for computing the French vehicles fleet from 1980 to 2030, for all vehicle categories and in view of pollutants emission calculation. Complementary in-situ video observations also demonstrate the sensitivity of fleet composition to local contexts.

22

Impact of FAME Content on Regeneration Frequency of Diesel Particulate Filters (DPFs)

Modern diesel vehicles utilize two technologies that have been motivated by recent European legislation: diesel fuel blends containing fatty acid methyl esters (FAME) and diesel particulate filters (DPF). Oxygenates, like FAME, are known to reduce PM formation in the combustion chamber and reduce the amount of soot that must be filtered from the engine exhaust by the DPF. This effect is also expected to lengthen the time between DPF regenerations and reduce the fuel consumption penalty that is associated with soot loading and regeneration. This study investigated the effect of FAME content, up to 50% v/v (B50), in diesel fuel on the DPF regeneration frequency by repeatably running a Euro 5 multicylinder bench engine over the NEDC until a specified soot loading limit had been reached. The results verify the expected reduction of engine-out PM emissions with increasing FAME content, and the reduction in fuel consumption penalty associated with reducing the frequency of DPF regenerations.

22.1. Introduction

Recent European legislation, such as the Renewable Energy Directive (RED) and the Fuel Quality Directive (FQD), set targets for increasing renewable energy and reducing greenhouse gas (GHG) emissions from road transportation by 2020. Meeting these targets has encouraged the use of bio-derived blending components in market fuels, such as ethanol from sugar fermentation for gasoline blending and fatty acid methyl esters (FAME) from the esterification of vegetable oils and animal fats for diesel fuel blending. Although there is considerable work in progress

Chapter written by Kenneth D. ROSE, Heather HAMJE, Dimitris KATSAOUNIS, Christos SAMARAS, Savas GEIVANIDIS and Zissis SAMARAS.

to develop more advanced products that utilize more of the plant's biomass, commercial volumes of these products are still quite small and are not expected to make a large contribution to transport fuels before 2020. At the same time, vehicle emissions limits for both CO_2 and other regulated pollutants will continue to tighten over this decade to further improve transport-related air quality.

Modern diesel vehicles utilize two technologies that have been motivated by this combination of European fuel and vehicle legislation: diesel fuel blends containing FAME and diesel particulate filters (DPF). DPFs are now widely used on European diesel vehicles and are responsible for removing filterable particulate matter (PM) and much of the ultrafine combustion particle numbers (PN) from the diesel engine-out exhaust.

The addition of FAME into diesel fuel is well-known to decrease the PM emissions of diesel engines [LAM 13, WIL 06, HAS 07, CZE 12, BHA 13, EPA 02]. This effect is largely attributed to the addition of oxygen into the fuel which increases the local oxygen concentration in the rich area of the diesel flame [LAM 13] as well as by dilution of polyaromatic hydrocarbons in the diesel fuel with an aromatics-free blending component.

The use of DPFs in modern vehicles results in a small but important increase in fuel consumption, mainly due to two factors. First, additional engine work is required to compensate for the increasing pressure drop across the DPF as the filter accumulates soot. As soot increases and the back-pressure also increases across the DPF, the engine must compress exhaust gases to a higher pressure which involves additional mechanical work. Less energy is also extracted by the exhaust turbine which can affect the intake manifold boost pressure [JAA 07, MIK 10].

Second, the DPF must be periodically regenerated to remove the accumulated soot. This is usually done by introducing a small amount of additional fuel through late cycle injection using the post injection system of the engine. This results in higher concentrations of hydrocarbons in the exhaust, which are oxidized in the diesel oxidation catalyst or the catalytic DPF. This exothermic process increases the temperature at the DPF to sufficient levels that the accumulated soot is burned away with the oxygen that is present in the exhaust. The fuel consumption penalty associated with this process depends on the rate of soot build-up and on the frequency of the DPF regeneration.

Although this process is generally understood, the nature of this effect and the role of FAME have not been systematically investigated. This study was designed to investigate in detail the effect of FAME content, ranging from 0 to 50 vol% (B0 to B50) in diesel fuel, on the DPF regeneration frequency. There were three objectives for this study: (1) develop a repeatable bench engine test protocol to evaluate the impact of FAME content on DPF regeneration frequency, (2) use this test protocol to relate the DPF regeneration interval to the FAME content in diesel, and (3) assess the possible benefits or determent of FAME content on fuel consumption.

22.2. Experimental setup

Measurements were performed on an engine bench with a Euro 5-compliant 1.4 liter turbocharged diesel engine (66kW at 3800RPM) installed on an AVL Dynoperform 350. Several parameters were constantly monitored including exhaust temperature at the DPF inlet and outlet, DPF pressure drop, O_2 and NO_x concentrations and engine data (speed, torque, acceleration pedal position, EGR, and inlet air flow rate). Fuel consumption was measured with an AVL 735 fuel meter.

The DPF was weighed at ambient temperature before and after each test to provide an accurate value for the soot loading. For some selected tests, PM mass and PN emissions at the DPF inlet were measured according to the legislated method in the CVS. The PM mass emissions were also monitored with the AVL Micro Soot Sensor (MSS). Additionally, gaseous emissions (CO, HC, NO_x, and CO_2) were measured with an AVL AMA i60 analyzer (Figure 22.1).

Fuels for this study were blended from a conventional diesel fuel complying with the European norm EN 590 and having a sulphur content less than 10 ppm. A single batch of rapeseed methyl ester complying with the European norm EN 14214 was used to produce the FAME/diesel blends. The FAME was additized with butylated hydroxytoluene (BHT) antioxidant after production in order to ensure acceptable oxidation stability throughout the study. The oxidation stability on the FAME/diesel blends was measured at the beginning and end of the study using the Rancimat method (EN 15751). Selected fuel properties are shown in Table 22.1.

Fuel Designation	FAME Content (% v/v) (EN 14078)	Density (kg/l) (EN ISO 12185)	Lower Heating Value (LHV) (MJ/kg) (IP 12)	Distillation Range (°C) (EN ISO 3405)	Initial Oxidation Stability (h) (EN 15751)
B0	0.0	0.8334	42.93	173.0–357.8	–
B10	10.2	0.8386	42.53	174.0–355.7	39.2
B30	28.0	0.8472	41.34	178.0–354.3	21.5
B50	48.7	0.8578	40.34	183.5–352.7	16.7

Table 22.1. *Fuel properties*

The experimental setup is shown in Figure 22.1. The exhaust gas could follow two paths according to the needs of the measurement. During soot loading and DPF regeneration, the exhaust gas went through path 1, exiting the tail pipe. For the emission measurements, the exhaust gas followed path 2 to the CVS. A ceramic diesel oxidation catalyst (DOC, cordierite substrate, 600cpsi/3mils, 1.1 liters) was installed at the exhaust line, upstream of the DPF. The conventional DPF (a SiC, 300cpsi/12mils, 16 segment, 2.5 liters) was installed downstream of the DOC. Four identical and initially unused DPFs were used, one for each test fuel.

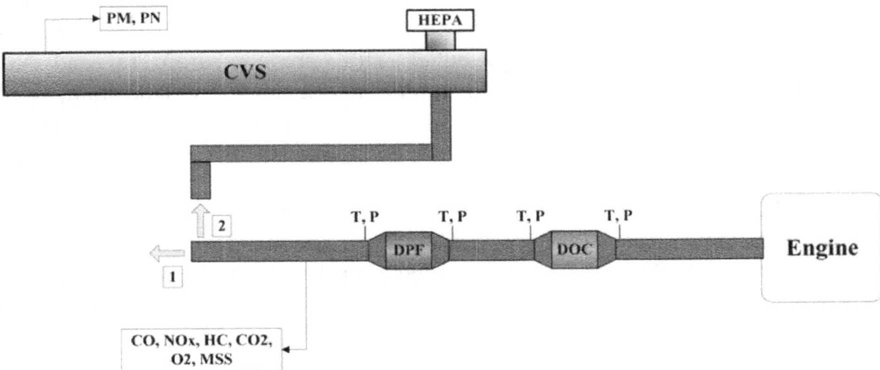

Figure 22.1. *Schematic representation of experimental setup and measured quantities*

22.3. Test procedure

A repeatable test procedure was first developed in a scoping study that was then used to evaluate the effects of FAME content in diesel fuel on fuel consumption and DPF regeneration. Before each test on a new diesel blend, the engine oil was changed and a sample of the fresh oil was taken. At the start of each test, the engine was warmed up with the DPF removed from the exhaust line, until a sufficient operating temperature had been reached. This excluded any repeatability problems that might occur during cold start and ensured that the DOC would be active from the beginning of each measurement. After this preconditioning, the DPF was installed in the exhaust to start the test. The engine simulated multiple repetitions of the New European Driving Cycle (NEDC) until a specified loading limit on the DPF had been reached at which point the DPF was regenerated using post injection.

The DPFs were unused at the start of testing so a conditioning procedure consisting of a loading/regeneration cycle was used to bring the filters to a state similar to that after some road use. The engine was operated over the NEDC for an equivalent distance of 100 km after which the DPF was fully regenerated at 2000 RPM/40 Nm using the active regeneration system of the engine. The post injection was adjusted to achieve at least 600°C temperature at the outlet of the DPF and ensure that the filter was completely cleaned. The duration of the regeneration was defined to be 5min after the pressure drop had been stabilized. Then, the DPF weight at the clean condition was measured and another lubricant sample was taken.

The soot loading procedure was then initiated over the NEDC with a continuous engine speed and torque demand from the dyno. The soot loading of the DPF was measured by weight measurements at the end of each day. The gaseous emissions after the DPF were measured during the first 6 cycles of the last loading day, to check engine repeatability. This procedure was repeated until a loading of 6 g/l had been achieved. The DPF was then fully regenerated according to the procedure described above and another lubricant sample was taken. This loading/regeneration cycle was repeated two more times to complete the testing on each test fuel. All of the recorded engine data were evaluated to check engine repeatability during the tests. The engine lubricant samples were analyzed to measure FAME and gasoil content.

After the end of all repetitions with all fuels, the filtration efficiency of all the DPFs was measured to verify proper condition and to exclude any damages caused during regenerations that would affect the collected amount of soot during loading. This was achieved by measuring the PM emissions at the outlet of the DPF with the AVL MSS during the NEDC loading procedure. The same fuel (a market diesel fuel) was used for all DPFs. The test started from a clean condition where the lowest filtration efficiency was observed. After two NEDCs, the filtration efficiency of all the DPFs had reached 99% and slowly increased during the next cycles. This confirmed that there were no filtration problems with the DPFs.

The next step included some specific tests to measure the PM mass and PN emissions at the inlet of the DPF. This was done by removing the DPF and connecting the exhaust line to the CVS. PM and PN emissions were measured in the CVS using the legislated method while the AVL MSS was measuring the raw exhaust at the same time. This procedure gave a more precise measurement of emissions for all fuels and a good comparison with the emissions measured by the DPF weight measurement. The volatile part of the PM measured with the legislated gravimetric method was measured by heating the PM filters in a furnace.

22.4. Results

22.4.1. *PM emissions*

As described above, the DPF was loaded until 6 g/l of soot loading had been reached. The soot loading values based on DPF weight measurements indicated that increasing FAME content in the fuel blend lengthens the time between DPF regenerations (Figure 22.2). For example, the required mileage to reach the 6 g/l regeneration limit was about two times greater for the B50 fuel compared with the B0 fuel.

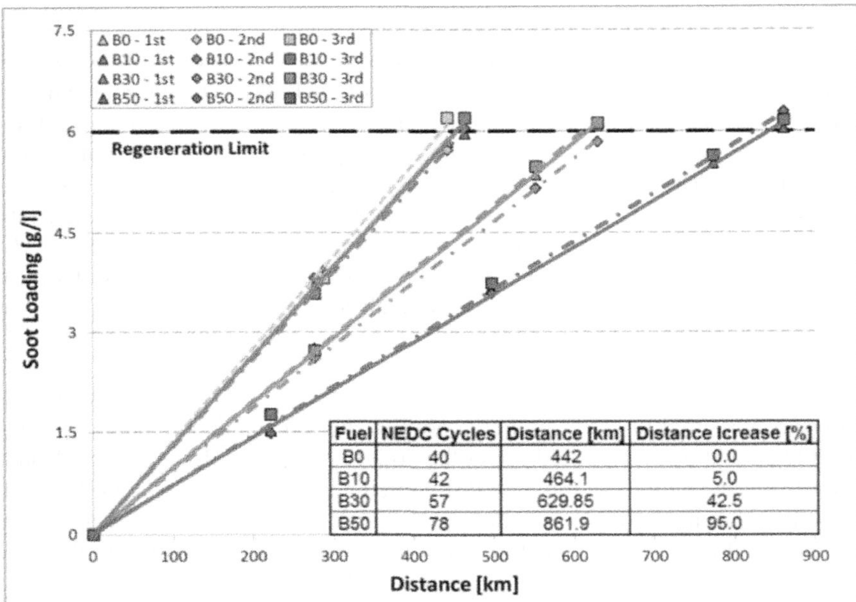

Fuel	NEDC Cycles	Distance [km]	Distance Icrease [%]
B0	40	442	0.0
B10	42	464.1	5.0
B30	57	629.85	42.5
B50	78	861.9	95.0

Figure 22.2. *Soot loading based on DPF weight measurements for all fuels. For a color version of the figure, see www.iste.co.uk/jacob/energy.zip*

As the soot accumulated on the DPF, the pressure drop across the DPF gradually increased (Figure 22.3) and all DPFs reached a similar ΔP level when 6 g/l loading had been reached. The pressure drop of the DPF at the end of a loading day was slightly different from that measured on the following morning. This might be attributed to humidity adsorbed overnight on the accumulated soot changing the soot's properties.

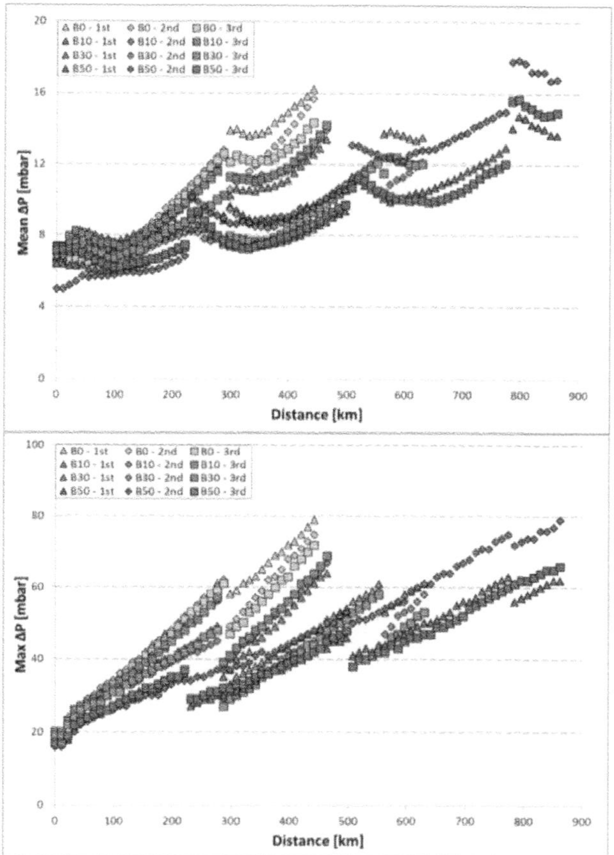

Figure 22.3. *a) Mean and b) maximum pressure drop over the DPF. For a color version of the figure, see www.iste.co.uk/jacob/energy.zip*

The specific PM emission measurements carried out after the end of the loading repetitions with all fuels provided additional information and a comparison with the results from the DPF weight measurement. The trend of decreasing PM emissions with increasing FAME content in the fuel from the DPF weights was confirmed by measurements of the PM emissions using the AVL MSS and the CVS gravimetric measurements (Figure 22.4). The PM emissions calculated after the thermal removal of the volatile part of the PM collected on the DPFs agreed with the MSS measurements, which also measureed only the solid part of the PM emissions. The trend was much steeper between B10 and B30, suggesting that a stronger effect on PM emissions occurs in this range of FAME content.

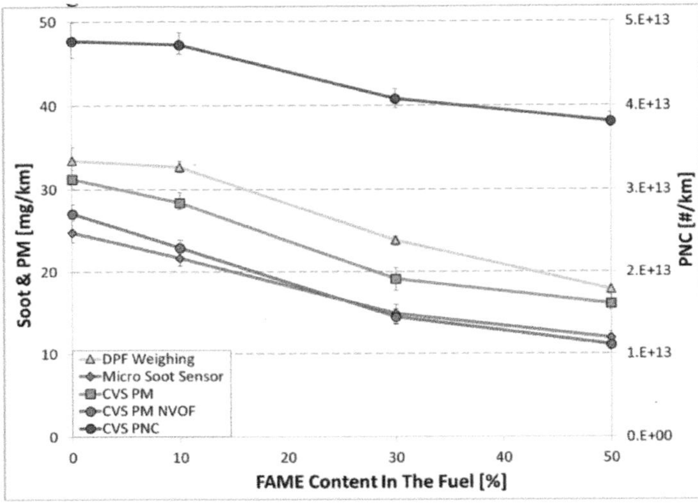

Figure 22.4. *Soot, PM emissions and particle number concentration (PNC) vs. FAME content. For a color version of the figure, see www.iste.co.uk/jacob/energy.zip*

22.4.2. *Fuel economy penalty*

The fuel economy penalty (FEP) attributed to the pressure drop over the DPF (FEP_p) can be expressed as [MUN 04]:

$$FEP_p \ (in \ \%) = 100 \cdot \Delta P / BMEP \qquad [22.1]$$

Where:

– FEP_p: fuel economy penalty due to increased backpressure [%];

– ΔP: pressure drop over the DPF [kPa];

– BMEP: brake mean effective pressure of the engine [kPa].

The BMEP of the engine can be calculated by the following formula [HEY 88]:

$$BMEP = 2\pi \cdot N_R \cdot T / V_d \qquad [22.2]$$

Where:

– N_R: number of crank revolutions for each power stroke per cylinder (which is two for four-stroke cycles and one for two-stroke cycles);

– T: engine torque [Nm];

– V_d: engine volume [dm^3].

The FEP due to the extra fuel consumed to regenerate the DPF (FEP_r) is calculated from the fuel used to actively regenerate the filter and the fuel consumed by the engine [SIN 09]:

FEP_r (in %) = 100 · Fuel injected during post injection/Total fuel consumed by the engine [22.3]

The FEP_{total} is then the sum of the individual fuel economy penalties due to backpressure and regeneration [MAJ 05]:

$$FEP_{total} = FEP_p + FEP_r \qquad [22.4]$$

The calculated FEPs for all fuel blends are shown in Figure 22.5. FEP_p is almost constant because it depends mainly on the DPF backpressure and the BMEP which is the same for all fuels. In general, the backpressure differences among the four test fuels are small (Figure 22.3) so the variation from fuel to fuel is also quite small. The FEP_p values (0.5–0.6%) are somewhat lower than the values found in previous literature [WIL 06, MIK 10, SIN 09, MAJ 05, JAC 05]. This may be attributed to the relatively low exhaust flow rates during the NEDC. The pressure drop of the filter depends on the exhaust velocity, so, if the engine operates at higher speeds and loads, the ΔP will be higher and, consequently, the FEP_p will increase.

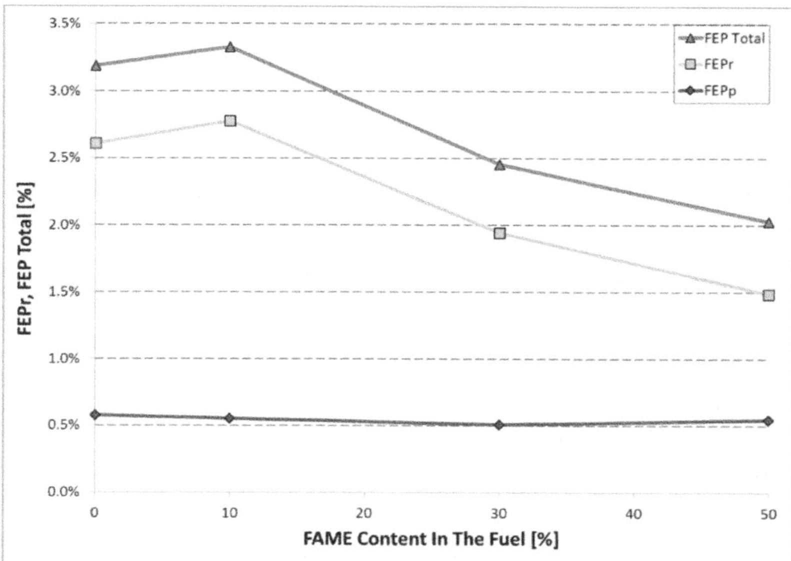

Figure 22.5. *Fuel Economy Penalty (FEP) factors for the same soot loading (6 g/l) vs. FAME content. For a color version of the figure, see www.iste.co.uk/jacob/energy.zip*

The higher FAME contents in the B30 and B50 fuels reduce the FEP_r as expected. This trend is the result of two opposite effects. First, the lower energy content (LHV) of the FAME/diesel blends (Table 22.1) means that slightly more fuel must be consumed during post injection in order to achieve the same exhaust temperature at the DPF. However, during engine operation over the NEDC, the final soot loading on the DPF was the same for all fuels (6 g/l), so the fuel quantity that must be consumed when the engine runs on B50 is much higher than with B0 due to its lower soot loading rate (Figure 22.2). From the definition of FEP_r (equation [22.3]), both the numerator and denominator increase with increasing FAME content. However, the B50 fuel consumed over the NEDC is almost double compared to the B0, while the corresponding fuel consumed during post injection is only 14% higher. The overall effect is a 43% reduction in FEP_r when the engine runs with B50 compared to B0. The calculated FEP_r values are similar to those found in the literature [SIN 05, SIN 09].

The FEP_{total} shows the same trend as FEP_r which is not surprising because the FEP_p is almost constant for all fuel blends. The FEP_{total} for B10 seems to be slightly higher compared to B0 although the difference is very small (3.3 vs. 3.2%). The FEP_{total} is within the range previously reported in the literature [SIN 09]. Higher FAME contents in diesel fuel clearly have a beneficial effect on FEP_{total}.

22.5. Conclusions

A repeatable procedure has been developed on a Euro 5-compliant 1.4 liter turbocharged diesel bench engine to evaluate the effect of FAME content in diesel fuel on the frequency of DPF regeneration. The DPFs were loaded over the regulatory NEDC until a specific loading limit had been reached and the filters were then regenerated.

The results confirmed that the addition of FAME in diesel fuel decreases the engine-out PM emissions. A larger decrease seems to occur for FAME contents higher than 10% v/v. For example, the distance needed to reach the same DPF loading limit was 42% longer with the B30 blend than with the B10 blend. The distance with the B50 blend was almost twice that with the B0 blend. This trend was confirmed with other measurements that showed a good agreement between the DPF weighing procedure, the PM measured gravimetrically in the CVS, and the solid PM measured with the MSS.

The FEP_p due to increased backpressure over the DPF was essentially constant at 0.5–0.6% for all four test fuels. The FEP_r due to DPF regeneration decreased with increasing FAME in the fuel, from 2.6-2.7% for the B0 and B10 blends reducing to 1.5% for the B50 blend. Since the FEP from backpressure is constant, the FEP_{total} for

DPF regeneration follows the same trend as FEP_r reaching 3.1–3.2% for B0–B10 and about 2% for B50.

22.6. Acknowledgments

The CONCAWE co-authors wish to acknowledge the members of its Diesel Fuel Special Task Force, who contributed substantially to this study with the support of CONCAWE's Member Companies.

22.7. Bibliography

[BHA 13] BHARDWAJ O., KREMER K., PISCHINGER S. *et al.*, "Impact of biomass-derived fuels on soot oxidation and DPF regeneration behavior", *SAE 2013-01-1551*, 2013.

[CZE 12] CZERWINSKI J., BÜRKI S., BONSACK P. *et al.*, "DPF's regeneration procedures and emissions with RME blend fuels", *SAE 2012-01-0844,* 2012.

[EPA 02] EPA, *A comprehensive analysis of biodiesel impacts on exhaust emissions*, Environmental Protection Agency, 2002.

[HAS 07] HASEGAWA M., SAKURAI Y., KOBAYASHI Y. *et al.*, "Effects of fuel properties (content of FAME or GTL) on diesel emissions under various driving modes", *SAE 2007-01-4041,* 2007.

[HEY 88] HEYWOOD J.B., *Internal Combustion Engine Fundamentals*, McGraw-Hill Inc, New York, 1988.

[JÄÄ 07] JÄÄSKELÄINEN H., *Engine exhaust back pressure*, available at: http://www.dieselnet.com/tech/diesel_exh_pres.php, 2007.

[JAC 05] JACQUE E., SAID Z., Catalyzed diesel particulate trap – components and engine controls integration for an optimum performance, *Aachen Colloquium for Automobile and Engine Technology,* 2005.

[LAM 13] LAMHARESS N., STARCK L., MILLET C.N. *et al.*, "Effect of biofuels on catalyzed diesel particulate filter regeneration", *Topics in Catalysis,* vol. 56, pp. 462–466, 2013.

[MAJ 05] MAJEWSKI, W.A., Diesel Filter Systems, available at: http://www.dieselnet.com/tech/dpf_sys.php, 2005.

[MIK 10] MIKULIC I., ZHAN R., EAKLE S., "Dependence of fuel consumption on engine backpressure generated by a DPF", *SAE 2010-01-0535*, 2010

[MUN 04] MUNTEAN G., "How exhaust emissions drive diesel engine fuel efficiency", *10th Diesel Engine Emissions Reduction Conference (DEER),* San Diego, 2004.

[OIN 05] OINUMA R., TAKUMA S., KOYANO T. *et al.*, "Effects of post injection on piston lubrication in a common rail small bore diesel engine", *SAE 2005-01-2166,* 2005.

[SIN 05] SINGH N., JOHNSON J., PARKER G. *et al.*, "Vehicle engine aftertreatment system simulation (VEASS) model: application to a controls design strategy for active regeneration of a catalyzed particulate filter", *SAE 2005-01-0970,* 2005.

[SIN 09] SINGH N., RUTLAND C., FOSTER D. *et al.*, "Investigation into different DPF regeneration strategies based on fuel economy using integrated system simulation", *SAE 2009-01-1275*, 2009.

[WIL 06] WILLIAMS A., MCCORMICK R.L., HAYES R.R. *et al.*, "Effect of biodiesel blends on diesel particulate filter performance", *SAE 2006-01-3280*, 2006.

23

Exhaust Aftertreatment Potential of Advanced Coupled NSC-SCR System

The upcoming increasingly stringent global emission legislation and CO_2 restriction are highly challenging for modern diesel engine and aftertreatment technology. Meeting all requirements, some conflicting, advanced combined aftertreatment systems are beneficial due to the use of synergistic effects. In this study, focusing on NO_X treatment, an aftertreatment system consisting of NSC plus exhaust bypass, passive SCR and integrated on-board exhaust fuel reforming was developed and investigated. Without the use of a urea dosing system, the passive SCR increases the total NO_X conversion by enforcing the intrinsic NH_3 formation during the NSC regeneration in comparison to conventional NSC-only application. Especially at medium temperature (200°C–300°C) with maximum NH_3 selectivity, the possible NO_X conversion shift from NSC to SCR offers significant fuel consumption savings. In addition, fuel reforming shows much higher reductant generation efficiency than engine enrichment. This study discusses the aftertreatment potentials of the combined system.

23.1. Introduction

The upcoming increasingly stringent global emission legislation and CO_2 restrictions are highly challenging for modern diesel engine and aftertreatment technology. In addition to the more restrictive emission limits, new emission evaluation test procedures, like WLTP, RDE and PEMS result in enlarged emission relevant engine operation areas. Beside the increased coverage of engine load and speed, the emission testing from cold start to high temperature cycles also demand an extended active temperature range of aftertreatment systems. Additionally, improved engine combustion technologies with increased fuel efficiency and thereby lowered exhaust gas temperature are even more challenging for exhaust aftertreatment functionalities. In response to the increased scope of demands,

Chapter written by Thomas WITTKA, Bastian HOLDERBAUM and Thomas KÖRFER.

specifications of aftertreatment systems have to be upgraded and alternative solutions and technologies are moving forward in development and technical realization. Meeting all requirements, some conflicting, advanced combined aftertreatment systems allow the best overall performance by the benefit of synergistic effects. In addition to the well-established DPF-technology, the combination of the two state-of-the-art NO_X removal technologies NSC and SCR is a suitable approach due to their specific single characteristics. Under cold-start conditions, the NSC offers appropriate low temperature activity by intermediate NO_X adsorption. After heat-up, the stored NO_X will be reduced at warm temperatures depending on the reductant species activity. Typically, the light-off temperature of the NSC regeneration by engine internal enrichment is around 250°C. In contrary to HC, the more reactive H_2 and CO reductant species might shift the light-off temperature of the rich NSC regeneration toward lower temperatures at around 180°C. However, at a higher temperature of around 400°C the NSC NO_X storage capacity drops due to thermodynamic reasons, which results in lower NO_X conversion and higher absolute fuel consumption by increased NSC regeneration frequency. Contrary to NSC, the SCR offers NO_X conversion activity especially at high temperature. Therefore, regarding the active temperature operational areas the combination of NSC and SCR seems to be advantageous. A further motivation for combining NSC and SCR is based on the intrinsic NH_3 formation during the NSC regeneration at reducing conditions. Typically, the undesired NH_3 emission of the NSC is prevented by operational and catalytic converter formulation measures. However, by the NH_3 storage capacity of a downstream positioned SCR the NH_3 emission can be eliminated and even utilized for additional NO_X conversion.

In general, NSC and SCR combination systems can be classified in 3 types:

– NSC based system NSC + passive SCR (operated without urea dosing system);

– SCR based system passive NO_X adsorber (PNA) + active SCR (operated with urea dosing);

– NSC + SCR based system NSC + active SCR (operated with urea dosing).

In an NSC based system, the SCR contributes passively to the total NO_X conversion by utilization of the desired or even enforced intrinsic NH_3 formation during rich NSC regeneration, without the disadvantages of a urea dosing system, such as additional hardware costs or urea refilling effort. However, the general total NO_X conversion performance is still mainly dominated by the NSC behavior, i.e. by NO_X adsorption capacity and NH_3 formation.

The intended task of a PNA in an SCR based system is intermediate NO_X storage during cold start and NO_X release after SCR light-off. The SCR is conventionally operated by active urea dosing control. However, the passive NO_X storage function is only sufficient in the case of low initial NO_X load at engine start. Therefore, a PNA could still require active control effort and additional fuel consumption [KOR 13].

The highest NO_X conversion potential is given by combination of two active systems, i.e. NSC including rich regeneration and urea based SCR. However, the NSC + SCR based system shows maximum packaging, costs and system complexity issues.

In this study, an NSC based system consisting of NSC plus exhaust bypass, passive SCR and on-board exhaust fuel reforming was developed and investigated. The concept is in line with the requirement specification of the 7th Framework Programme of the European Union, demonstrating NO_X aftertreatment without urea dosing and with maximum additional fuel consumption of 2%.

23.2. Description of investigated exhaust system and fuel reformer performance

The concept of the investigated exhaust system combines three technical approaches, with the overall objective of researching potentials regarding NO_X conversion performance and CO_2 emission. First, the combination of NSC and passive SCR was investigated in terms of synergy effects by interaction of the specific $DeNO_X$ technology properties. Second, the NSC regeneration was realized by hydrocarbon free synthesis gas (H_2 + CO), which was generated on-board by an integrated fuel reformer. The use of this highly reactive reformate gas for NSC enrichment offers enhancement of NSC regeneration efficiency, especially at low temperature. Third, an exhaust bypass of the NSC enables engine independent operation.

The exhaust system layout and simplified method of operation are shown schematically in Figure 23.1.

The investigated combination system is positioned under-floor downstream closed-coupled DOC-DPF. The system is discontinuously operated in two operation modes.

During the NO_X adsorption mode, the total system is operated under an oxidizing exhaust atmosphere (air/fuel ratio $\lambda > 1$). The NSC path is opened and the NSC bypass is closed. Similar to conventional NSC application, NO_X is stored on the NSC by forming nitrate and nitrite surface adsorbates. Mainly dependant on SCR temperature and ammonia load, NSC NO_X slip is continuously reduced on the SCR by stored NH_3, which was formed and adsorbed during previous NSC regeneration. The integrated on-board fuel reformer is inactive.

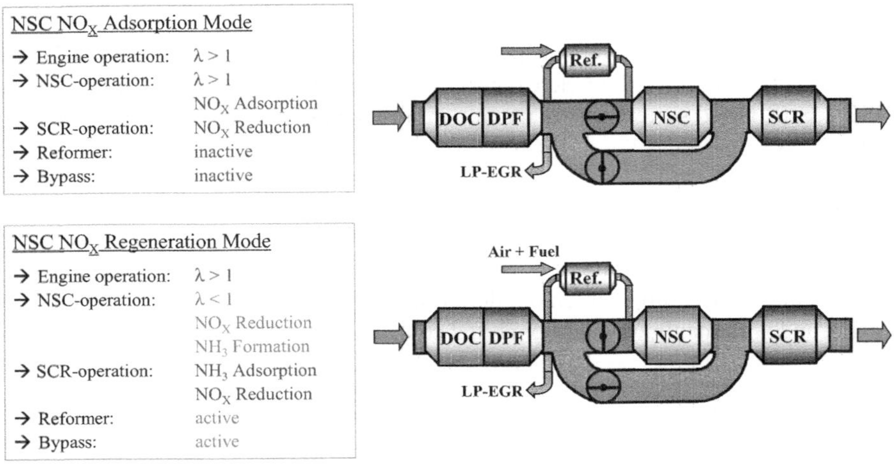

Figure 23.1. *Simplified method of operation of NSC based combination system, with under-floor NSC plus exhaust bypass, passive SCR and integrated fuel reformer*

Similar to conventional NSC application, the NSC needs to be regenerated regularly due to limited NO_X storage capacity and decreasing NO_X adsorption efficiency at increased NO_X load. Therefore, the NSC is operated under a reducing exhaust atmosphere ($\lambda < 1$) during the regeneration mode. NO_X is released by decomposition of the nitrate and nitrite surface adsorbates, and subsequently reduced by redox reactions with hydrocarbon (HC), carbon monoxide (CO) and hydrogen (H_2). Besides the target product N_2, NH_3 is intrinsically formed and emitted as a by-product. Conventional NSC application usually need to avoid this NH_3 emission during the rich NSC regeneration. In contrast, NH_3 can be stored on the downstream positioned SCR in the combination system. The consumed SCR NH_3 load by NSC NO_X slip during NO_X adsorption mode is refilled. Thus, the SCR is operated discontinuously. Conventionally, the reductant supply for NSC regeneration is realized by engine enrichment. This non-typical diesel engine operation causes drawbacks, e.g. highly decreased engine internal combustion efficiency, oil dilution, and drivability and noise issues. In contrast, in the investigated system the NSC regeneration is carried out independently of the engine by closing the NSC exhaust gas path, opening the NSC bypass and feeding in synthesis gas at low space velocity via the active fuel reformer. The exhaust paths are actuated by two serial exhaust flaps. Only the NSC is locally enriched. The diesel engine is continuously operated at efficient lean operation. The integrated on-board fuel reformer is predominantly operated by exhaust gas, which is taken downstream DPF. Diesel is injected by an air-assisted, low-pressure injector at a significant under-stoichiometric air/fuel ratio ($\lambda_{Ref} = 0.35–0.80$) and is catalyzed to a H_2- and CO-rich synthesis gas. The completely newly developed on-board fuel reformer is a coaxial design, and is shown in more detail in Figure 23.2.

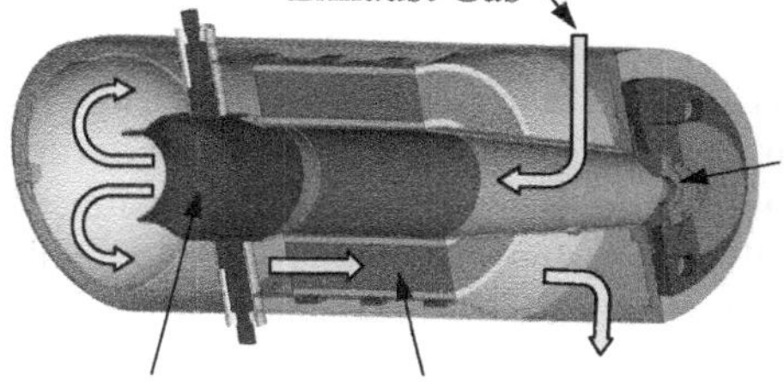

Figure 23.2. *CAD design of the fuel reformer*

The reaction enthalpy, which is released in the ring-shaped reformer catalyst, supports fuel vaporization via internal heat transfer. The reforming reaction process can be described by superposition of exothermic partial oxidation (POx) by the residual oxygen in the exhaust gas as well as the secondary air, and endothermic steam (SR) and CO_2 reforming (DR) which occurs via H_2O and CO_2 introduced by the exhaust gas. Additionally, the reforming products in temperature-dependent chemical equilibrium, mainly according to the water-gas shift (WGS) and methanation reaction. In comparison to pure POx, exhaust gas reforming exhibits favorable thermal behavior as well as benefits regarding H_2 and CO yield, due to the endothermic part of the reaction. The reformer operation is strongly dependent on the exhaust gas composition. Fuel is injected by feed-forward control of λ_{Ref} depending on the fraction of molar oxygen to the total amount of reactant oxygen sources (O_2, H_2O und CO_2) and space velocity. Additionally, fuel supply is corrected by closed-loop control of reformer temperature in terms of maximizing H_2 and CO as well as minimizing HC concentration in the reformate, while complying with temperature limits.

Figure 23.3(a) shows a comparison of H_2 and CO molar concentrations achieved under steady-state conditions with those attainable under stoichiometric conditions as function of λ_{Ref}. The concentrations are close to the maximum yield and demonstrate high fuel conversion. Compared to rich diesel engine operation, substantially higher concentrations can be achieved.

Figure 23.3(b) shows a comparison of the achieved ratio of heating values (HVR) as well as the reductant yield Y_{Red} (ratio of oxygen consumption potential of the reducing agent generated in the exhaust gas relative to the additional fuel introduced to reach rich operation) in contrast to corresponding theoretically attainable pure POx values. At lean engine operation and consequently predominantly exothermic reactant composition, due to thermal reasons a low λ_{Ref} must be set. With decreasing engine lambda and consequently increasing endothermic reactant composition, λ_{Ref} must on one hand be enleaned to support endothermic reaction enthalpy. On the other hand, it offers high reformate quality. In contrast to HVR, Y_{Red} is an important indicator for NSC regeneration in terms of CO_2 emissions. In addition, Figure 23.4 shows Y_{Red} by fuel reforming in comparison to diesel engine enrichment of various engine calibrations as function of brake mean effective pressure (BMEP). Compared to actual engine calibrations for rich operation, the reformer generally exhibits considerably higher efficiency and thus represents a potential decrease in CO_2 emissions for NSC regeneration. The reductant loss of the reformer is based on partial oxidation of fuel by molar oxygen. The reductant loss of engine enrichment is caused by internal combustion efficiency loss at $\lambda_{Engine} = 1$ due to intake throttling and post injection.

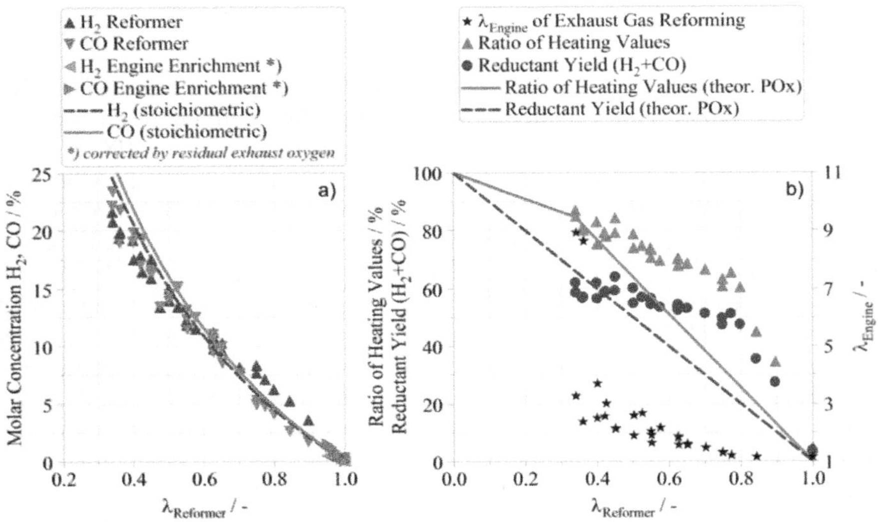

Figure 23.3. *Reformer output as a function of relative reformer air/fuel ratio: a) measured and stoichiometric H_2 and CO molar concentrations in reformate; b) measured and theoretical POx values for ratio of heating values (HVR) and reductant yield (Y_{Red}, only H_2 + CO are considered) of the reformer, engine relative air/fuel ratio of the exhaust gas supply to the reformer. For a color version of the figure, see www.iste.co.uk/jacob/energy.zip*

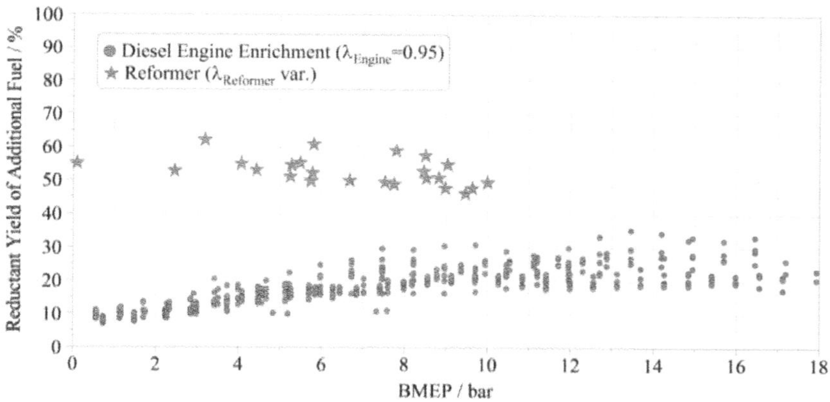

Figure 23.4. *Reductant yield of additional fuel for NSC enrichment by fuel reforming in comparison to diesel engine enrichment of various engine calibrations as function of brake mean effective pressure (BMEP)*

23.3. Discussion of results and aftertreatment potentials of investigated exhaust system

The measurements, whose results are discussed here, were performed on a dynamic engine test bench. The NO_X conversion behavior of the combination system was investigated in steady-state operation. For each measuring point, the system was operated at least by 10 adsorption-regeneration-cycles to stabilize both NSC NO_X load and SCR NH_3 load. In a first project phase, the system performance was initially investigated by reformate substitute instead of on-board generated reformate. The reformate substitute is a gas mixture with 20% H_2, 20% CO and 60% N_2 and equivalent to $\lambda_{Ref} = 0.4$.

As already mentioned, usually the emission of the intrinsic NH_3 formation during rich NSC regeneration needs to be avoided. The avoidance of NH_3 emission is typically realized by catalyst formulation and operational issues. However, in an NSC based combination system, for sufficient operation and high NO_X conversion share of the passive SCR, the NH_3 formation and emission needs to be enforced. The ammonia formation yield (molar ratio of cumulated NH_3 emission during NSC regeneration to initial NSC NO_X load at NSC regeneration start) is an important parameter for operation of the combined system. According to the state of knowledge [MUL 08, PIH 06, KOC 08, CLA 08, PAR 09], inside the reaction front, which axially propagates through the NSC during the rich regeneration, NH_3 is intermediately formed by over-reduction of NO_X. The formed NH_3 re-oxidizes by reaction with surface NO_X or oxygen (OSC: oxygen storage capacity) downstream the reaction front. At the end of the NSC regeneration, when the rich reaction front reaches the catalyst outlet, the intermediate NH_3 can no longer be re-oxidized and is typically emitted as NH_3 spike.

In Figure 23.5, the ammonia yield of the NSC regeneration as a function of the NSC NO_X load is shown, exemplarily for an NSC temperature of 250°C. Additionally, the regeneration duration is varied and represented by the relative fuel consumption penalty.

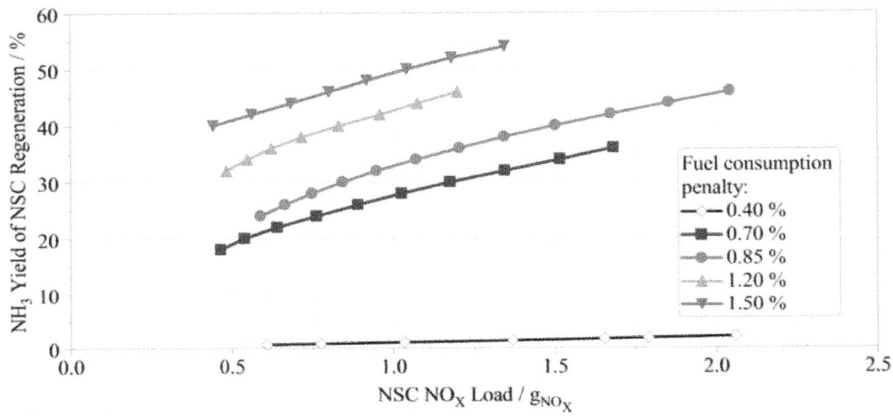

Figure 23.5. *Ammonia yield of NSC regeneration as a function of NSC NOX load and fuel consumption penalty (due to regeneration duration); NSC temperature = 250°C, operation by reformate substitute (20% H_2, 20% CO, 60% N_2). For a color version of the figure, see www.iste.co.uk/jacob/energy.zip*

At low total NSC NO_X load, most of the NO_X is stored at the NSC inlet and the intermediately formed NH_3 mostly re-oxidizes at the NSC outlet. However at high total NO_X load, NO_X is also stored at the NSC outlet. NH_3, which is intermediately formed at the outlet, is only partially re-oxidized and gets emitted. Therefore, beside insufficient short regeneration duration (0.40% fuel consumption penalty), the relative NH_3 yield increases with increasing NSC NO_X load. In parallel, in line with the literature, elongating the duration length and reduction of the complete catalyst surface minimizes the process of NSC internal NH_3 re-oxidation and therefore results in increased NH_3 emission.

In Figure 23.6, the NO_X conversion performance as well as the related fuel consumption penalty and NO_X conversion share of NSC with bypass and passive SCR are shown as function of NSC NO_X load, exemplarily for an NSC temperature of 250°C.

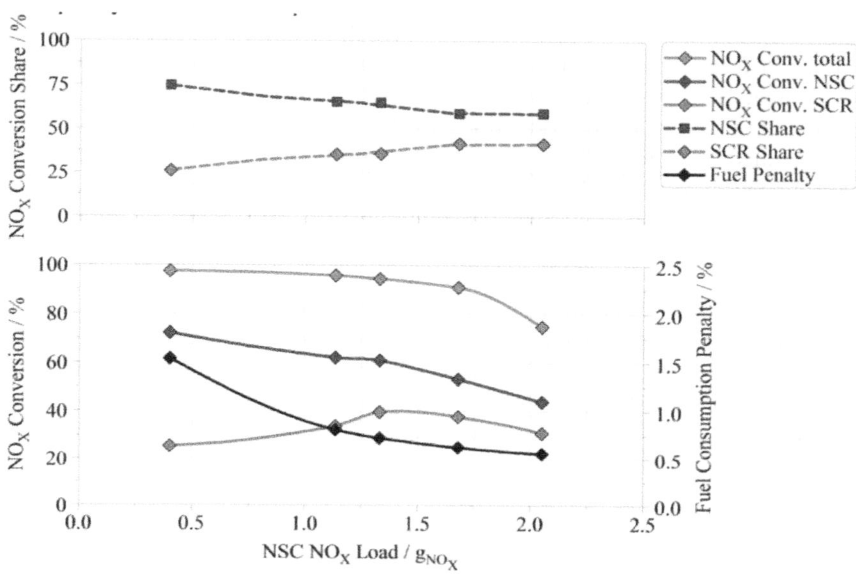

Figure 23.6. *NO_X conversion, related fuel consumption penalty and NO_X conversion share of NSC with bypass and passive SCR as function of NSC NO_X load; NSC temperature = 250°C, operation by reformate substitute (20% H_2, 20% CO, 60% N_2). For a color version of the figure, see www.iste.co.uk/jacob/energy.zip*

The total NO_X conversion is almost constant for a NO_X load of 0.4–1.4 g. However, with increasing NO_X load by elongating NO_X adsorption duration the NO_X conversion distribution is shifting from NSC to SCR. As shown in the lower diagram, the loss of NSC conversion is almost compensated by increasing SCR conversion, which achieves a maximum of 40%. The NO_X conversion share between NSC and SCR is changing from 75% / 25% at 0.4 g NO_X load to 60% / 40% at 1.4 g NO_X load. For NO_X adsorption durations up to the NO_X load level of 1.4 g, the increasing NO_X slip through NSC and bypass can be compensated by the increased NH_3 formation on the NSC as shown in Figure 23.5. The cycle integrated NH_3-NO_X-ratio upstream SCR (α) is ≥ 1 (not displayed) for NSC NO_X load below 1.4 g. For longer NO_X adsorption duration resulting in NSC NO_X loads above 1.4 g, α is dropping below the stoichiometry of 1, which results in a corresponding decrease of SCR NO_X conversion and therefore also of the total NO_X conversion. A great potential of the conversion shifting between NSC and SCR is the impact on the fuel consumption penalty, shown in the bottom diagram of Figure 23.6. By expanding the NO_X adsorption duration toward higher NSC NO_X load, the fuel penalty can be lowered from 1.5% at 0.4 g to 0.75% at 1.4 g. The higher the NO_X load, the more reducing agent is consumed for NO_X reduction. Reasons for this are the lower regeneration frequency as well as the higher NH_3 yield of the NSC regeneration for higher NOx loads. A further reason for the benefit of high NO_X loads is caused by the OSC of the NSC, which has to be reduced in parallel to the NO_X reduction. In conventional NSC application, one functionality of the OSC is the avoidance of emission of the intrinsic NH_3 formation by catalyst internal re-oxidation. In contrary, in a combined NSC and SCR system the NH_3 formation on the NSC is desired and does not need to be avoided. Hence, an additional potential of a combined NSC and SCR system with respect to fuel consumption resp. CO_2 emission minimization might be the reduction of OSC.

In summary, contrary to conventional NSC application, the NSC NO_X slip caused by decreased adsorption efficiency at increased NO_X storage level does not automatically drop the system NO_X conversion but can be converted by the passive SCR. Therefore the NSC NO_X adsorption duration can be elongated, resulting in benefits of lowered fuel consumption penalty by NSC enrichment as well as minimized secondary emissions.

Increasing NSC NO_X conversion causes increased fuel consumption by higher regeneration frequency and reductant demand. Another strong influencing factor on this relation is the absolute NO_X mass resp. NO_X raw emission level. Figure 23.7 shows a correlation between NO_X conversion of NSC only and NSC + SCR and fuel consumption penalty for various NO_X raw emission levels, exemplarily for an NSC temperature of 230°C.

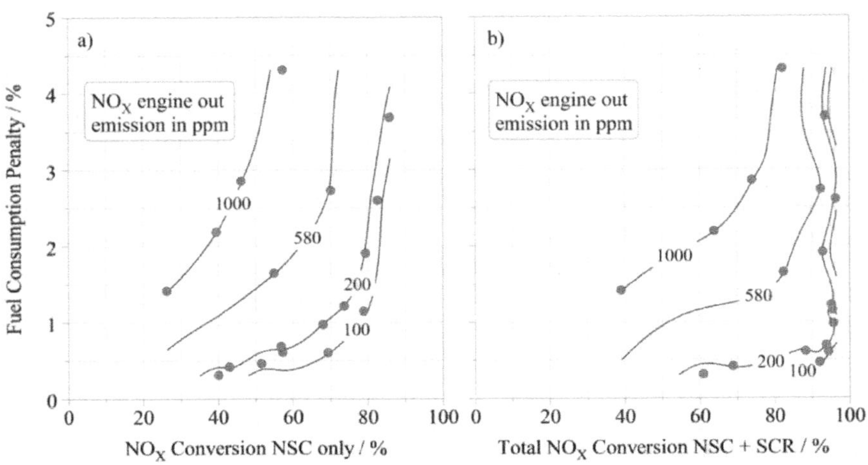

Figue 23.7. *Correlation between NOX conversion of a) NSC only and b) NSC + SCR and fuel consumption penalty as function of NO_X raw emission level; NSC temperature = 230°C, operation by reformate substitute (20% H_2, 20% CO, 60% N_2)*

Independently of NO_X raw emission, the fuel consumption penalty progressively increases for higher NSC NO_X conversion. The NSC NO_X conversion converges asymptotically against a maximum at strong increase of fuel consumption. Furthermore, higher NO_X raw emission causes decreasing NO_X conversion for given fuel consumption and lower achievable maximum NO_X conversion. In comparison to the NSC-only, the combined system shows benefits regarding fuel consumption and NO_X conversion. High maximum total NO_X conversion of up to > 95% is achievable at low fuel consumption penalty of < 1% for reasonable NO_X raw emissions. At a high NO_X raw emission level, the total NO_X conversion decreases but is significantly higher than for NSC only. For given NO_X conversion, significantly lower fuel consumption is attainable by the combination system.

As an overview, Figure 23.8 shows the steady-state NO_X conversion and fuel consumption penalty of the combination system for two variants with reformate and reformate substitute as function of NSC temperature. The system variants are defined as follows:

– system A: 1.4 L NSC A and 2.9 L Cu-SCR;

– system B: 1.4 L NSC B (+50% precious metal loading compared to NSC A) and 2.9 L Cu-SCR.

The results of system B indicate that the attained NO_X conversion is comparable for both operation with the integrated reformer as well as operation with reformate substitute. Compared to system A, system B achieves a slight increase in conversion at T < 250°C due to higher NSC precious metal loading. For both variants, almost complete conversion is achieved at 250°C with a corresponding fuel consumption increase of roughly 1.5%. The operation strategy of NO_X adsorption and regeneration is optimized with regard to maximum total NO_X conversion and feasible fuel consumption penalty. The SCR significantly contributes to the overall NO_X conversion up to 35%. At T < 250°C the total steady-state NO_X conversion drops due to reduced NSC activity as well as NH_3 yield. The fuel consumption penalty increases due to higher NSC regeneration frequency as well as lower engine fuel consumption due to lower engine load. Nevertheless, 65–70% steady-state NO_X conversion at 185°C resp. 30% steady-state NO_X conversion at 150°C are still achieved. In transient operation, higher conversion rates can be achieved due to pure NO_X adsorption at the NSC as well as SCR conversion with previously stored NH_3. At T > 350°C the steady-state NO_X conversion drops due to the decreasing (for thermodynamic reasons) NO_X storage capacity of the NSC as well as a lowered NH_3

yield of NSC regeneration. The decreased NSC NH_3 yield causes limited steady-state SCR conversion in contrast to urea based SCR operation. However, during transient operation, complete conversion can still be maintained by the SCR for a short time, e.g. during DPF regeneration, by consuming previously stored NH_3 depending on SCR NH_3 load and NO_X mass flow. This represents high NO_X performance benefits compared to conventional NSC application.

Finally, Figure 23.9 shows lean oxidation conversion of synthesis gas (reformate substitute) as a function of NSC substrate temperature for various synthesis gas mass flows. At low synthesis gas mass flow, the lean oxidation reaction shows a light-off temperature of 120°C. In case of necessary NSC heating, e.g. during cold started emission test procedures, the low oxidation light-off temperature offers an efficient local heating with only short delay after engine start and represents an interesting alternative solution instead of engine internal or electric heating.

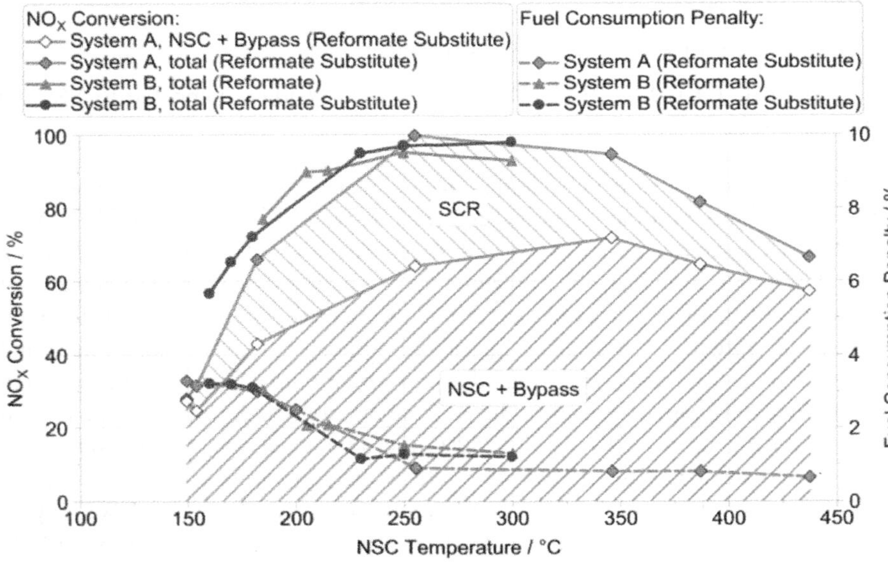

Figure 23.8. *Steady-state NO_X conversion and fuel consumption penalty of the combination system with reformate and reformate substitute (20% H_2, 20% CO, 60% N_2) as a function of NSC temperature; System A: 1.4 l NSC A + 2.9 l Cu-SCR, System B: 1.4 l NSC B + 2.9 l Cu-SCR; EURO4 NO_X raw emission level. For a color version of the figure, see www.iste.co.uk/jacob/energy.zip*

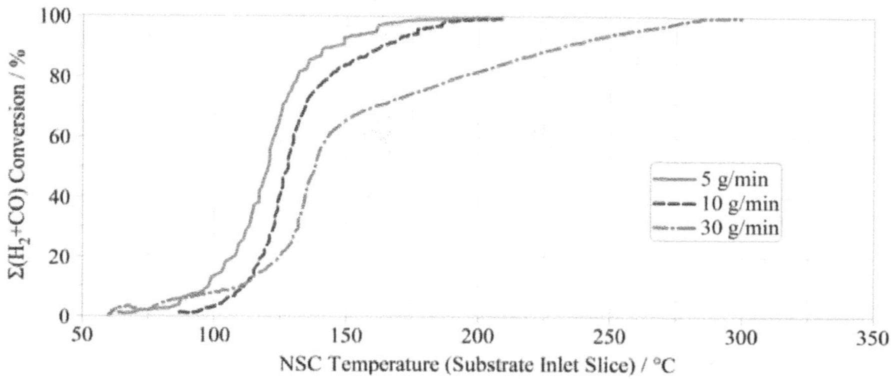

Figure 23.9. *Lean oxidation conversion of synthesis gas (20% H_2 + 20% CO + 60% N_2) as function of NSC substrate temperature for various synthesis gas mass flows; steady-state operation: λ = 1.4, space velocity NSC = 35000 h^{-1}. For a color version of the figure, see www.iste.co.uk/jacob/energy.zip*

23.4. Summary

The upcoming increasingly stringent global emission legislation and CO_2 restriction are highly challenging for modern diesel engine and aftertreatment technology. Meeting all requirements, some conflicting, advanced combined aftertreatment systems are beneficial due to the use of synergistic effects. In this study, focusing on NO_X treatment, an aftertreatment system consisting of NSC plus exhaust bypass, passive SCR and integrated on-board exhaust fuel reforming was developed and investigated. The discussion is based on steady-state measurements on engine test bench and point out potentials regarding NO_X performance and fuel consumption respectively CO_2 emission savings.

Conventional NSC application usually needs to avoid emission of the intrinsic NH_3 formation during NSC regeneration. By enforcing NH_3 emission, the downstream positioned SCR can be activated and significantly contributes to the total NO_X conversion. Almost 100% NO_X is converted at 250°C with fuel consumption increase of roughly 1.5%. Steady-state NO_X conversion drops toward lower temperature, but still reaches 65% at 185°C resp. 30% at 150°C. Due to the passive and discontinuous SCR operation, the steady-state SCR NO_X performance is strongly dependent on NSC NO_X adsorption and NH_3 formation. However, during transient operation, complete conversion can still be maintained by the SCR for a short time, e.g. during DPF regeneration, by consuming previously stored NH_3. In contrary to conventional NSC application, the NSC NO_X slip caused by decreased adsorption efficiency at increased NO_X storage level does not automatically drop the system NO_X conversion but can be converted by the passive SCR. Therefore the NSC NO_X adsorption duration can be extended resulting in benefits of lowered fuel consumption penalty by NSC enrichment.

Additionally, the fuel consumption penalty was reduced significantly by use of engine independent NSC regeneration with reformate gas in bypass operation. Compared to diesel engine enrichment, reducing agent generation by fuel reforming is more efficient. Furthermore, reformate gas is almost hydrocarbon free and enhances NSC NO_X performance and secondary hydrocarbon emission. Avoiding un-typical rich diesel engine operation minimizes drawbacks, like oil dilution and drivability and noise issues.

In summary, the combination of NSC, passive SCR and fuel reformer significantly increases the NO_X performance at lower fuel consumption penalty in comparison to conventional NSC application, and represents an attractive technology for future emission aftertreatment.

23.5. Acknowledgments

The research leading to these results has received funding from the European Union 7th Framework Programme [FP7/2007-2011] under grant agreement No. 234032.

23.6. Bibliography

[CLA 08] CLAYTON R.D., HAROLD M.P., BALAKOTAIAH V., "NO_X storage and reduction with H_2 on $Pt/BaO/Al_2O_3$ monolith: Spatiotemporal resolution of product distribution", *Applied Catalysis B: Environmental*, vol. 84, pp. 616–630, 2008.

[KOČ 08] KOČÍ P., PLÁT F., ŠTĚPÁNEK J. *et al.*, "Dynamics and selectivity of NO_X reduction in NO_X storage catalytic monolith", *Catalysis Today*, vol. 137, pp. 253–260, 2008.

[KÖR 13] KÖRFER T. *et al.*, Potential of advanced, combined aftertreatment systems for diesel engines to meet upcoming EU and US emission regulation, *SAE 2013-24-0163*, 2013.

[MUL 08] MULLA S.S., CHAUGULE, S.S., YEZERETS A. *et al.*, "Regeneration mechanism of $Pt/BaO/Al_2O_3$ lean NO_X trap catalyst with H_2", *Catalysis Today*, vol. 136, pp. 136–145, 2008.

[PAR 09] PARTRIDGE W.P., CHOI J.-S., "NH_3 formation and utilization in regeneration of Pt/Ba/Al_2O_3 NO_X storage-reduction catalyst with H_2", *Applied Catalysis B: Environmental*, vol. 91, pp. 144–151, 2009.

[PIH 06] PIHL J.A., PARKS II J.E., DAW C.S. *et al.*, Product selectivity during regeneration of lean NO_X trap catalysts, *SAE 2006-01-3441*, 2006.

24

Power Controlled Microwave Reactor for the Removal of NOx and SOx from the Exhaust of Marine Diesel Engines

Microwave energy in combination with AC corona initiated non-thermal plasma is a promising method for removing NOx and SOx from the exhaust of diesel engines. Recently, experimental work shows that a microwave oven (1.8 kW) can be used as an experimental system to research this subject. However, microwave ovens are often based on cyclic mode operation for controlling the power. This type of control is not suitable for the purpose of continuously cleaning the exhaust gas. This study investigates the use of a phase control circuit as a regulator for controlling the power in the microwave reactor. The system was tested and its performance was evaluated. It was observed that phase-controlled power regulator could be successfully used for quasi-continuous (fast-switching) power regulation with the maximization of power efficiency. The developed system is used to reduce the emission gases from a marine diesel engine. Different power settings were used to investigate the effect of the power on the reduction efficiency. The same controlled microwave source-based non-thermal plasma reactor will be used to investigate the pollutant removal efficiency of the exhaust gas emitted from a 40 kW marine diesel engine.

24.1. Introduction

Air pollutants generated by ships in both gaseous and particulate forms have a long-term effect on the quality of the environment, and cause a significant exposure risk to people living in the proximity of harbors or in neighboring coastal areas

Chapter written by Maysam ABBOD, Radu BELECA, David PEIRCE, Lionel GANIPPA, Nadarajah MANIVANNAN and Wamadeva BALACHANDRAN.

[EYR 09]. It was recently estimated that ships produce at least 15% of the world's NOx (more than all of the world's cars, buses and trucks combined), between 2.5 and 4% of greenhouse gases, 5% black carbon (BC) and between 3 and 7% of global SO_2 output [NEE 09]. Estimation of the contribution of maritime shipping to global emissions of VOC and CO is not yet available. In the EU, 80% of international freight is carried by sea and 40% of intra-European freight uses short sea shipping, while European countries must face ever-fiercer competition from foreign competitors who are favoured by more flexible regulations, cheaper labor or government support. In this controversial climate, the shipping industry faces an unprecedented level of new regulations aimed at improving the environmental performance of the shipping industry. Compared to other industries in the developed world, the shipping industry is the last to be regulated for emissions to air. In order to reduce the environmental footprint of ships, the International Maritime Organization recently issued the legislation of Marpol Annex VI guidelines to be enforced in 2012 for the Emission Control Areas on sulphur fuel content (1%) and NOx emissions based on ship engine speed. The Kyoto Protocol also calls for the reduction of greenhouse gas emissions from ships (EU Commission, 2006/944/EC). All these limits can be reached, in practice by burning higher cost, low sulphur fuel or by the use of exhaust scrubber systems as 'zero-waste, zero-emission' maritime transport is a priority for the EU (EU Commission, [COM 02] 595). At the moment, fuel switching is still the option most commonly adopted by ship-owners to comply with environmental prescriptions and only about 30 scrubbers are in operations or under commissioning (EGCSA handbook). The decision to retro-fit exhaust gas cleaning systems will be based largely on the cost-effectiveness and a short payback period and will be compared to the cost of compliance by simply using a low sulphur fuel. The current merchant marine business model presents a further challenge to the investment in capital equipment to enable compliance, as the owner of the ship may not be adequately reimbursed for the investment for the saving in fuel cost accruing to the vessel charterer.

One of the proposed solutions toward marine diesel emission control is the non-thermal plasma process. A non-thermal plasma reactor (NTPR) was designed and built using a combination of AC corona and microwave (MW) for treatment of marine diesel exhaust gas [BEL 12]. The reactor performance in NOx and SOx removal was tested initially with a Ford Duratorq (Puma) high speed direct injection, 4 cylinders, 2 l diesel engine in various operating conditions and fuelled with ultra-low sulphur diesel and with a 40 kW two-stroke marine engine fuelled with low/high sulphur diesel. This study is a part of the DEECON (Innovative After-Treatment System for Marine Diesel Engine Emission Control) FP7 European project.

24.2. Non-thermal plasma experimental set-up using diesel car engine

The DEECON project is being developed in two stages. In the first stage, we are building a laboratory scale NTPR reactor as a combination of microwave and AC corona discharge presented in this chapter. The second stage involves development of a pilot-scale NTPR reactor using a combination of electron beam and microwave plasma. The Brunel laboratory scale NTPR reactor consists of a modified commercial Samsung CM1919 microwave oven, 1850 W total power with cavity dimensions of 370 × 190 × 370 mm, 1 mm thick stainless steel, (Figure 24.1(b)). Diesel exhaust gas can pass through a custom AC corona injection assembly and a glass tube, 5 mm thick of 0.0023 m^3 volume. All components were chosen to sustain the estimated car engine gas flow rate of 0.04 kg/s, exhaust gas temperatures at the sampling point of 70–100°C and to contain the non-thermal plasma and its complex chemistry. A set of specially designed grounded copper wire meshes were placed at both inlet and outlet of the reactor to avoid any plasma leaks. This laboratory scale reactor performance in NOx and SOx removal was tested using a Ford Duratorq (Puma) high speed direct injection (HSDI) 4 cylinder, 2 l diesel engine in various operating conditions including high level of soot – engine system available at Brunel University (Figure 24.1(a)). The gas was analyzed using a Testo 350 portable emission analyzer (Testo AG, Germany) with gas sensors for NO, NO_2, O_2, SO_2 and CO. In parallel, Testo data measurements were confirmed by Horiba data obtained at the outlet of the engine. The AC corona source is an Electro-Technic Product (Chicago, USA) high-frequency generator which produces a high-voltage, high-frequency spark at the tip of the electrode (Figure 24.2). Power is supplied by a transformer, housed separately for a better heat dissipation with an output voltage between 10–48 kV and a 4–5 MHz frequency. AC corona in air has two different current modes associated with the positive and negative polarity of applied voltage. At positive half–period, the AC corona is not uniform because it operates under streamer regime, while the negative half-period corresponds to homogeneous glow regime. The special electrode design will concentrate a strong electric field around it, which will provoke the streamers. Electron density at the extremities of the electrodes varies around 10^{10} and agrees with the experimentally measured value. Primary radicals are created directly by these collisions, e.g. O, H, OH and N. They may react rapidly with surrounding gas molecules to form secondary radicals such as HO_2 or O_3. If contaminations such as SO_2 or NO are present O, HO_2 and O_3 can oxidize them to acids. The N radical can reduce NO to N_2. The extent to which this happens will depend on the gas composition. NTPR based on a combination of MW and AC corona discharge at atmospheric pressure, if it proves an effective unit, is simple, clean and highly cost-effective.

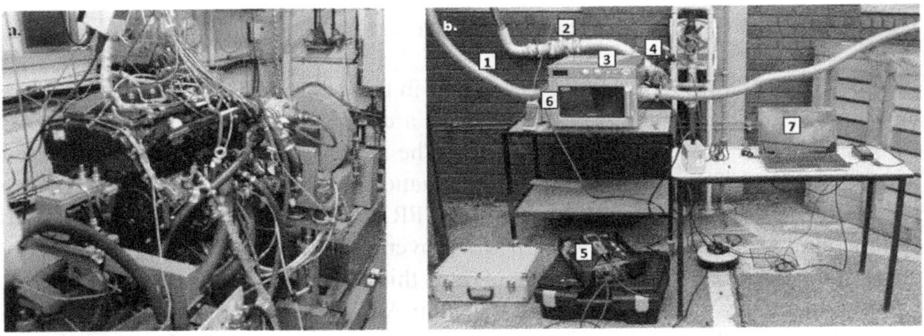

Figure 24.1. *a) Car diesel engine; b) Brunel laboratory scale NTPR experimental set-up: 1) diesel exhaust sampling point, 2) gas temperature sensor, 3) laboratory scale NTPR reactor, 4) Testo sampling probe, 5) Testo 350 gas analyzer, 6) AC corona gun and 7) Laptop for data acquisition*

Figure 24.2. *AC corona discharges in air using two different electrode designs at 48 kV, 5 MHz. For a color version of the figure, see www.iste.co.uk/jacob/energy.zip*

The temperature at the exit of the diesel engine can vary based on the operating parameters, between 360 and 500°C so that at the NTPR sampling point the gas temperature varies between 70 and 100°C. The diesel engine was housed inside the Engine Laboratory and the NTPR had to be located in open air outside. Tests were performed outdoors in the winter season at temperatures varying between –5 and 0°C, and humidity values of 30–40%.

24.2.1. *Experimental results*

For the results presented below, we have used the following car diesel engine conditions: 80 Nm torque (applied load); 2,000 rpm engine speed; start of injection at 9° before top dead centre, smoke number approx. 1.2–1.4, ultra low sulphur diesel fuel, air flow rate at engine output 0.02 kg/s. The engine was operated to produce higher levels of NOx. Testo gas analyzer was monitoring the SO_2 and NOx reduction levels with a dilution factor x5 in ppm at the outlet of NTPR, while Horiba was monitoring CO(H)/ppm; CO_2/Vol%; CO(L)/ppm; O_2/Vol%; THC/ppm; NOx/ppm; A/F; Lambda; EGR/% at the outlet of the diesel engine. Figure 24.3 shows 85% removal of NOx and 100% removal of SO_2 when using a combination of MW and AC corona within the NTPR reactor (see also Table 24.1). In this experiment, only 370 W of the total MW power was used. These early preliminary results also show significant promise from power requirements to treat exhaust gases in the configuration where 25% of MW available total power was used while the AC corona was at ≈ 40 kV. In this preliminary experimental stage, the NTPR feasibility was also tested to perform these experiments and successfully maintain the non-thermal plasma at various power intensities and engine operating conditions. The short duration of a few seconds for each result (Figure 24.3) is because of the technical features of the commercial microwave oven, as it uses pulse width modulated power into the magnetron, and as soon as the temperature inside the cavity exceeds certain limits the thermistor sensors will cut off the magnetron power and as the temperature goes down, the magnetron restarts.

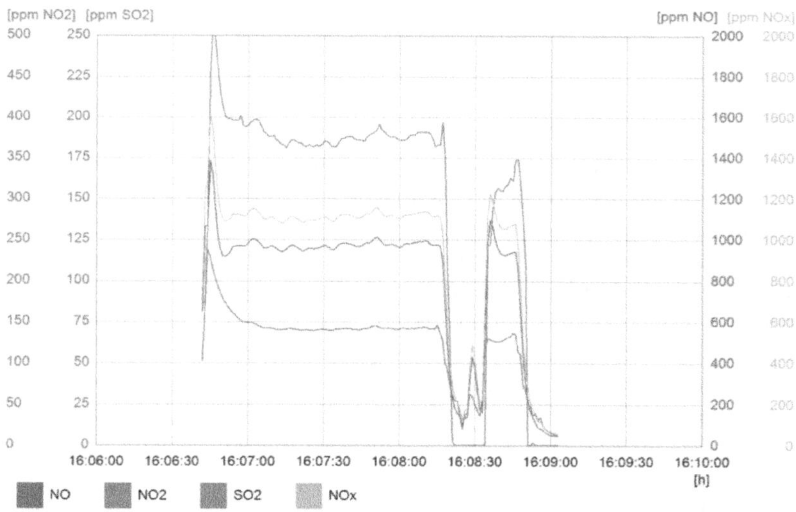

Figure 24.3. *NOx and SO_2 reduction using MW 370 W power and AC corona at ≈ 40 kV. For a color version of the figure, see www.iste.co.uk/jacob/energy.zip*

Horiba average values – measured at engine output								
CO(H)/ ppm	CO_2/Vol%	CO(L)/ ppm	NOx/ ppm	THC/ ppm	O_2/Vol%	A/F	Lambda	EGR/%
295.72961	9.340661	280.74581	1286.552	112.5045	7.7123527	22.778564	1.5570704	86.619274
Testo average values – measured at NTPR input								
SO_2/ ppm	NO/ ppm	NO_2/ ppm	NOx/ ppm					
198	1004.5	140.4	1232.3					
Testo average values – measured at NTPR output with MW+AC Corona on								
SO_2/ ppm	NO/ ppm	NO_2/ ppm	NOx/ ppm					
0	101.4	36.2	177.3					
MW+AC Corona reduction %								
SO_2/ ppm	NO/ ppm	NO_2/ ppm	NOx/ ppm					
100%	90%	75%	85%					

Table 24.1. *Engine diesel exhaust measured parameters and non-thermal plasma reduction efficiency*

24.3. MW power control

Throughout this experimental work, the Brunel NTPR reactor showed the importance of being able to control the magnetron power to be able to stabilize the MW plasma and obtain NOx and SOx reduction for longer periods. The MW unit comes with dual magnetron set-up and a top feeding microwave distribution to ensure an even MW distribution pattern within the cavity. Several options have been explored and tested. Typical microwave ovens operate on cyclic mode (on/off) to control the power inside the oven. The magnetron operates with two voltages, DC voltage (4 kV typical) for the magnetron, and AC signal for the filament. The filament signal is offset by the DC signal. There is a transformer with two secondary coils; the first is high voltage, which is converted into DC using a capacitor and a high-voltage diode, and the other secondary is used to supply AC voltage to the filament. Reducing the input voltage to the oven for the purpose of reducing the power usually does not work, as the oven will cut off. In order to provide continuous power at a variable rate, a pulse width modulation (PWM) circuit can be used to reduce the power and keep the voltage at the same level. Load dimmers are a good choice for this purpose, however, standard types do not work due to the inductive load type. For a non-resistive load dimmer, either a leading edge dimmer (for inductive load) or trailing edge dimmer (for capacitive load) can be used for the microwave oven. An MK K1400 1 kW Universal Leading Edge Dimmer Module which can support master/slave operation mode was selected, that can control both magnetrons of Samsung CM1919 equally and simultaneously (Figure 24.4). The control operations are performed using the input microcontroller. These features were implemented for the marine engine experimental work.

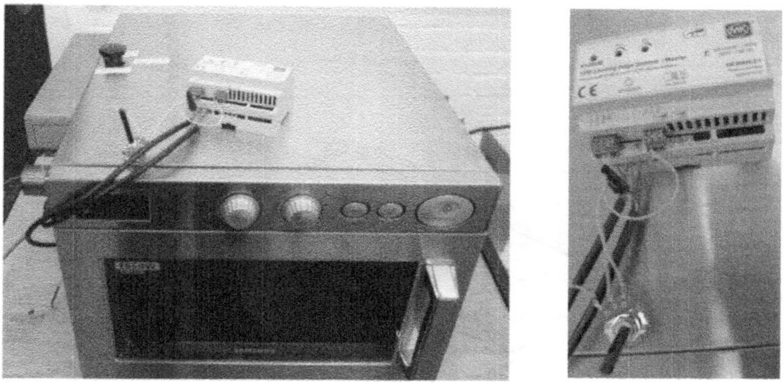

Figure 24.4. *K1400 Universal Dimmer Module for NTPR magnetron control*

Microwave power output measurement is made with the microwave oven supplied at its rated voltage and operated at its maximum microwave power setting with a load of (1,000 ± 5) g of potable water. The initial temperature (T1) of the water is (10 ± 2)°C. It is measured immediately before the water is added to the vessel. After addition of the water to the vessel, the load is immediately placed on the centre which is in the lowest position and the microwave power switched on. The time T for the temperature of the water to rise by a value ΔT of (10 ± 2)° is measured, where T is the time in seconds and ΔT is the temperature rise. The initial and final water temperatures are selected so that the maximum difference between the final water temperature and the ambient temperature is 5°. The microwave power output P in watts is calculated from the following formula:

$$P=[4187 \times (\Delta T)] / T \quad [24.1]$$

This power is measured while the microwave generator is operating at full power. Magnetron filament heat-up time is not included (about 3 sec). Figure 24.5 shows a linear trend between installed dimmer control and MW oven supplied power following the procedure described above.

MW leakages were not exceeding CE Directive 2004-40 of maximum acceptable microwave leakage level of 50 W/m^2, i.e. 5 mW/cm^2 measured at 5 cm from the equipment. These were continuously monitored during MW operation using a Sairem (France) 2450 MHz microwave leakage meter.

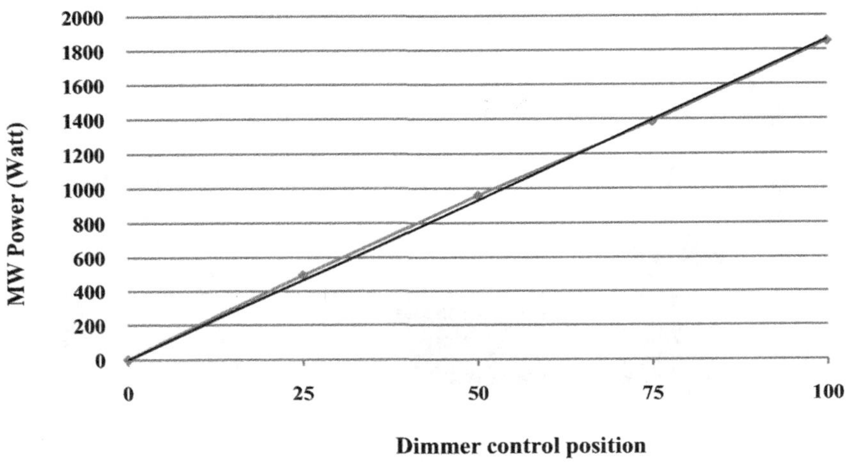

Figure 24.5. *MW calibration using a K1400, 1 kW universal dimmer module*

24.4. Non-thermal plasma experimental set-up using 40 kW marine engine

In the second stage of the DEECON project, the NTPR system was integrated together with other modules developed within the project and was tested using a 40 kW marine engine available through our project partners at Gdynia Maritime University, Poland, as seen in Figures 24.6(a) and 24.5(b). Throughout this experimental process, the NTPR unit was improved and upgraded focusing on the AC corona electrode design to maximize the ionization capabilities of the unit and inlet/outlet mesh set-up, to avoid soot clogging which would block the gas path throughout the system and create back pressure for the engine.

Figure 24.6. *a) 40kw Marine diesel engine and b) Brunel laboratory scale NTPR integrated with DEECON modules*

In the results shown in Figure 24.7 and Table 24.2, the gas flow rate through the NTPR was 57 l/s, with an exhaust gas temperature of 40°C. Engine power was at 7 kW (≈10%) with a torque of 25%. Gas analysis was performed at the outlet of the NTPR unit using the Testo portable gas analyzer. The MW power used was 1.85 kW, with AC corona at 48 kV and 5 MHz. The engine was supplied with a mixture of heavy fuel oil (HFO) with light fuel (MGO). Physical and chemical analysis of the fuel is presented in Table 24.3. The results show > 90% reduction of NOx and SO_2.

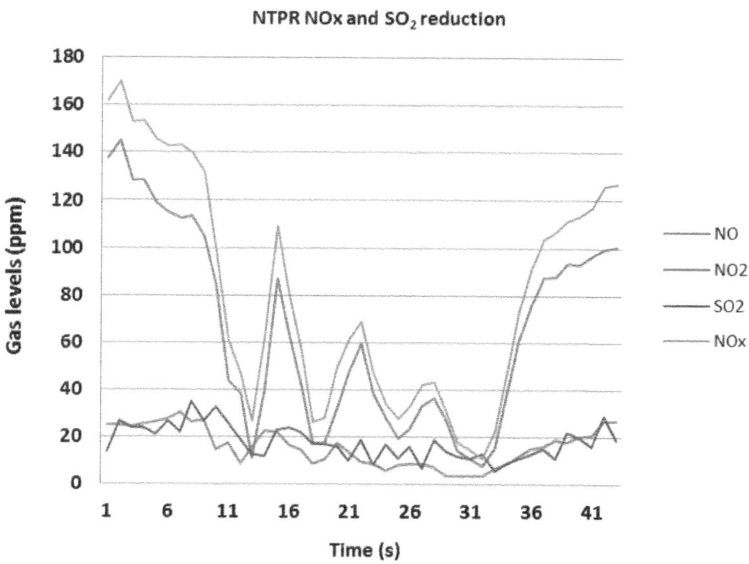

Figure 24.7. *NTPR, NOx (NO+NO_2) and SO_2 reduction results using MW & AC Corona at full power. For a color version of the figure, see www.iste.co.uk/jacob/energy.zip*

Testo average values – measured at NTPR input				
SO_2/ppm	NO/ppm	NO_2/ppm	NOx/ppm	
70	240.8	26	271.6	
Testo average values – measured at NTPR output with MW+AC Corona on				
SO_2/ppm	NO/ppm	NO_2/ppm	NOx/ppm	
4	18	1.2	17.9	
MW+AC Corona reduction %				
SO_2/ppm	NO/ppm	NO_2/ppm	NOx/ppm	
94%	93%	95%	94%	

Table 24.2. *Measured marine engine diesel exhaust parameters and non-thermal plasma reduction efficiency*

Parameter @ Temperature	Value, unit		Fractional analysis:	start at 142°C
Viscosity @ 50°C	17.71 mm^2/s		5 vol.% evaporated at	205°C
Viscosity @ 100°C	4.719 mm^2/s		10 vol.% evaporated at	219°C
Density @ 50°C	0.9031 g/cm^3		30 vol.% evaporated at	287°C
Density @ 100°C	0.8690 g/cm^3		50 vol.% evaporated at	341°C
Sulphur content	0.6677 wt.%		End of distillation	360°C
Heat value	41 115 kJ/kg		Remnants	38 vol.%
PM content	0.03 wt.%		Distillation efficiency	62 vol.%
Elemental analysis:				
Carbon	88.16 wt.%			
Hydrogen	11.16 wt.%			
Nitrogen	t.%			

Table 24.3. *Analysis of the HFO+MGO fuel (LOTOS Laboratory, Gdansk, Poland)*

24.5. Practical implementation

The proposed laboratory scale NTPR system based on AC corona and MW presents itself as an efficient tool in understanding the complex science behind non-thermal plasma process for exhaust gas abatement. The simple design enables easy cleaning and maintenance of each component. Advanced modeling is implemented in corona electrode design and positioning in MW field using COMSOL software package. The design also allows implementation of DC corona or dielectric barrier discharge as an alternative to AC corona. Each component/material selected supports the non-thermal plasma environment, elevated temperatures of the diesel engine exhaust and variable flow rates (up to 40 l/s), while the optimum exhaust temperature through the NTPR was controlled to vary between 70 and 90°C, based on engine running conditions. The corona injection system and gas path components (e.g. flanges, meshes and glass tube) are custom built. The power requirements are modeled based on the amount of gas to be treated and the energy required to break down the exhaust gas into radicals. It is easy to transport, install and adapt with any diesel car engine or small marine engine set-up (e.g. 40 kW). The total cost of this experimental NTPR unit and spare parts was € 3000.

24.6. Conclusions

– Brunel laboratory scale NTPR reactor based on a combination of AC corona and MW was successfully tested in several engine operating conditions, monitoring several key parameters: inlet and outlet gas temperature, exhaust gas humidity, flow rate, NOx and SOx reduction.

– Although within the NTPR, the pollutants reduction time is of a short duration (30 s), early preliminary results showing a promising (>90)% removal of NOx and SO_2.

– MW control assured a continuous supply of magnetron power in the second stage of experimental work using the marine engine. However, in this scenario, the random distribution of MW plasma within the commercial cavity of the MW oven ensured a short and random energy enhancement of the AC ionization area.

– This laboratory scale NTPR unit has provided us with very useful data for the next stage of the project, where the pilot scale system has a fully controlled MW system that will allow us to focus and concentrate the MW energy in the electron density area of the NTPR to achieve NOx and SO_2 removal for a longer periods.

24.7. Bibliography

[BEL 12] BELECA R., MANIVANNAN N., GANIPPA L. *et al.*, Microwave induced plasmas for marine diesel engine emission control, *Proceedings of the Electrostatic Joint Conference*, Ontario, Canada, June 12–14, 2012.

[COM 10] COMMISSION DECISION 2006/944/EC of determining the respective emission levels allocated to the Community and each of its Member States under the Kyoto Protocol pursuant to Council Decision 2002/358/EC, Official Journal L 358 of 16.12.2010, 14 December 2006.

[COM 02] COMMUNICATION FROM THE COMMISSION TO THE EUROPEAN PARLIAMENT AND THE COUNCIL, of 20 November 2002, "A European Union strategy to reduce atmospheric emissions from seagoing ships", COM (2002) 595 final, Volume I, 2002.

[EYR 07] EYRING V., STEVENSON D.S., LAUER A. *et al.*, Multi-model simulations of the impact of international shipping on atmospheric chemistry and climate in 2000 and 2030, *Atmospheric Chemistry and Physics*, vol. 7, pp. 757–780, 2007.

[GRE 12] GREGORY D., A practical guide to exhaust gas cleaning systems for the maritime industry, *EGCSA Handbook*, London, UK, available at: www.egcsa.com, 2012.

[NEE 09] NEEF D., The development of a global maritime emissions inventory using electronic monitoring and reporting techniques, *Proceedings of the 18th Annual International Emission Inventory Conference "Comprehensive Inventories – Leveraging Technology and Resources"*, Baltimore, Maryland, April 14–17, 2009.

25

Quantification of Non-Exhaust Particulate Matter Emissions from Road Transport

This work deals with the quantification of non-exhaust particulate matter emissions in terms of emission from road transport, factors including tyre wear, brake wear, road wear and resuspension as derived by COPERT 4 and a supplementary methodology. The emission factors derived correspond to urban, rural and highway driving conditions. Non-exhaust EFs corresponding to tyre, brake and road wear components are given in detail (according to vehicle technologies), whereas the resuspension component is given separately for passenger cars, light duty vehicles and heavy duty vehicles and buses, and not in the same detailed form. The pollutants include PM2.5, PM10, particle number, elemental carbon, organic matter, seven metals and 28 PAHs.

25.1. Introduction

While EU regulations have resulted in substantial reductions in the exhaust emissions from road traffic, this is not the case for non-exhaust emissions. It is estimated that today emissions of PM10 from wear processes are about equal to exhaust emissions. If the reduction of exhaust emissions progresses as predicted, wear emissions will dominate traffic-related PM emissions in the near future, increasing by 80–90% by the end of this decade [DER 12].

Many studies have sought to distinguish between the different non-exhaust particulate emissions by identifying particular chemical components [HAR 12]. While the constituents of brake material may vary among manufacturers and over time, iron, copper, antimony and barium have been associated with the particulate

Chapter written by Ilias Vouitsis, Leonidas Ntziachristos, Christos Samaras and Zissis Samaras.

matter (PM) released from brake operation. Road surfaces are likely to result in PM of mineral origin. Tyre wear is likely to result in predominantly carbonaceous particles, although small quantities of metals, in particular zinc, may be present. Resuspension of material from the land or road surface by high wind speeds tends to result in PM with aerodynamic diameters toward the larger end of abrasive size spectra, although forces generated at the interface between the tyre and the road surface contribute substantially to particle resuspension.

There are two general approaches for determining EFs from non-exhaust sources:

1) direct measurements from the sources, either under real-world test conditions or in the laboratory;

2) receptor modeling where ambient pollutant concentration data are subdivided according to their different sources and EFs are derived using mass-balance techniques.

Both methods have advantages and limitations. In particular, the first method provides emission factors (EFs) for a relatively small number of vehicles, but for very well-controlled conditions. However, tyre, brake and road surface wear are difficult to simulate in controlled tests. Speed, acceleration/deceleration, tyre material, temperature and parameters such as road surface construction and curvature are all important considerations. Further, resuspension fluxes are difficult to measure directly because the space around a vehicle does not form a “closed” system with obvious air inlets and outlets. The receptor modeling method requires knowledge of source composition and assumes that the sources specified are responsible for the species measure at the receptor. Beyond that, source apportionment and quantification of airborne PM measured in the vicinity of roads is a rather complex task, as road traffic contribution can be small compared with the background concentrations.

This chapter deals with the quantification of road transport non-exhaust PM-emissions in terms of EFs. Section 25.2 presents the methodology applied for the determination of the EFs of various key pollutants and the vehicle categories taken into account and section 25.3 discusses the database produced. Finally, section 25.4 summarizes the conclusions.

25.2. Methodology

The database of exhaust EFs developed in Vouitsis *et al.* [VOU 14] was extended to include non-exhaust EFs. It includes non-exhaust EFs (in g/km) of various vehicle categories and technologies for several pollutants (see Table 25.1 of

Vouitsis *et al.* [VOU 14]) derived by COPERT 4 and from a combined methodology explained below. The COPERT approach is based on the methodology of the calculation of emissions for tyre, break and asphalt wear that has been included in the EMEP/EEA air pollutant emission inventory guidebook [NTZ 12]. It does not account for: (1) resuspension, as it is intended for use by inventory developers and resuspension is not a primary emission source, (2) road surface wear – studded tyres and (3) weather effects. In this chapter, we made an attempt to include (1) and (2) of the above parameters after appropriate analysis of the corresponding literature. Hence, the non-exhaust EFs have the following contributions:

$$\text{EFnon-exhaust=tyre wear + brake wear + road wear and resuspension} \quad [25.1]$$

In general, the EFs derived correspond to two different urban conditions (peak and off-peak) and to rural and highway conditions. For vehicle technologies, Euro 5, 6 and 7 as well as Euro V, VI and VII, however, the EFs derived correspond to only one urban condition. EFs included in the database correspond to tyre, brake and road wear components of equation [25.1], whereas resuspension and studded tyre components are given separately for PCs, LDVs, HDVs and buses and not in the same detailed form. In addition, along with COPERT methodology, a supplementary methodology was applied for the derivation of EFs including:

– literature review to extract non-exhaust EFs reported during the last 5 years;

– comparison with EFs and their relations derived from the literature with the EFs provided by COPERT;

– combination of the above two to derive trends and to estimate EFs by applying the most appropriate (reasonable) fitting of data. An example of this procedure is the derivation of EFs for studded tyres (see section 25.3.4).

Only one EF for each pollutant is given for all EU-27 countries and for all vehicle technologies. This procedure does not take into account the share of vehicle cars in different countries, but it is based on averages extracted from data of the largest countries (specifically, Germany, France and Italy). The pollutants include:

– PM2.5, PM10, OM and EC (PM2.5 and PM10 include exhaust emissions as well);

– Metals: Pb, Cd, Cu, Cr, Se, Zn and Ni;

– PAHs: Anthracene, Benzo(α)anthracene, Benzo(α)pyrene, Benzo[α] fluorene, Benzo[b]fluoranthene, Benzo[e]pyrene, Benzo[g,h,i[perylene, Benzo[g,h,i] fluoranthene, Benzo(k)fluoranthene, Chrysene, Dibenzo[α,h]anthracene, Dibenzo [α,h]anthracene, Dimethyl anthracene, Dimethyl Phenanthrene, Dimethyl fluoranthene, Dimethyl pyrene, Fluoranthene, Indeno[1,2,3-c,d]pyrene, Methyl anthracene, Methyl fluoranthene, Methylbenz[α]anthracene, Methylcrysene,

Methyltriphenylene, Methyl phenanthrene, Methyl pyrene, Naphthalene, Phenanthrene, Pyrene, and Triphelyne.

EFs for technologies Euro 5 and 6 (PC and LDV), Euro V and VI (HDV and Buses) and Euro 3 and 4 (Motorcycles) are taken equal to those of Euro 4 (PC and LDV), Euro IV (HDV and Buses), Euro 2 (Motorcycles) assuming no improvement. However, EFs for future technologies Euro 7 (PC and LDV), Euro VII (HDV and Buses) and Euro 5 (Motorcycles) are taken to be 30% lower. This is based on the assumption of the future maximum technically feasible reductions, presupposing (Borken-Kleefeld and Ntziachristos, 2012): (1) better brake pads and disks, (2) reformulated rubber mixtures for tyres, (3) appropriate adjustment of pavement properties and (4) use of dust suppressants to keep road surfaces wet. Note, however, that this assumption is to a great extent arbitrary; so the user has to be cautious when using it. It would be better to examine two alternative future scenarios: with constant EFs and with 30% lower.

25.3. Database of road non-exhaust EFs

The space here is insufficient for a detailed presentation of the database. For this reason, we focus on current and future technologies for which the availability of data is limited.

25.3.1. *PM10, PM2.5, organic matter and elemental carbon*

PM10, PM2.5, organic matter (OM) and elemental carbon (EC) were calculated from COPERT as described above but without resupsension (see section 25.3.4 resuspension). Tables 25.1, 25.2, 25.3 and 25.4 are the examples of such EFs (in g/km) corresponding to current and future technologies of various vehicles. Similar tables corresponding to other vehicle categories are also included in the database.

(a) PCs (1.4 l–2.0 l, all fuels)				(b) HDVs (Rigid 28–32 t, all fuels)				(c) MCs (250–750 cm^3)			
Vehicle technol.	Driving condition			Vehicle technol.	Driving condition			Vehicle technol.	Driving condition		
	U	R	H		U	R	H		U	R	H
Euro 5	0.021	0.014	0.007	Euro V	0.082	0.054	0.034	Euro 3	0.010	0.007	0.003
Euro 6	0.021	0.014	0.007	Euro VI	0.082	0.054	0.034	Euro 4	0.010	0.007	0.003
Euro 7	0.015	0.010	0.005	Euro VII	0.057	0.038	0.024	Euro 5	0.007	0.005	0.002

Table 25.1. *PM10 EFs (g/km) (exhaust plus non-exhaust) in different driving conditions (U = Urban, R = Rural, H = Highway) for a) current and future passenger cars (PCs) (1.4–2.0 l, all fuels), b) heavy duty vehicles (HDVs) (Rigid 28–32 t, all fuels) and c) motorcycles (MCs) (250–750 cm^3)*

(a) PCs (1.4 l–2.0 l, all fuels)				(b) HDVs (Rigid 28–32 t, all fuels)				(c) MCs (250–750 cm^3)			
Vehicle technol.	Driving condition			Vehicle technol.	Driving condition			Vehicle technol.	Driving condition		
	U	R	H		U	R	H		U	R	H
Euro 5	0.011	0.008	0.005	Euro V	0.041	0.029	0.019	Euro 3	0.005	0.003	0.002
Euro 6	0.011	0.008	0.005	Euro VI	0.041	0.029	0.019	Euro 4	0.005	0.003	0.002
Euro 7	0.008	0.005	0.003	Euro VII	0.029	0.020	0.014	Euro 5	0.0035	0.002	0.0015

Table 25.2. *PM2.5 EFs (g/km) (exhaust plus non-exhaust) in different driving conditions (as in Table 25.1)*

(a) PCs (1.4 l–2.0 l, all fuels)				(b) HDVs (Rigid 28–32 t, all fuels)				(c) MCs (250–750 cm^3)			
Vehicle technol.	Driving condition			Vehicle technol.	Driving condition			Vehicle technol.	Driving condition		
	U	R	H		U	R	H		U	R	H
Euro 5	0.008	0.005	0.003	Euro V	0.029	0.020	0.012	Euro 3	0.004	0.002	0.001
Euro 6	0.008	0.005	0.003	Euro VI	0.029	0.020	0.012	Euro 4	0.004	0.002	0.001
Euro 7	0.005	0.004	0.002	Euro VII	0.021	0.014	0.009	Euro 5	0.0028	0.0014	0.0007

Table 25.3. *Non-exhaust OM EFs (g/km) in different driving conditions (as in Table 25.1)*

(a) PCs (1.4 l–2.0 l, all fuels)				(b) HDVs (Rigid 28–32 t, all fuels)				(c) MCs (250–750 cm^3)			
Vehicle technol.	Driving condition			Vehicle technol.	Driving condition			Vehicle technol.	Driving condition		
	U	R	H		U	R	H		U	R	H
Euro 5	0.003	0.002	0.001	Euro V	0.013	0.008	0.005	Euro 3	0.0015	0.001	0.0005
Euro 6	0.003	0.002	0.001	Euro VI	0.013	0.008	0.005	Euro 4	0.0015	0.001	0.0005
Euro 7	0.002	0.0014	0.0007	Euro VII	0.009	0.006	0.004	Euro 5	0.00105	0.0007	0.00035

Table 25.4. *Non-exhaust EC EFs (g/km) in different driving conditions (as in Table 25.1)*

25.3.2. *Metals*

Table 25.5 (Cu) is an example of such EFs (in g/km) corresponding to current and future technologies of various vehicles. Similar tables corresponding to other vehicle categories are also included in the database.

(a) PCs (1.4 l–2.0 l, all fuels)				(b) HDVs (Rigid 28–32 t, all fuels)				(c) MCs (250–750 cm^3)			
Vehicle technol.	Driving condition			Vehicle technol.	Driving condition			Vehicle technol.	Driving condition		
	U	R	H		U	R	H		U	R	H
Euro 5	2.51	1.34	0.28	Euro V	10.9	7.06	3.60	Euro 3			
Euro 6	2.51	1.34	0.28	Euro VI	10.9	7.06	3.60	Euro 4	1.24	0.66	0.14
Euro 7	1.76	0.94	0.19	Euro VII	7.65	4.94	2.52	Euro 5	0.87	0.46	0.09

Table 25.5. *Non-exhaust Cu EFs (x10^{-4} g/km) in different driving conditions (U = Urban, R = Rural, H = Highway) for a) current and future passenger cars (PCs) (1.4–2.0 l, all fuels), b) heavy duty vehicles (HDVs) (Rigid 28–32 t, all fuels) and c) motorcycles (MCs) (250–750 cm^3)*

25.3.3. *PAHs*

There are a limited number of studies and reports available on PAHs in non-exhaust aerosol. EFs were averaged on the basis of published data on chemical composition of tyre, brake and abrasion wear particulate matter [WAL 12, BOU 07]. Table 25.6 depicts the derived averages of concentrations of 28 PAHs in tyre, brake and road wear in total, whereas Table 25.7 is an example of such EFs (benzo[a]pyrene) corresponding to current and future technologies of various vehicles. Similar tables for all vehicle technologies and for all PAHs reported previously are included in the database.

PAH	μg/g	PAH	μg/g
Anthracene	2,1	Dimethyl Pyrene	19
Benzo[a]anthracene	8	Fluoranthene	15,8
Benzo[a]pyrene	5,4	Indeno[1,2,3-c,d]pyrene	2
Benzo[a]fluorene	1,6	Methyl anthracene	24,3
Benzo[b]fluoranthene	16,8	Methyl fluoranthene	25,3
Benzo[e]pyrene	6,9	Methylbenz[a]anthracene	20,9
Benzo[g,h,i]perylene	15,2	Methylcrysene	20,9
Benzo[g,h,i]fluoranthene	6,6	Methyltriphenylene	20,9
Benzo[k]fluoranthene	9,7	Methyl phenanthrene	24,3
Chrysene	17.8	Methyl pyrene	25,3
Dibenzo[a,h]anthracene	1,7	Naphthalene	7,2
Dimethyl anthracene	39.1	Phenanthrene	13,2
Dimethyl phenanthrene	39,1	Pyrene	55,2
Dimethyl fluoranthene	19	Triphelyne	8,2

Table 25.6. *Total concentration of PAHs in tyre, brake and road wear particles*

(a) PCs (1.4 l–2.0 l, all fuels)				(b) HDVs (Rigid 28–32 t, all fuels)				(c) MCs (250–750 cm^3)			
Vehicle technol.	Driving condition			Vehicle technol.	Driving condition			Vehicle technol.	Driving condition		
	U	R	H		U	R	H		U	R	H
Euro 5	1.73	1.16	5.84	Euro V	8.16	5.41	3.41	Euro 3	0.74	0.49	0.24
Euro 6	1.73	1.16	5.84	Euro VI	8.16	5.41	3.41	Euro 4	0.74	0.49	0.24
Euro 7	1.21	0.81	0.41	Euro VII	5.71	3.79	2.39	Euro 5	0.74	0.49	0.24

Table 25.7. *Non-exhaust benzo[a]pyrene EFs (x10^{-7} g/km) in different driving conditions (U = Urban, R = Rural, H = Highway) for a) current and future passenger cars (PCs) (1.4–2.0 l, all fuels), b) heavy duty vehicles (HDVs) (Rigid 28–32 t, all fuels) and c) motorcycles (MCs) (250–750 cm^3)*

25.3.4. *Resuspension*

Entrainment of particles into suspension is a complex process and depends on a number of environmental and meteorological factors. Large particles with diameters of between 500 and 1,000 μm roll along the ground (surface creep) or move with small bounces (saltate), whereas smaller particles below 100 μm can become suspended [NIC 88]. Particles with settling velocities smaller than vertical velocities in the turbulent boundary layer can remain in airborne suspension for long periods.

Vehicle-induced resuspension.– driving speed and aerodynamic properties of the vehicle affect its turbulence and resuspension of the deposited material from surfaces [KAR 98]. Furthermore, the tyres resuspend particles due to turbulence generated as the air squeezes from beneath the rolling tyre as well as through the shearing action generated by the rotation of the tyre [GIL 05] showed that large vehicles contributed several times more resuspended road dust than light duty vehicles.

Sources of resuspendable material.– surface properties and amount of loose material are other important factors affecting resuspension. Resuspension is high from surfaces that have much loose material of suitable size to be entrained into the air. Good examples of such situations are unpaved roads, which have several orders of magnitude higher emissions than paved roads [VEN 99]. In modern cities, unpaved roads have become scarce, but resuspension of particles from paved streets also affects urban air quality [PAK 01]. In paved road environments, sources of resuspendable material are to a large extent wear products from tyres, brakes and road surface. In subarctic regions of the world, enhanced road dust resuspension has been linked to the use of traction control methods [GER 06]. Resuspendable material can also be delivered from outside the street environment, for example from nearby construction works. Pavement properties may affect the resuspension levels but this factor has been studied very little [DÜR 03].

Surface moisture also affects resuspension. It has been observed that road dust emissions are low during periods with wet surfaces [KUH 03]. During rainy and melting periods, the dust deposit may decrease; some studies indicate that intensive rain events can reduce the surface loadings of roads significantly [VAZ 02]. However, this process may also work the other way if the rain and melting waters relocate loose material from outside onto the street surface and its surroundings. When the surfaces dry out, the smaller PM is released and may be resuspended. This process can give rise to high concentrations during melting periods in subarctic cities during spring. Nicholson and Branson [NIC 90] suggested that resuspension is greatest immediately after the road becomes dry when the dust accumulated and translocated onto the road surface during the wet period is lifted. The impact of road surface wetness is shown in Table 25.8 [JOH 12].

Month	January	February	March	April	May	June	July	August	September	October	November	December
Ratio	4.5	5.5	5	2.8	1.6	1.15	3	2	3.3	1.8	4.45	4.5

Table 25.8. *Dry/wet ratio of PM10 emissions during the year [JOH 12]*

Traction sanding has both direct and resuspension emission characteristics. Traction sand can act as a source of resuspendable material if it contains significant amounts of fine material (for example, grain sizes below 63 μm). Dust is also formed as the sand grains are crushed under tyres into smaller pieces, some of which are small enough to become airborne [VAZ 02]. The interaction between the tyre, sand and pavement materials may wear the pavement surface and result in dust formation from all three sources.

Emission factors measured in road conditions – studies and emission models give EFs varying over several orders of magnitude depending upon which fraction is considered, the traffic speed, type of road, type of tyres, etc. [KUP 07].

Study	*Road type/Traffic conditions*	*PM10*	*PM2.5*
LDVs			
Abu-Allaban *et al.*, 2003	Two lane road/low-speed	125	0.55
Abu-Allaban *et al.*, 2003	Four lane road/high-speed	323	0.7
Luhana *et al.*, 2004	Tunnel/high speed	0.8	
Boulter *et al.*, 2006	Modeling	0.8	
Stocker and Carruthers, 2007	Modeling	0.02	
HDVs			
Abu-Allaban *et al.*, 2003	Four lane road/low-speed	675	3.4
Abu-Allaban *et al.*, 2003	Four lane road/high-speed	3820	4
Lohmeyer *et al.*, 2004	Tunnel/high speed	14.4	
Boulter *et al.*, 2006	Modeling	141	
Stocker and Carruthers, 2007	Modeling	116	
Mixed Fleets			
Fitz. 2001	Freeway	95	
Venkatram and Fitz. 1998	Freeway	200	
Cahill *et al.*, 1995	Intersection	300	
Claiborn *et al.*, 1995	Freeway	500-34000	
Harding and Lawson, 1996	Freeway	30-18000	

Table 25.9. *Resuspension EFs (mg/km) reported in various studies*

Studies carried out in the United States in the early 1990s report PM10 emission factors of about 1,000 mg/km or higher for paved roads with traffic, probably also including some heavy duty vehicles. The range is still wide for light duty traffic as well. For heavy duty vehicles, the average PM10 emission factor is estimated about 675 mg/km for low speed conditions and 3,820 mg/km for high-speed conditions (an average of 2,247 mg/km). In recent years, European measurements have also become available (Table 25.9). Recently, mobile vehicles or trailers with road dust measurement systems have also been introduced [PIR 10]. Table 25.10 shows the PM10 and PM2.5 resuspension EFs for PCs, LDVs, HDVs and buses derived after analyzing published data from studies in Europe and taking into account the speed dependency [BRI 97] to allocate between urban, rural and highway-driving conditions.

Vehicle	*PM10*		
	Urban	Rural	Highway
PCs and LDVs	0.04 (0.05)	0.4 (0.5)	0.8 (1.0)
HDVs and buses	0.06 (0.08)	0.51 (0.65)	1.16 (1.40)
Vehicle	*PM2.5*		
	Urban	Rural	Highway
PCs and LDVs	0.0035 (0.004)	0.012 (0.015)	0.013 (0.015)
HDVs and buses	0.0052 (0.007)	0.016 (0.022)	0.018 (0.025)

Table 25.10. *PM10 and PM2.5 resuspension EFs for PCs, LDVs, HDVs and buses (g/km). Numbers in parentheses correspond to standard deviation values*

Tyre–road interface: effect of studded tyres.– in regions where studded tyres and traction material are used during winter, mechanically generated particles from traffic are the main reason for high particle concentrations in busy street- and road environments. In many Nordic municipalities, the quality standard for inhalable particles is exceeded due to these particles (i.e., [DEN 12]). Table 25.11 shows the studded tyres EFs for PM10 and PM2.5 derived after analyzing data from the above studies and taking into account the speed dependency [GUS 09a] to allocate between urban, rural and highway-driving conditions. EFs for HDVs were taken five times larger than those of PCs [DEN 13], whereas EFs PM2.5 were taken equal to 7.5% of PM10 values [DEN 12].

Vehicle	*PM10*		
	Urban	Rural	Highway
PCs and LDVs	0.12 (0.03)	0.4 (0.06)	0.6 (0.2)
HDVs and buses	0.6 (0.15)	2 (0.30)	3 (1.02)
Vehicle	*PM2.5*		
	Urban	Rural	Highway
PCs and LDVs	0.009 (0.002)	0.03 (0.004)	0.045 (0.015)
HDVs and buses	0.045 (0.010)	0.15 (0.020)	0.23 (0.08)

Table 25.11. *PM10 and PM2.5 studded tyres EFs for PCs, LDVs, HDVs and buses (g/km). Numbers in parentheses correspond to standard deviation values*

25.3.5. *Particle number*

Only a few articles report on airborne tyre particles smaller than 1 mm. Dahl *et al.* [DAH 06] reported the number of EFs between 4×10^{11} and 3×10^{12} particles/km at speeds of 50 and 70 km/h for studded and non-studded tyres. Gustafsson *et al.*

[GUS 08] estimated the particle number concentrations in the 15–700 nm size range from the road–tyre interaction on a road simulator between 1.81–2.65, 0.72–0.82 and 0.14–0.17 ($\times10^4$) cm^{-3} for different road surfaces at a vehicle speed of 70, 50 and 30 km/h, respectively. The PN concentration from the 30 km/h test was in the same range as the background level (0.12–0.17 $\times$ 10^4 cm^{-3}). Particle number emissions increased strongly with increasing speed for all combinations of tyres and pavements. Mathissen *et al.* [MAT 11] reported that for steady straight driving, the upper limit was below 1×10^{11} particles/km. Under extreme conditions, ultrafine particles were measured with mean particle sizes between 30 and 60 nm and peak concentrations up to 1×10^7 particles/cm^3. Both the acceleration test and the cornering tests reveal that non-exhaust particles were formed at the tyre/road interface for extreme conditions due to significant tyre slip. Table 25.12 shows the derived EFs for studded tyres based on EFs and speed dependency given by Gustafsson *et al.* [GUS 09b].

Vehicle	Urban	Rural	Highway
PCs and LDVs	3 (2.9)	160 (180)	500 (650)
HDVs and buses	15 (15)	800 (920)	2500 (3200)

Table 25.12. *Studded tyres PN EFs for PCs, LDVs, HDVs and buses (1010 particles/km). Numbers in parentheses correspond to standard deviation values*

25.4. Conclusions

A database of road transport non-exhaust EFs was developed. The database includes EFs for vehicle technologies pro-Euro and Euro1 – Euro 7 (PCs and LDVs), Euro I – Euro VII (HDVs, buses and coaches) and Euro 1 – Euro 5 (mopeds and motorcycles). For vehicle technologies pro-Euro – Euro 4, Euro IV and Euro 1 – Euro 2 (mopeds and motorcycles), the database includes EFs for the EU-27 and for EE separately, whereas for current and future technologies, only one EF for each pollutant is given for all EU-27 countries. All EFs correspond to urban, rural and highway conditions. The pollutants include PM2.5, PM10, PN, EC, OM, seven metals and 28 PAHs. Attempts were made to include resuspension, road surface wear and studded tyre effects. EFs for technologies Euro 5 and 6 (PC and LDV), Euro V and VI (HDV and buses) and Euro 3 and 4 (motorcycles) are taken equal to those of Euro 4 (PC and LDV), Euro IV (HDV and buses), Euro 2 (motorcycles) assuming no improvement. On the other side, EFs for technologies Euro 7 (PC and LDV), Euro VII (HDV and buses) and Euro 5 (motorcycles) are taken as 30% lower. This is based on the assumption of the maximum technically feasible future reductions.

25.5. Acknowledgments

This work was funded by the 7th European Framework project TRANSPHORM (ENV.2009.1.2.2.1 – Transport-related air pollution and health impacts – Integrated Methodologies for Assessing Particulate Matter).

25.6. Bibliography

[BOR 12] BORKEN-KLEEFELD J., NTZIACHRISTOS L., The potential for further controls of emissions from mobile sources in Europe, Service Contract on Monitoring and Assessment of Sectorial Implementation Actions (ENV.C.3/SER/2011/0009), *TSAP Report* #4 Version 1.0, 2012.

[BOU 06] BOULTER P., THORPE A., HARRISON R. *et al.*, Road vehicle non-exhaust particulate matter: final report on emission modelling, published project report PPR110, TRL, limited, Wokingham, 2006.

[BOU 07] BOULTER P.G., A review of emission factors and models for road vehicle non-exhaust particulate matter, *TRL report for Defra*, Scottish Executive, Welsh Assembly Government, DoENI, 2007.

[BRI 97] BRINGFELT B., BACKSTRÖM H., KINDELL S. *et al.*, Calculations of PM10 concentrations in Swedish cities – modelling of inhalable particles, *RMK* no. 78, SMHI, 1997.

[BOR 12] BORKEN-KLEEFELD J., NTZIACHRISTOS L., The potential for further controls of emissions from mobile sources in Europe, Service Contract on Monitoring and Assessment of Sectorial Implementation Actions (ENV.C.3/SER/2011/0009), *TSAP Report* #4 Version 1.0, 2012.

[CAH 95] CAHILL T.A., SPERLING D., CHANG D.P.Y. *et al.*, PM10 "Hot Spot" emissions from California Roads, *Report submitted to the California Department of Transportation from Air Quality Group*, University of California, 1995.

[CLA 95] CLAIBORN C., MITRA A., ADAMS G. *et al.*, Evaluation of PM10 emission rates from paved and unpaved roads using tracer techniques, *Atmospheric Environment*, vol. 29, no. 10, pp. 1075–1089, 1995.

[DAH 06] DAHL A., GHARIBI A., SWIETLICKI E. *et al.*, Traffic-generated emissions of ultrafine particles from pavement-tyre interface, *Atmospheric Environment*, vol. 40, no. 7, pp. 1314–1323, 2006.

[DEN 12] DENBY B.R., SUNDVOR I., JOHANSSON C. *et al.*, NORTRIP model development and documentation, *NILU OR* 23/2012, 2012.

[DEN 13] DENBY B.R., Personal communication, 2013.

[DEN 12] DENIER VAN DER GON H., JOZWICKA M., CASSEE F. *et al.*, The policy relevance of wear emissions from road transport, now and in the future, *TNO report*, TNO-060-UT-2012-00732, 2012.

[DÜR 03] DÜRING I., BÄCHLIN W., LOHMEYER A., Quantifi zierung der PM10-Emissionen durch Staubaufwirbelung und Abrieb von Strassen auf Basis vorhandener Messdaten, *Projekt* 1772, Achim Lohmeyer, Germany, 2003.

[ETY 03] ETYEMEZIAN V., KUHNS H., GILLIES J. *et al.*, Vehicle-based road dust emission measurement: I – methods and calibration. *Atmospheric Environment*, vol. 37, pp. 4559–4571, 2003.

[FIT 01] FITZ D.R., Measurements of PM10 and PM2.5 emission factors from paved roads in California, report 02-AP-18381-03-FR, Center for Environmental Research and Technology, University of California, 2001.

[GEH 04] GEHRIG R., HILL M., BUCHMANN B., Separate determination of PM10 emission factors of road traffic for tailpipe emissions and emissions from abrasion and resuspension processes, *International Journal of Environment and Pollution*, vol. 22, no. 3, pp. 312–325, 2004.

[GEH 06] GERTLER A., KUHNS H., ABU-ALLABAN M. *et al.*, A case study of the impact of winter road sand/salt and street sweeping on road dust re-entrainment, *Atmospheric Environment*, vol. 40, pp. 5976–5985, 2006.

[GIL 05] GILLIES J.A., ETYEMEZIAN V., KUHNS H. *et al.*, Effect of vehicle characteristics on unpaved road dust emission, *Atmospheric Environment*, vol. 39, pp. 2341–2347, 2005.

[GUS 08] GUSTAFSSON M., BLOMQVIST G., GUDMUNDSSON A. *et al.*, Properties and toxicological effects of particles from the interaction between tyres, road pavement and winter traction material, *Science of the Total Environment*, vol. 393, pp. 226–240, 2008.

[GUS 09a] GUSTAFSSON M., BLOMQVIST G., GUDMUNDSSON A. *et al.*, Factors influencing PM10 emissions from road pavement wear, *Atmospheric Environment*, vol. 43, pp. 4699–4702, 2009.

[GUS 09b] GUSTAFSSON M., BLOMQVIST G., BRORSTRΦM-LUNDIN E. *et al.*, NanoWear – Nanopartiklar frεn slitage av dδck och vδgbana, *VTI Rapport* 660 (In Swedish), available at: http://www.vti.se/sv/publikationer/pdf/nanowear–nanopartiklar-fran-slitage-av-dack-och-vagbana.pdf, 2009.

[HAR 96] HARDING & LAWSON ASSOCIATES, *Final Report for the 1993-1994 ADEQ Paved Road Emissions Research Study*, Maricopa Association of Governments, 1996.

[HAR 12] HARRISON R.M., JONES A.M., GIETL J. *et al.*, Estimation of the contributions of brake dust, tire wear. and resuspension to nonexhaust traffic particles derived from atmospheric measurements, *Environmental Science and Technology*, vol. 46, pp. 6523–6529, 2012.

[HOS 04] HOSIOKANGAS J., VALLIUS M., RUUSKANEN J. *et al.*, Resuspended dust episodes as an urban air-quality problem in subarctic regions, *Scandinavian Journal of Work, Environment & Health*, vol. 30, no. 2, pp. 28–35, 2004.

[JOH 12] JOHANSSON C., DENBY B.C., SUNDVOR I. *et al.*, NORTRIP, Non-exhaust road traffic induced particle emissions: development of a model for assessing the effect on air quality and exposure, ITM-report 212, 2012.

[KAR 98] KARIM M.M., MATSUI H., GUENSLER R. *et al.*, A mathematical model of wind flow, vehicle wake, and pollutant concentration in urban road microenvironments. Part II: model results, *Transport research Part D: Transport and Environment*, vol. 3, no. 3, pp. 171–191, 1998.

[KUH 03] KUHNS H., ETYEMEZIAN V., GREEN M. *et al.*, Vehicle based road dust emission measurement–Part II: Effect of precipitation, wintertime road sanding, and street sweepers on inferred PM10 emission potentials from paved and unpaved roads, *Atmospheric Environment*, vol. 37, pp. 4573–4582, 2003.

[KUM 13] KUMAR P., PIRJOLA L., KETZEL M. *et al.*, Nanoparticle emissions from 11 non-vehicle exhaust sources – a review, *Atmospheric Environment*, vol. 67, pp. 252–277, 2013.

[KUP 07] KUPIAINEN K., Road dust from pavement wear and traction sanding, *Monographs of the Boreal Environment Research*, vol. 26, 2007.

[LOH 04] LOHMEYER A., BÄCHLIN W.M., DÜRING I., Modelling of vehicle induced non-exhaust PM10 emissions, *Proceedings of the PM Emission Inventories Scientific Workshop*, Lago Maggiore, Italy, p. 14, 18 October 2004.

[LUH 04] LUHANA L., SOKHI R., WARNER L. *et al.*, Characterisation of Exhaust Particulate Emissions from Road Vehicles, FP5 Particulates Project, 2004.

[MAT 11] MATHISSEN M., SCHEER V., VOGT R. *et al.*, Investigation on the potential generation of ultrafine particles from the tyre-road interface, *Atmospheric Environment*, vol. 45, pp. 6172–6179, 2011.

[NIC 88] NICHOLSON K.W., A review of particle resuspension, *Atmospheric Environment*, vol. 22, pp. 2639–2651, 1988.

[NIC 90] NICHOLSON K.W., BRANSON J.R., Factors affecting resuspension by road traffic, *Science of the Total Environment*, vol. 93, pp. 349–358, 1990.

[NTZ 12] NTZIACHRISTOS L., SAMARAS Z., *Exhaust emissions from road transport, EMEP/EEA emission inventory guidebook 2009*, updated May 2012.

[PAK 01] PAKKANEN T.A., LOUKKOLA K., OJANEN C. *et al.*, Sources and chemical composition of atmospheric fine and coarse particles in the Helsinki area, *Atmospheric Environment*, vol. 35, pp. 5381–5391, 2001.

[PIR 10] PIRJOLA L., JOHANSSON C., KUPIAINEN K. *et al.*, Road dust emissions from paved roads measured using different mobile systems, *Journal of Air & Waste Management Association*, vol. 60, pp. 1422–1433, 2010.

[STO 07] STOCKER J., CARRUTHERS D., Road vehicle non-exhaust particulate matter: Initial air quality model development and application, and model uncertainty analysis and further model improvements, *TRL Report PPR224*, TRL Limited, Wokingham, 2007.

[VAZ 02] VAZE J., CHIEW H.S., Experimental study of pollutant accumulation on an urban road surface, *Urban Water*, vol. 4, pp. 379–389, 2002.

[VEN 99] VENKATRAM A., FITZ D., BUMILLER K. *et al.*, Using a dispersion model to estimate emission rates of particulate matter from paved roads, *Atmospheric Environment*, vol. 33, pp. 1093 –1102, 1999.

[VOU 14] VOUITSIS I., NTZIACHRISTOS L., SAMARAS C. *et al.*, Quantification of road transport particulate matter using detailed technology emission factors, *Transport Research Arena*, pp. 14–17, France, April 2014.

[WAL 12] WALKER H., Road transport emission factors from 2010 NAEI, Roadtransport EFs NAEI10.v2.xls, available at: http://naei.defra.gov.uk/datawarehouse/3_9_323_136259_roadtransportefs_naei10_v2.xls, 2012.

26

COPERT Micro: a Tool to Calculate Vehicle Emissions in Urban Areas

This chapter discusses the main characteristics of COPERT Micro software, an Excel/VBA-based tool that calculates the road transport emissions from urban areas. The main points of the COPERT Micro methodology are presented, focusing on input data, traffic model interconnection, vehicle classification, emission factors and the calculation procedure. The software application and the user interface are also described. The results from two typical case-studies are presented in order to demonstrate the capabilities of the tool. The main function of COPERT Micro is to assist experts to compile city-scale and other microinventories. However, the case studies revealed that COPERT Micro can also be used to run scenarios in order to investigate the effect of policy measures and the introduction of new technologies on vehicle emissions of a specific area.

26.1. Introduction

Road transport is a significant source of air pollution, especially in urban areas. According to a review of the CLRTAP emission data by the European Environment Agency [EEA, 13], road transport is an important contributor to NO_x, non-methane volatile organic components (NMVOCs), CO, PM_{10} and $PM_{2.5}$ emissions. In 2011, road transport alone was responsible for 40.3% of total NO_x, and 16.5% of total $PM_{2.5}$ in the 27 European Union countries (EU27). Due to their large contribution of total emissions, road transport emissions should be calculated as detailed and precisely as possible.

Road transport exhaust emissions are the aftermath of fuel combustion in internal combustion engines. Gasoline, diesel, liquefied petroleum gas, biofuels and natural

Chapter written by Christos SAMARAS, Leonidas NTZIACHRISTOS and Zissis SAMARAS.

gas are the main fuels used in this sector. The total emissions from road traffic can be divided into three main categories [NTZ 13]:

– exhaust emissions, subdivided into "hot" and "cold start" emissions;

– evaporative emissions;

– emissions due to road, vehicle tyre and brake wear.

Several methods and tools have been developed to calculate road transport emissions [BOU 07, BAR 09, ALK 13, KOU 13]. However, in most European countries (22 out of the EU27), COPERT is used usually for country-level emission inventories [NTZ 09]. Based on methodology and data from COPERT 4, a new methodology and tool was developed. COPERT Micro is an Excel/VBA-based tool that calculates the vehicle pollutant emissions in urban areas. Provided with the necessary road and traffic data, it is able to calculate the fuel consumption, the solid and total particles number, as well as the emissions of the following pollutants: CO, NO_x, VOC, PM exhaust, CO_2, NH_3, B(a)P, EC, OM, NO and NO_2. It is a bottom-up model; hence, it is able to calculate the emissions from a single traffic link up to a whole town, taking into account the vehicle fleet composition of the simulated region.

This chapter discusses the main points of COPERT Micro methodology, focusing on input data, vehicle classification, emission factors and the calculation procedure. The chapter also presents the software and its user interface and demonstrates two typical case studies that were handled with COPERT Micro.

26.2. Methodology

26.2.1. *Input data: interaction with the traffic model*

COPERT Micro is an average speed emissions model, so the average vehicle velocity is the main input in all calculations. However, contrary to other average speed models [FAL 12], it is used in "micro level" (per traffic link). Consequently, it can be used for microinventories (road, region and city-scale inventories). COPERT Micro was initially developed focusing primarily on hot exhaust emissions. The hot emissions are calculated following a quite straightforward approach. If the road and activity data are known, the output emissions are the result of multiplying road and activity data with the corresponding emission factors. The data required to run COPERT Micro are the following:

– Road data: GIS network of the simulated area. The major parameter here is the link length. The GIS network is required to visualize how the output emissions are distributed throughout the area. Moreover, the GIS network is necessary for traffic modeling.

– Activity data: number of vehicles, average speed and vehicle fleet composition per traffic link. The average speed is necessary for calculating the emission factors of each vehicle category, while the number of vehicles and the vehicle fleet composition are used to calculate the total emissions per link and for the entire area.

COPERT Micro was designed to calculate hourly and daily emissions, so the activity data should be given for each hour of the day and for every traffic link. The required input is provided by the traffic model. There are several traffic models available on the market [JON 04, KOK 11]. In general, traffic simulation modeling tools can be divided into two broad categories: micro and macro simulation models [KOT 09]. Both are used by traffic engineers and planning professionals in order to examine traffic flow, congestion and to understand the interaction of vehicles on the roadway. Micro simulation tools model the individual vehicles in the traffic stream and consider the features and characteristics of the individual vehicles and use car-following logic and algorithms to predict and model the movement of each vehicle in the traffic stream. On the other hand, macro simulation models evaluate the traffic flow as a whole without consideration of the characteristics and features of individual vehicles in the traffic stream. These macroscopic models usually require less data input and simpler coding efforts, but provide lower levels of output detail [HEL 02].

Although COPERT Micro is not limited to a specific traffic model/software, the macro simulation models seem to be more suitable for interaction, as long as they provide the necessary input to the software. Macro simulation models calculate the volume of demand and travel patterns (i.e. origin-destination, mode share and routes taken) in the simulated area. They are used to analyze measures that impact travel behavior at the regional level. Example software include AIMSUN, EMME, CUBE and PTV VISUM.

26.2.2. *Vehicle classification*

COPERT Micro contains 230 vehicle categories and respective vehicle technologies, including:

– mopeds and motorcycles (NFR code 1.A.3.b.iv);

– passenger cars (NFR code 1.A.3.b.i);

– light-duty vehicles (NFR code 1.A.3.b.ii);

– heavy-duty vehicles (NFR code 1.A.3.b.iii);

– buses (NFR code 1.A.3.b.iii).

These main categories are further split into classes specific to each vehicle category. Passenger cars are distinguished by those which use gasoline, and further divided into three capacity classes (<1.4 l, 1.4 – 2.0 l and >2.0 l), diesel cars are split into two capacity classes (<2.0 l and >2.0 l), and, then, some additional subcategories, such as LPG-driven cars, 2-stroke engines and hybrids are also included. Light duty vehicles contain vans and small trucks. They are divided into diesel and gasoline vehicles. Heavy duty vehicles consist of 15 subcategories in total, with distinction according to type of vehicle (rigid or articulated), gross vehicle weight (>3.5 t to 60 t) and the fuel (gasoline and diesel). Buses include urban buses and coaches, in different weight classes. Finally, two-wheelers include mopeds (<50cc) and three subcategories of motorcycles (<250cc, 250 – 750cc and >750cc).

The emission level of different vehicles depends on their technological level, i.e. their emission standard. COPERT Micro covers all the relevant EU vehicle technologies for each vehicle category. The technology steps considered range from 1970s technologies (PRE-ECE) up to more recent technologies (Euro 6–2015). The technology availability for passenger cars and light-duty vehicles is shown in Table 26.1, while Table 26.2 shows the technologies for heavy duty vehicles, buses and two-wheelers. COPERT Micro contains emission factors for all these technologies and for all the main pollutants.

The allocation of vehicle stock to the different categories is a key parameter in the development of an accurate emissions inventory regardless of its scale (road, region, city and country). So, before running COPERT Micro, the vehicle fleet of each traffic link must be split into the corresponding vehicle categories according to the fuel used, engine size, weight and technology level of the vehicle.

Passenger cars	Light duty vehicles
PRE ECE	Conventional
ECE 15/00-01	LD Euro 1 – 93/59/EEC
ECE 15/02	LD Euro 2 – 96/69/EEC
ECE 15/03	LD Euro 3 – 98/69/EC Stage2000
ECE 15/04	LD Euro 4 – 98/69/EC Stage2005
Improved Conventional	LD Euro 5 – EC 715/2007
Open Loop	LD Euro 6 – EC 715/2007
PC Euro 1 – 91/441/EEC	
PC Euro 2 – 94/12/EEC	
PC Euro 3 – 98/69/EC Stage 2000	
PC Euro 4 – 98/69/EC Stage 2005	
PC Euro 5 – EC 715/2007	
PC Euro 6 – EC 715/2007	

Table 26.1. *Vehicle technologies for passenger cars and light duty vehicles*

Heavy duty vehicles/buses	Mopeds/motorcycles
Conventional	Conventional
HD Euro I – 91/542/EEC Stage I	Euro 1 – 97/24/EC
HD Euro II – 91/542/EEC Stage II	Euro 2 – 97/24/EC
HD Euro III – 2000 Standards	Euro 3 – 2002/51/EC
HD Euro IV – 2005/55/EC	
HD Euro V – 2005/55/EC	
HD Euro VI – COM(2007)851	

Table 26.2. *Vehicle technologies for heavy duty vehicles/buses and mopeds/motorcycles*

26.2.3. *Emission factors*

Similarly to COPERT, the calculation of emission factors is the core of the COPERT Micro methodology. Emission factors describe the emission performance of a vehicle and depend on the vehicle operation (e.g. momentary speed), particular processes (e.g. acceleration and deceleration), ambient conditions (most importantly temperature) and environment (e.g. altitude). COPERT Micro was built around the average-speed approach; emission factors are expressed as functions of the mean travelling speed over a complete driving cycle.

For the calculation of emission factors, a sample of vehicles is driven over several well-defined speed profiles (driving cycles) on a chassis dynamometer. The mean emission level of each vehicle (in g/km) is then linked to the mean driving speed over each cycle. This produces a matrix of emission level values (g/km) over various mean speeds (km/h). By testing a large number of cars, representative emissions for the fleet can be derived as a function of mean speed. An example of this approach is shown in Figure 26.1, where each data point corresponds to the average emission level of the sample of vehicles measured over the driving cycles. The relatively large scatter of measurements is due to the variable emission level of different cars, even when they share the same characteristics (size, fuel and emission standard). In addition, the mean emission level over some driving cycles differs from the level estimated by the trendline. This is a limitation of using the average speed as the only determinant of driving kinematics. However, the trendline has been derived as an unbiased predictor of the mean emission performance, so real-world deviations from the predicted values should, in principle, be cancelled out.

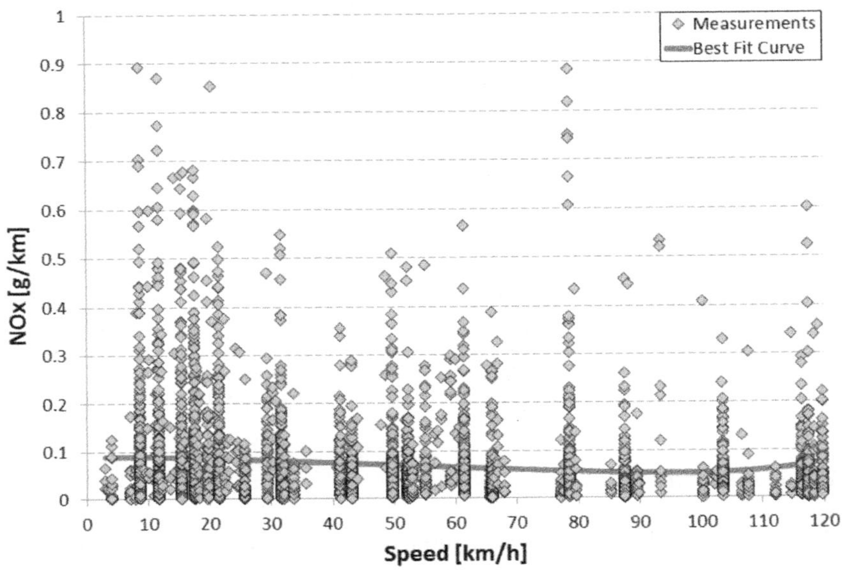

Figure 26.1. *NOx emission factor for gasoline Euro 3 passenger cars*

COPERT Micro incorporates the most up-to-date emission factors database of COPERT 4 complemented by emission factors of TRANSPHORM's database [SAM 13]. COPERT 4 functions were derived in the framework of several research projects conducted by the Laboratory of Applied Thermodynamics and other research institutes in Europe. Major sources of information, between others, have been the MEET [MEE 99], the ARTEMIS [ART 08] and PARTICULATES [SAM 05] projects. In the EMEP/EEA emission inventory guidebook [NTZ 13], more information and details on the emission factors can be found.

26.2.4. *Equations: calculation procedure*

The equations below summarize the main calculations performed by COPERT Micro during hot exhaust emissions calculation. First, based on the average speed of each traffic link the corresponding emission factors are calculated for each of the 230 vehicle categories and for every pollutant:

$$EF_i(V)=f_{EF}(V) \qquad [26.1]$$

where:

$EF_i(V)$: Emission factor per vehicle category, for average speed V and for pollutant i [g/km].

i: Pollutant of interest (CO, NO_x, VOC, PM exhaust, CO_2, NH_3, B(a)P, EC, OM, NO and NO_2, fuel consumption, solid and total particles number).

f_{EF}: Polynomial function derived from measured data (as trendline); unique for every vehicle category and for each pollutant.

V: Average vehicles speed on the traffic link [km/h].

After calculating the emission factors for all pollutants, the hot exhaust emissions of the traffic link j are calculated by the formula:

$$Emissions_{i,j} = L_j \times N_j \times P_{vehcile\ category,\ j} \times EF_i(V) \qquad [26.2]$$

where:

$Emissions_{i,\ j}$: Hot exhaust emissions of pollutant i produced by N_j vehicles that circulate on the link j [g].

L_j: Length of the traffic link j [km].

N_j: Number of vehicles circulating on the link j.

$P_{vehcile\ category,\ j}$ = Percentage of vehicles of the specific category in the overall vehicle fleet (230 categories), e.g. passenger cars - gasoline <1.4 l, Euro 3 = 5%, light duty vehicles – diesel Euro 3 = 2% etc.

Finally, the total hot exhaust emissions of the pollutant i from the entire area ($Emissions_{i,\ area}$) are calculated by summing the emissions of individual traffic links:

$$Emissions_{i,\ area} = \sum Emissions_{i,j} \qquad [26.3]$$

If the hourly traffic activity data are available, the calculation is repeated for each hour of the day and the hourly and daily emissions are calculated for all pollutants and for every traffic link. The emission modeling of an entire city with thousands of traffic links involves an enormous amount of calculations. However, COPERT Micro was designed to keep computational time as minimum as possible. Hence, the hourly and daily emissions of a road network that consists of 40,000 traffic links can be calculated in less than 15 min.

26.3. Software application

26.3.1. *Software development*

COPERT Micro methodology has been integrated into an Excel application, the latest version of which is 1.7 (September 2012). The programming language used for the development of COPERT Micro is Microsoft Visual Basic for Applications (VBA). Microsoft Excel was selected due to its user-friendly interface, which is quite convenient for entering data. Furthermore, the majority of available traffic software can export their results to Excel, making the preparation of the input traffic data to COPERT Micro a much easier process. VBA on the other hand is the standard programming language of Microsoft Office suite, so it can be used to connect different Office applications, a quite useful feature when creating reports in Word based on data from Excel. It should be highlighted that COPERT Micro runs only on the latest Excel versions (2010 and 2013), since its source code incorporates advanced Excel features – like sparklines and conditional formatting – that are not available in older Excel versions (2007, 2003 and earlier).

The source code of COPERT Micro is constituted of several modules and interface forms that are interconnected. The main process is consisted of nine modules and is responsible for calculating the hourly and daily emissions from all traffic links (Figure 26.2). There are also modules for updating the vehicle fleet split and the emission factors, as well as modules responsible for exporting the calculated emissions in a new Excel workbook and to a PDF report, respectively. All necessary data, such as input data, calculation parameters and final results are stored in the same file, so the user can work on the file on different machines and send or receive files from other users for further processing.

26.3.2. *User interface*

The user interface of COPERT Micro follows the standard Excel interface. It has one main sheet, two input data sheets, two sheets for intermediate calculations, 14 sheets for emissions results (one for each pollutant), one overview sheet, which contains the daily emissions from each traffic link and one sheet containing some representative charts (emissions per hour of the day, emissions vs. number of vehicles etc.).

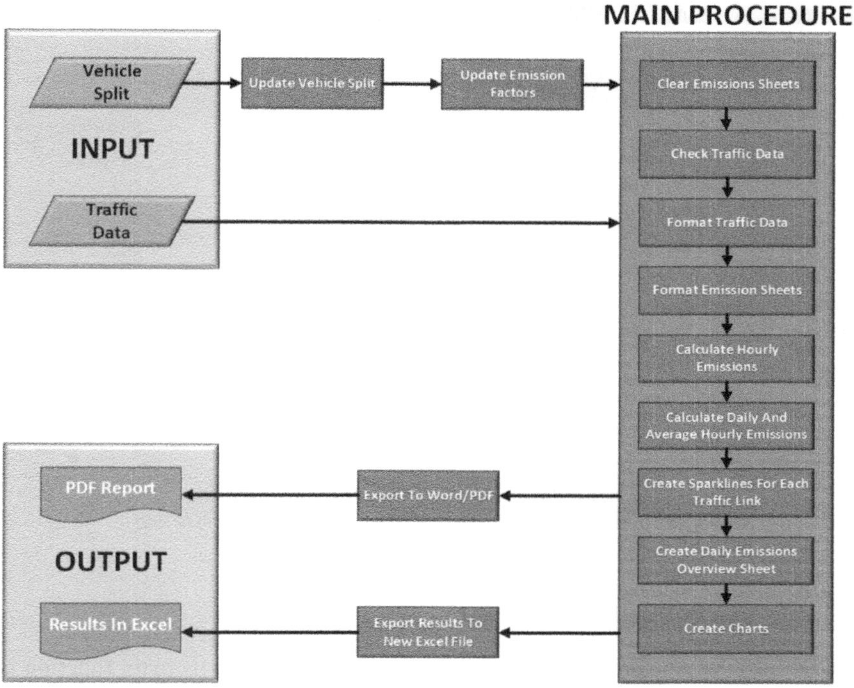

Figure 26.2. *COPERT Micro main modules*

The main sheet (Figure 26.3) of the application consists of seven buttons, each of which runs a different process. There are four discrete steps to calculate emissions. The user first should update the vehicle split according to his/her own input data (step 1). Then, he/she must update all the emission factors (step 2). Then, the traffic/link data must be entered in sheet "Link Data", according to the sheet headings (step 3). By pressing the "Emissions" button, all the emissions are calculated (step 4). The computational time of the last step depends on the amount of traffic data entered. For the first three steps, special interface forms exist to help the user enter the input data to the corresponding sheets. After finishing the calculations, the user has two options regarding the results: he/she can either export all emission results to a new workbook (step 5), or he/she can create a short PDF report (step 6) by pressing the corresponding buttons on the main sheet. Finally, if the user needs to use COPERT Micro with new input data he/she can press the "Clear All Data" button, and all the data from all the sheets will be deleted.

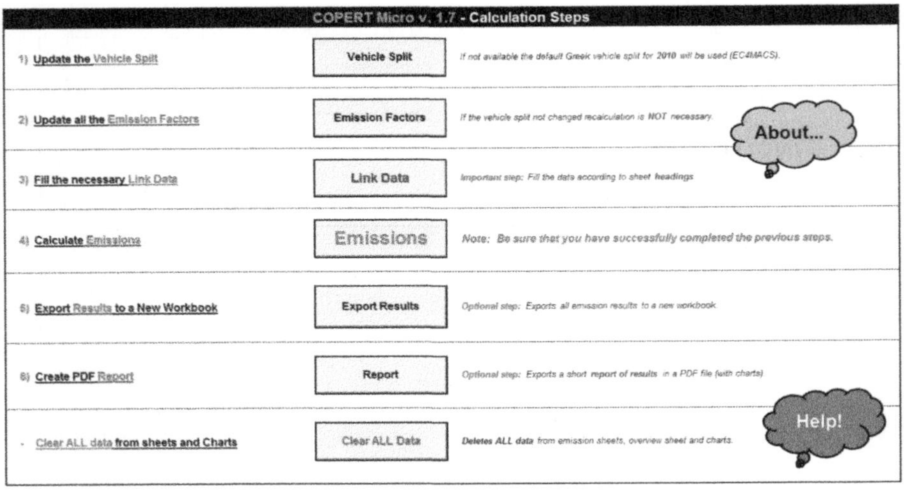

Figure 26.3. *COPERT Micro user interface (main sheet)*

26.4. Results and applications

COPERT Micro can be used in a number of applications. Its main function is to assist experts to compile city-scale and other microinventories. Since the core methodology of COPERT Micro – the emission factors – is based on detailed reports and experimental data, it can also be used for studies on emission calculations for specific fleets (e.g. captive fleets), as well as an input to air quality modeling. Moreover, the user can use COOPERT Micro to run scenarios in order to investigate the effect of policy measures and the introduction of new technologies on emissions produced from a specific area.

26.4.1. *Athens emissions inventory*

In the emissions inventory of Athens, the hourly volume data measured by inductive loop detectors installed at 557 locations across the city were imported to PTV VISUM, as well as the vehicle fleet composition and a detailed representation of the urban and regional road network of the metropolitan region of Athens, based on open-source GIS, fused with traffic-related parameters The output of traffic modeling was the hourly vehicle load, the average speed and the vehicle fleet composition of the 81,880 links of the network for a typical weekday of October 2010. In the study of Samaras *et al.* [SAM 12], more information can be found regarding the traffic modeling.

Using the output of PTV VISUM as an input to COPERT Micro, the hourly and daily emissions were calculated for all the links of the network. The emission values from COPERT Micro were imported to a GIS grid in order to visualize how the various emissions are distributed throughout the city (Figure 26.4). It is evident that the contribution of large boulevards – such as Leoforos Kifisou, Leoforos Kifisias and Attiki Odos – on total CO, NO_x, as well as on the other emissions is quite high (marked with red colour on the GIS grid).

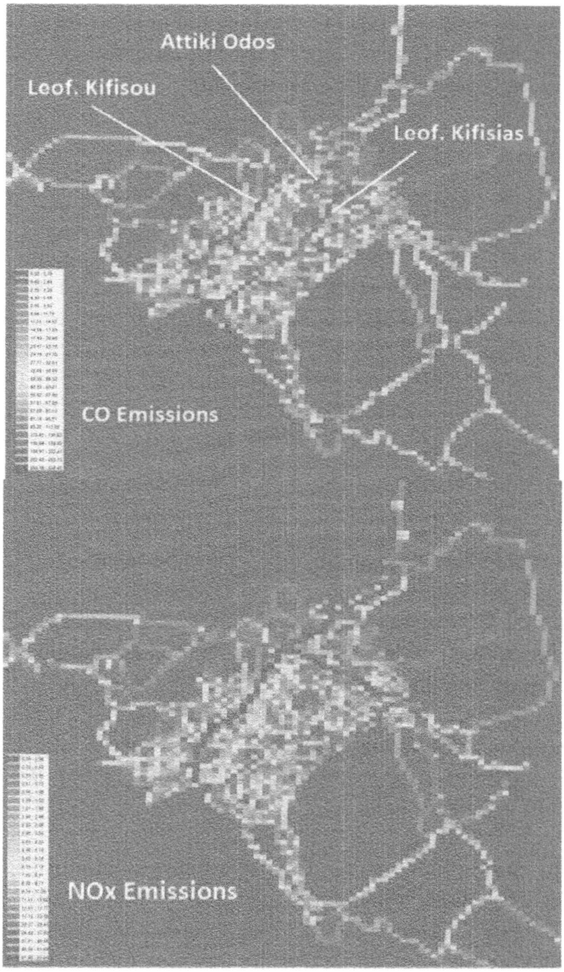

Figure 26.4. *Daily a) CO and b) NO_x emissions (kg) on a 100 x 100 (500 x 500 m^2) grid for a typical weekday of October 2010 in the Greater Athens Area. For a color version of the figure, see www.iste.co.uk/jacob/energy.zip*

Under the framework of the TRANSPHORM project [TRA 13], a low emissions zone (LEZ) scenario was examined for the city of Athens, but for the 2020 (projection). The scenario involves a shift from <Euro 5 vehicles to vehicles with the emission standard Euro 5 and higher. The emissions for 2020 (without LEZ) were calculated using the traffic activity data of 2010. However, the vehicle fleet was updated for 2020 using data from the EC4MACS project (EC4MACS, 2013). In the LEZ scenario, apart from the renewal of vehicle fleet, the vehicle-km of Euro 5 and Euro 6 vehicless was increased, while the vehicle-km of older vehicles reduced, respectively, using the formulas below. The idea behind the formulas is that with the application of LEZ, "cleaner" vehicles will travel more than the old ones, due to the circulation restriction imposed by the LEZ measure.

$$vkm_{Euro\,i\,(LEZ)} = 0.25 \times vkm_{Euro\,i},\ i = 0\ (conventional)\ to\ 4 \qquad [26.4]$$

$$vkm_{Euro\,j\,(LEZ)} = vkm_{Euro\,j} + (0.75 \times vkm_{Euro\,j} \times \Sigma vkm_{Euro\,i})\ /\ (vkm_{Euro\,5} + (vkm_{Euro\,6}),\ i = 0\ to\ 4, j = 5, 6 \qquad [26.5]$$

The emissions for both 2020 cases (with and without LEZ) were calculated with COPERT Micro. The results indicating that in 2020 the road transport emissions in Athens are expected to be less than 2010. Furthermore, if the authorities decide to apply an LEZ measure, emissions will be further reduced. For CO emissions, an overall reduction of about 25 and 43% (with LEZ) is expected to take place in 2020, compared with the base case scenario (2010). NO_x reduction is expected to be even higher – 43 and 69% (with LEZ),

26.4.2. *Emissions inventory of the Municipality of Pilea-Hortiatis*

A similar approach was followed in the inventory of Municipality of Pilea-Hortiatis, which is located on the south-east side of Thessaloniki. PTV VISUM was used for traffic modeling, the output of which was the hourly vehicle load, the average speed and the vehicle split composition of the 137,804 links of the network for a typical weekday of October 2011. In the report of Municipality of Pilea-Hortiatis [MUN 13], more information can be found regarding the traffic modeling (although the report is written in Greek).

The particular study focused on fuel consumption and CO_2 emissions. The results from COPERT Micro showed that about 93% of the daily fuel consumption is due to private vehicles and the rest 7% due to public transport vehicles. Moreover, 83% of the daily CO_2 emissions are caused by gasoline combustion and the rest 17% by diesel. The contribution of LPG on the total CO_2 emissions is close to zero, since the population of vehicles fuelled with LPG – mainly taxis – is substantially small compared to gasoline and diesel.

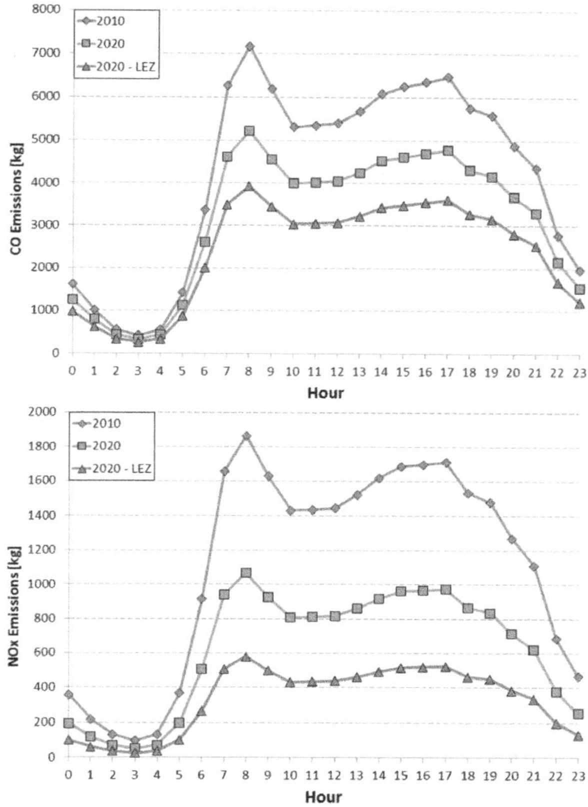

Figure 26.5. *Hourly a) CO and b) NOx calculated emissions for a base case (2010), a future projection (2020) and a low emissions zone measure applied in the future (2020 – LEZ) in the city of Athens. For a color version of the figure, see www.iste.co.uk/jacob/energy.zip*

26.5. Conclusions

COPERT Micro is an Excel/VBA user-friendly software developed to calculate road transport emissions at micro level (per traffic link). It is an average speed model and the core of its methodology is based on emission factors. It can calculate the emissions from a variety of pollutants, while its database contains emission factors (as a function of average speed) for 230 different vehicle categories.

Application of the software to calculate the total emissions from road transport in Athens (in 2010) showed how the different pollutants are distributed throughout the road network. The results indicated that the large boulevards are responsible for the majority of the pollutants emitted. In 2020, Athens is expected to have lower road transport emissions, due to a shift toward "cleaner" vehicle technologies (Euro 5 and Euro 6). Moreover, a possible application of a low emissions zone would be beneficial for the city since it would further reduce emissions.

In the emissions inventory of Municipality of Pilea-Hortiatis, the results of COPERT Micro demonstrated that the majority of daily CO_2 emissions (83%) produced by gasoline vehicles. Furthermore, in the particular area, the private vehicles are responsible for the 93% of the total daily fuel consumption.

26.6. Bibliography

[ALK 13] ALKAFOURY A., BADY M., ALY M. *et al.*, "Emissions modeling for road transportation in urban areas: state-of-art review", *23rd International Conference on Environmental Protection is a Must*, Alexandria, 2013.

[ART 08] ARTEMIS, Assessment and reliability of transport emission models and inventory systems, Retrieved from ARTEMIS, available at: http://www.trl.co.uk/artemis, 2008.

[BAR 09] BARLOW T.J., BOULTER P.G., Emission factors 2009: Report 2 – a review of the average-speed approach for estimating hot exhaust emissions, TRL Limited, 2009.

[BOU 07] BOULTER P.G., MCCRAE I.S., BARLOW T.J., A review of instantaneous emission models for road vehicles, Transport Research Laboratory, 2007.

[EC4 13] EC4MACS, Retrieved from European Consortium for Modelling of Air Pollution and Climate Strategies, available at: http://www.ec4macs.eu/, 2013.

[EEA 13] EEA, European Union emission inventory report 1990–2011 under the UNECE Convention on Long-range Transboundary Air Pollution (LRTAP), European Environment Agency, Luxembourg, 2013.

[FAL 12] FALLAHSHORSHANI M., ANDREA M., BONHORNRNEB C. *et al.*, "Coupling traffic, pollutant emission, air and water quality models – technical review and perspectives", *Procedia – Social and Behavioral Sciences*, vol. 48, pp. 1794–1804, 2012.

[HEL 02] HELBING D., HENNECKE A., SHVETSOV V. *et al.*, "Micro- and macro-simulation of freeway traffic", *Mathematical and Computer Modelling*, vol. 35, pp. 517–547, 2002.

[JON 04] JONES S.L., SULLIVAN A.J., CHEEKOTI N. *et al.*, Traffic simulation software comparison study, report, University Transportation Center for Alabama, 2004.

[KOK 11] KOKKINOGENIS Z., PASSOS L., ROSSETTI R. *et al.*, *Proceedings of the 6th Doctoral Symposium in Informatics Engineering*, pp. 77–90, 2011.

[KOT 09] KOTUSEVSKI G., HAWICK K., A review of traffic simulation software, technical report, Massey University, 2009.

[KOU 13] KOUSOULIDOU M., FONTARAS G., LONZA L. *et al.*, Review and evaluation of emission models and vehicle simulation tools, technical report, Publications Office of the European Union, 2013.

[MET 99] MEET, Methodology for calculating transport emissions and energy consumption, Office for official publications of the European communities, Luxembourg, 1999.

[MUN 13] MUNICIPALITY OF PILEA-HORTIATIS, Sustainable energy aciton plan, Thessaloniki: Municipality of Pilea Hortiatis, 2013.

[NTZ 09] NTZIACHRISTOS L., GKATZOFLIAS D., KOURIDIS C. *et al.*, "COPERT: A European Road Transport Emission", *Proceedings of the 4th International ICSC Symposium*, pp. 491–504, Thessaloniki, 2009.

[NTZ 13] NTZIACHRISTOS L., SAMARAS Z., Exhaust emissions from road transport, In EMEP/EEA emission inventory guidebook 2013, EMEP, 2013.

[SAM 05] SAMARAS Z., NTZIACHRISTOS L., THOMPSON N. *et al.*, Characterisation of Exhaust Particulate Emissions from Road Vehicles (PARTICULATES) report, Retrieved from Transport Research & Innovation Portal, available at: http://www.transport-research.info/Upload/Documents/201003/20100310_132356_74824_Particulates%20Final%20Publishable%20Report.pdf, 2005.

[SAM 12] SAMARAS Z., MOUSSIOPOULOS N., DOUROS I. *et al.*, "Transport Emissions and their Impact on Air Quality in Athens: A Case Study in the Framework of TRANSPHORM", *19th International Transport and Air Pollution Conference*, Thessaloniki, 2012.

[SAM 13] SAMARAS C., VOUITSIS E., SAMARAS Z., TRANSPHROM's emission factors database, Retrieved from Laboratory of Applied Thermodynamics, available at: ftp://transphorm:trans123@155.207.23.130, 2013.

[TRA 13] TRANSPHORM, Transport related Air Pollution and Health impacts, Retrieved from TRANSPHORM, available at: http://www.transphorm.eu/, 2013.

27

Traffic and Vehicle Fleet Statistics for the Calculation of Air Pollutant Emissions from Road Transport in France

Following European projects for calculating pollutant emissions from road transports, this chapter aims to works aims to question the French road-vehicle fleet and traffic-related data. Significant updates and improvements are conducted for the different vehicle categories. Tools are built-up to manage these data and to compute the French fleet from 1980 to 2030 that will be used as input of the HBEFA software for pollutant emission estimation.

Different estimations of the French fleet – in particular data used at the European level – led to significant differences, which justify such in-depth works and demonstrate the need of explicit approaches and assumptions. Such gaps also show the fragility of the emission estimation, highly sensitive to the vehicle fleet.

Data and assumptions are described, as well as the resulting French fleet composition. Other data needed for pollutant emission estimation are discussed in particular the driving conditions. Finally, local video observations demonstrate the sensitivity of fleet composition to local contexts.

27.1. Introduction

Air pollution is one of the main concerns in urban areas where regulatory limits are frequently overpassed, and where health impacts are significant. Road traffic is a major contributor to this pollution. Various measures (transport management, traffic regulation, public transports, depollution and technological improvements, etc.) are

---

Chapter written by Michel ANDRÉ, Marion CARTERET and Anaïs PASQUIER.

proposed to tackle such a problem and to limit the pollution due to transport, but their assessment remains difficult.

The elaboration of reliable pollutant emission models is thus important, and several European research projects (MEET, ARTEMIS) and European or international working groups (COST319, COST346, ERMES) have been conducted or are still active on that topic. These works have enabled the elaboration of well-accepted methods and tools (COPERT, HBEFA, etc.), as these rely on a large consensus and compile most available emission measurements and data from the European laboratories.

Today, these tools are considered quite satisfactory at least to estimate the main regulated pollutants (CO, NOx, total HC, etc.) from exhaust emissions. Indeed, there are not that many data and measurements for non-regulated pollutants (NO_2, NH_3, individual hydrocarbons, fine and ultrafine particulates and their chemical composition, PAH, heavy metals, etc.) and for non-exhaust emissions, and the tools remain weak for their estimation, although these pollutants appear to be the most important regarding their health impacts, for instance. Nevertheless, apart from these weaknesses, the inventory tools were built to enable the estimation of most emission phenomenon and for most components that need to be considered in impact studies.

However, for calculating the pollutant emission from road traffic, there is a main concern around the vehicle fleet composition – as emissions are highly dependent on the vehicle technologies, motorizations and fuels, and above all they are strongly influenced by the emission regulation (i.e. Euro standards). Other data, such as driving or traffic conditions are also important.

If a lot of effort has been dedicated to the refinement of emission models, we should observe that the necessary input data (vehicle fleet and traffic data) were not studied in detail, although their elaboration is very difficult. Although consistent works are conducted around vehicle fleets [KOL 12, HIR 07, GRE 96], the scientific literature regarding their modeling for air pollution is rather sparse, lacks information regarding assumptions and relies on old data [ZAC 95]. Furthermore, local applications (in urban or regional contexts) generally rely on national default data (fleet composition for instance) that could be questionable at these scales as different living standards or consumer behavior can strongly influence the purchase and choice of vehicles.

In this chapter, we describe the elaboration of French fleet data, we analyze and set up critical assumptions and discuss the resulting fleet composition in comparison with other existing estimations. Finally, we examine local observations that demonstrate the sensitivity of fleets to local contexts.

27.2. Data needed for estimating pollutant emissions from road transports

Typically, emission calculation tools require detailed information regarding (1) the fleet composition and vehicle characteristics, (2) the driving conditions (i.e. speeds and traffic situations) and (3) the vehicle use conditions (vehicle load and configuration, fuels, ambient conditions, etc.).

Regarding fleet description, the declination of the fleet (in number of vehicles) and traffic (in vehicles x km) by vehicle categories (passenger cars, light and heavy duty vehicles – of which lorries and articulated trucks, urban busses and coaches – motorized two-wheelers) is required. Furthermore, a detailed declination is needed according to motorization (diesel, petrol, alternative fuels and technologies, 2- or 4-strokes engine, etc.), “size” (engine size for the light vehicles and vehicle weight for the heavy ones), and finally according to the pollutant emission regulation (pre-Euro, and then Euro1 to 5 and the future ones). Complementary vehicle characteristics can be useful such as air conditioning, catalyst or depollution equipments, etc.

27.3. A typical approach for determining fleet composition

A typical approach for the determination of the French fleet composition for emission calculation is given in Figure 27.1. The base data (a) consist of the yearly vehicle registration data (number of vehicles), over a long period and that should be developed according to the necessary structure when possible (i.e. engine size and type, vehicle weight, auxiliary equipment, etc.). In France, such data were collected from 1945 up to 2012, but only 10–20 years are known in detail. Survival probability laws are applied to these registration data to determine the remaining vehicles after one or several years. We thus obtain the detailed fleet composition for a given year, taking into account the vehicles sold/registered each past years. By combining this fleet composition with the annual mileage experienced by each vehicle categories and according to their age (old vehicles are less driven than the newest), we can have a weighted fleet composition that corresponds to the in-use fleet (on the road), i.e. the traffic composition. That is the information required for emission computation for a given case study (i.e. a given year, a traffic flow and its spatial distribution, etc.) providing complementary information such as driving speeds, vehicle loads, etc. (d).

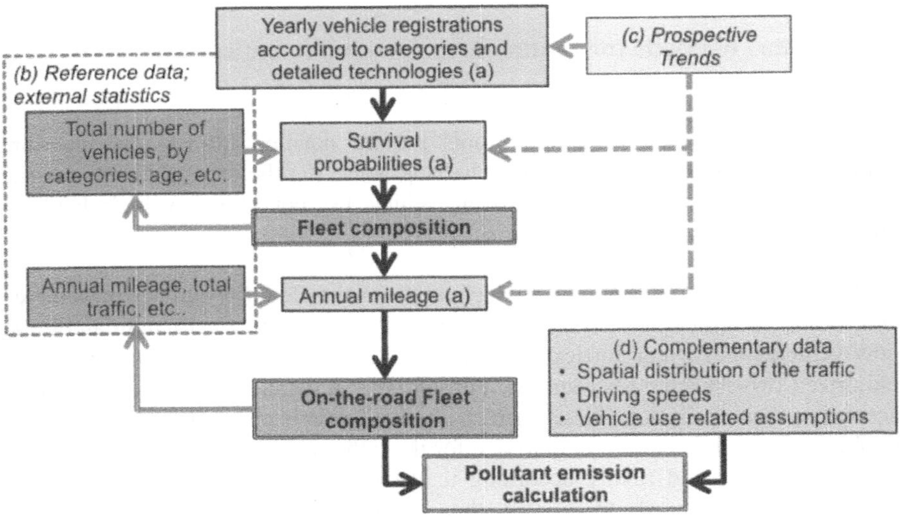

Figure 27.1. *Overall scheme for determining vehicle fleet composition. a) Base data, b) reference statistics, c) prospective trends and d) other data needed for emission calculation*

External statistics (set (b)) can be used to adjust the calculation. For instance, the total number of vehicles, their average age, annual mileage, which can be observed through complementary surveys and other works can be used to validate/adjust the two processes for the determination of the fleet composition.

Furthermore, as most applications concern impact studies of measures for a given term, prospective assumptions should be set-up (c), which applies to registration (i.e. what level of vehicle acquisition/renewal and for which vehicle types and technologies), survival (lifetime and renewal of the vehicles), mileage (mobility trends) and even to the complementary data (speeds, usages, etc.).

27.4. Detailed assumptions and data for France

A detailed description of the data collected and assumptions established for France and for each vehicle category is provided in [AND 13]. We will discuss these data in a more general way, highlighting specific difficulties, approaches or results.

27.4.1. *Registration data, engine and vehicle size, and equipment*

The main difficulty is to get detailed registration data to satisfy the required structure of the emission models and over a sufficient period to enable the validity of the survival process (at least 15–20 years are necessary to get a satisfying description of the oldest existing vehicles). In France, overall registration data without itemization were collected for the period 1945 up to the 1970s. Data which were more or less classified according to the COPERT structure were collected between 1980–1999, and finally we obtain almost fully disaggregated data (or rough data) for the last decade and up to 2012. While the oldest dataset enabled us to get a good approximation of the overall yearly numbers of vehicle registrations (which contribute to the total fleet), this last dataset enabled us to build an exact picture of the range according to engine sizes and types, vehicle weight categories, technologies and fuels of which the emerging ones (hybrid and electric vehicles, CNG, LPG and ethanol). This dataset also enabled a good description of the two-wheeler fleet (loosely known in the past), of the bus and coach categories (articulated, standard-, midi- and articulated- or 3-axle buses and coaches), and the distinction between lorries and tractors.

Several difficulties should be noted for this topic:

– Engine sizes: for passenger cars (and two-wheelers), engine sizes are evolving with time. A clear trend toward a decrease in engine size is observed that should be taken into account in the fleet composition estimation, and in particular for the prospective issue (Figure 27.2). However, the statistics are quite weak to establish a durable trend.

– Particulate trap and air conditioning for light vehicles, as well as exhaust gas recirculation (EGR) and selective catalyst reduction (SGR) for heavy vehicles: these characteristics are currently not documented through the registration, however they are crucial regarding the determination of the particulate, nitrogen and greenhouse gas-related pollutions. Assumptions should thus be established using external statistics to determine their equipment rate.

– Light commercial vehicles should be distinguished according to their regulatory type (M and N1-I, N1-II, N1-III), which is assumed to be linked to the vehicle weight. However, a strong evolution of the weight distribution is observed (strong increase in the heavier vehicles), which could not be directly linked to a change of vehicle type.

– For heavy-duty vehicles, the determination of the weight classes is possible using the recent dataset. However, vehicles around 40 tonnes are not documented properly (all are set up to be 44 tonnes) and an ambiguity may remain regarding the exact weight information (authorized or total weight). Furthermore, information regarding “configurations” (i.e. lorry, lorry + trial, tractor, tractor + trial) and their exact rolling weight capacity (independent of the exact weight) is still missing for France. Specific surveys (*in situ* monitoring or on-board diary) should be conducted to document such statistics.

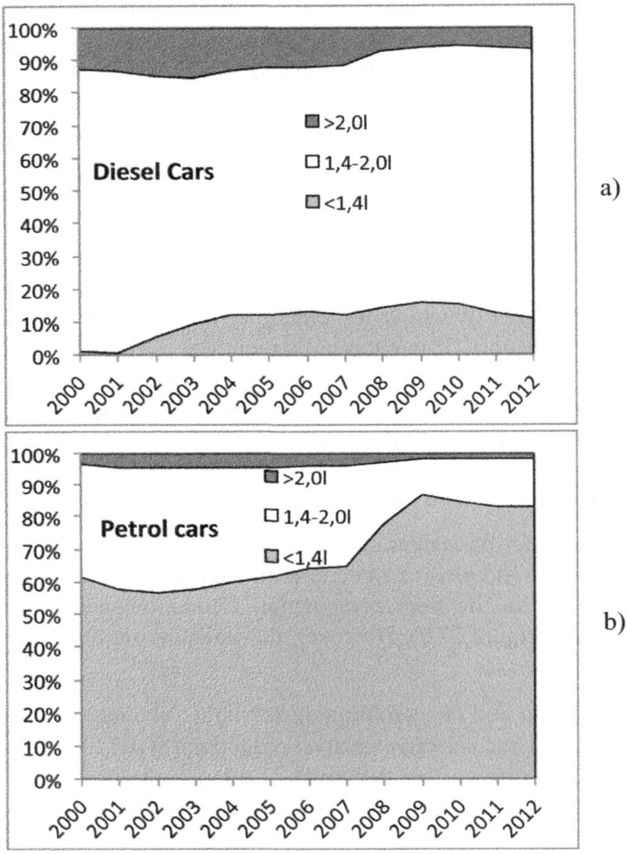

Figure 27.2. *Evolution of the vehicle distribution as regards engine capacity, over 2000–2012; Diesel a), Petrol b)*

– For buses and coaches, the classification envisaged in the emission calculation tools is not satisfying: indeed, those which are articulated are envisaged to be above 18 tonnes, while they are generally above 25 tonnes. In this case, the emission tools should evolve for a correct taking into account of the vehicles.

– The two-wheelers declination as regards engine size is rather envisaged through the engine power (good correspondence). The distinction between 2- and

4-strokes engines is not documented from the registration, and it was approached through sale statistics.

– Emission standard: this information has only been available for a few years in the vehicle registration file. It was approximated by the year of registration and a theoretical scheme of implementation of the European regulation in France. However, it is known that the transition between two regulations (i.e. Euro 4 – Euro 5) is progressive and the two (or even three) standards can coexist for a few years. Due to the lack of accurate statistics on this topic, it was approximated using rough estimations.

27.4.2. *Crucial statistics: survival probability and annual mileage*

As shown in Figure 27.1, survival probability and usage (annual mileage) functions are crucial data as they are combined successively with the yearly registration data to generate fleet compositions. These functions are generally derived from surveys monitoring the vehicles and their use (in France, a National mobility survey and observation in the big conurbations, and surveys dedicated to goods transport, light-duty vehicles, and two-wheelers). However, these functions are established for broad categories of vehicles and long periods.

In France, for instance, survival functions were established for diesel and petrol cars, and depend roughly on engine/vehicle size. Log-normal or Weibull forms were envisaged over large periods (before and after 1978, in 2–3 periods at best). This is clearly not sufficient as vehicle survival/demography strongly depends on their reliability, origins and technologies, as well as on sales which are influenced by the conjuncture as well as incentives, and the use of constant functions can lead to large divergences in the estimation of the total vehicle number. To solve this difficulty, the survival functions can be adjusted yearly, by considering external reference statistics derived from surveys (see, for instance, transport and mobility surveys by [SOE 13]). This process is highlighted in Figure 27.3, where diesel and petrol car fleets were estimated with and without adjustments of the survival to the estimation of the actual fleets through periodic survey. It can be observed that the drift can reach several billions of vehicles after a few years, and it was different for petrol and diesel cars.

A similar adjustment process was applied to mileage functions, which are exponential forms decreasing with vehicle age, and differing for petrol and diesel cars as a function of their engine size. Currently, a single set of functions is available, although mileage is supposed to evolve. In this case, the statistical annual estimation of the diesel and petrol car mileage derived from the "ParcAuto TNS Sofres" periodic survey [PAR 13] was used to adjust the mileage function. This adjustment is shown in Figure 27.3 (b) where we can also observe the variability of the annual mileage.

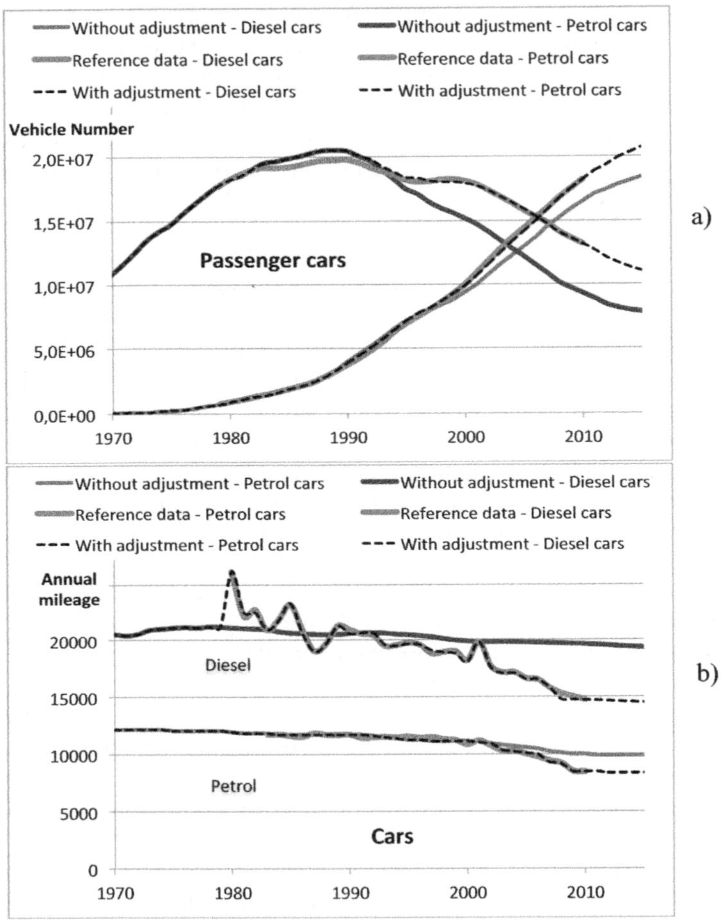

Figure 27.3. *Adjustments of the survival functions a) and the annual mileage functions b) and incidence on the determination of diesel and petrol car fleets and resulting mileage. For a color version of the figure, see www.iste.co.uk/jacob/energy.zip*

For the other vehicle categories, available statistics are poorer (the total vehicle numbers are available over a few decades or less and rarely declined in categories, the annual mileage is available for a few recent years), and the adjustment is more difficult, but it was realized using all the available statistics.

27.4.3. *Prospective trends*

Prospective trends constitute certainly the most crucial assumption regarding fleet and traffic. They concern on one hand vehicles (i.e. number of vehicles and their renewal) and mobility (i.e. traffic volume), and on the other hand the assumptions regarding the vehicle characteristics (size, fuels, etc.) and technology development (emerging or new ones). Generally, we consider that robust macroeconomic scenarios should be taken as guidelines to derive the necessary assumptions for the fleet and traffic estimation. For instance, assumptions regarding the daily and long-distance mobility should be considered to estimate the traffic travelled by cars, urban sprawl can be considered as an indicator for bus and coach traffic development, goods transport volumes, combined with modal distribution can be considered as indicators to set up the traffic by lorries, etc. Nevertheless, this exercise is difficult as numerous parameters are concerned (vehicle number and mileage, etc.).

In a first approach for France, the registration trends observed over 2002–2011 (except the year 2009 that was linked to the economic conjuncture) were considered and prolonged up to 2030: increase rates were thus 0.9%/year for cars, 0.5% for light commercial vehicles, 1 and 1.5% for lorries and articulated vehicles, and around 2% for buses and coaches. A stability of the annual mileage (by vehicle) was assumed.

Regarding passenger car (PC) technologies, we used the technological prospective study by [ADE 12] in cooperation with the car manufacturers. A clear decrease in the number of diesel and petrol cars is assumed (16 and 15% of the registrations in 2030 against 73/25% today), while electric vehicles reach 10%, diesel hybrids 9% petrol hybrids; 28% and plug-in hybrids; 22% of the annual registrations. Liquid petroleum gas (LPG) and ethanol remain at a very low level. For light commercial vehicles (LCVs), similar trends were adapted in lack of other data (high share of hybrids and electric in 2030). For lorries and buses, trends have been established on the basis of current years, plus assumptions regarding the development of compressed natural gas (CNG) (5, 10% of the registrations in 2030, respectively), hybrids (3 and 5%) and electric vehicles (3 and 5%). Electric two-wheelers were considered as 10% of the registrations in 2030.

Despite the significant figures above, the vehicle fleet should, however, include quite a limited number of these alternative technologies in 2030, due to its high inertia.

27.5. Comparing different estimations for France

Several estimations of the French vehicle fleets exist for the pollutant emission estimation, of which the national one (used for yearly national inventories, http://www.citepa.org/en/; this estimation, referred to hereafter as national, relies roughly on the same approach as described in section 27.3 with several simplified assumptions, and it is also adjusted according to the National Transports Accounts for the overall fleet number and annual traffic volume), and a fleet composition derived from European research works (FLEETS project, available on the COPERT/EMISIA website http://www.emisia.com/copert/Download.html, and referred to here as Emisia; this estimation is based on different survival and prospective data relying on European estimations). In the following, we compare these estimations with our current works. From Table 27.1, we observe that passenger cars are somewhat reliably estimated in number by the three works (3% of variation between the estimations). However, the estimation in total traffic diverges by 16%. For the other vehicle categories, the gaps are much higher, even omitting that Emisia estimation mistook buses and coaches.

2015	Cars	LCV	HCV	Bus	Coach
Total number of vehicles					
National estimation	32278447	5457637	562757	20920	62002
Emisia estimation for F	33065638	6134132	655164	(70384)	(22373)
Current works	32017544	6293469	515104	27566	68614
Differences (min-max) %	*3,3*	*15,3*	*27,2*	*31,8*	*10,7*
Traffic volume veh x km					
National estimation	418964	92955	27073	768	2276
Emisia estimation for F	458237	111793	33789	(992)	(2706)
Current works	396538	84951	16344	1067	2194
Differences (min-max) %	*15,6*	*31,6*	*106,7*	*39,0*	*23,3*

Table 27.1. *Comparison of different estimations of the French vehicles fleets (year 2015)*

Such differences on the absolute national figures highlight the strong variability that can be obtained for such data and its fragility (the overall weighted uncertainty on traffic volumes should be around 21%). Such differences led obviously to similar errors for the related pollutant estimations, as these ones – when using similar methods – are roughly directly proportional to the traffic volumes.

Table 27.2 compares in detail the car fleets regarding the diesel share (left) and the percentage of quite recent vehicles (Euro 5 and 6) in 2015. The French estimations are quit coherent regarding the share of the diesel vehicle in number, but tend to differ when considering traffic distribution (weighing by annual mileages). The gap is very high with the Emisia diesel estimation for France, and these differences should induce a strong divergence of pollution related to diesel (PM and NOx).

The share of recent vehicles is more coherent with our works, but national estimation underestimates it to a very large extent. Here, as recent vehicles (less polluting) replace older ones (more polluting), we can also expect big differences in the emission estimation due to these fleet composition divergences.

In all, these observed divergences demonstrate the need of harmonizing the approaches to elaborate the fleet composition, and to make explicit each underlying data and assumptions.

2015 – Cars	Diesel	Petrol	Others	%Euro5	%Euro6
Total number of vehicles					
National estimation	65,2	33,7	1,1	28,5	0,0
Emisia estimation for F	48,5	44,3	7,2	32,3	7,1
Current works	64,8	34,4	0,8	34,1	2,2
Differences (min-max) %	*34,4*	*31,5*		*19,6*	
Traffic volume (veh x km)					
National estimation	70,6	28,7	0,7	33,7	0,0
Emisia estimation for F	57,6	33,5	8,9	37,8	8,8
Current works	75,8	23,3	0,9	40,6	2,4
Differences (min-max) %	*31,6*	*43,8*		*20,5*	

Table 27.2. *Comparison of different estimations of the French vehicle fleets (year 2015) regarding diesel (left) and recent (right) vehicle share*

27.6. The French vehicle fleet composition: some highlights

Figure 27.4 shows the evolution of the resulting French fleet over 1980–2030, in number of vehicles and traffic volume. The total fleet increases from 32 million (M) vehicles in 1980 (19 M cars, 2.7 M LCVL, 580,000 lorries/tractors, 23,600 buses, 49,000 coaches, 9.4 M 2W) to 41.5 M in 2010, 42.5 M in 2015, 43.9 M in 2020, 47.9 M in 2030. The structure does not evolve significantly, a stronger increase in LCV is excepted. The overall increase rate over 2013–2030 is 0.8%.

Figure 27.5 highlights the evolution of the car fleet composition and the progressive emergence of alternative technologies according to the retained assumptions. Diesel cars increase from 58% of the total number and 71% of the traffic in 2010 up to 65/74% in 2020, and then decrease to 44/49% in 2030 (–1.3%/year over 2013–2030). Petrol cars are 41% in number and 29% of the traffic in 2010, 31/22% in 2020, 29/20% in 2030 (a decrease of about –0.45%/year). Hybrids (diesel, mainly petrol) should be 2.6/3.1% in 2020, 19/22% in 2030 (14/15% for the petrol ones), i.e. an increase rate of 25–30% per year. Electric and plug-in vehicles should represent 8% of the fleet and 10% of the traffic in 2030, with high but delayed increase rates (around 30 and 100%).

The EURO 4, 5 and 6 emission standards should represent 64% of the total car fleet and 72% of the traffic in 2015, 83 and 87% in 2020 and 85–86% in 2025. In 2030, more than 72% of the fleet and 67% of the traffic should be at least Euro 6, completed by alternative technologies (27% of the fleet and 32% of the traffic).

In conclusion, a strong evolution of the fleet is expected, with a significant share of alternative vehicles. However, the emission models are not ready for such an evolution as they do not deal with these technologies.

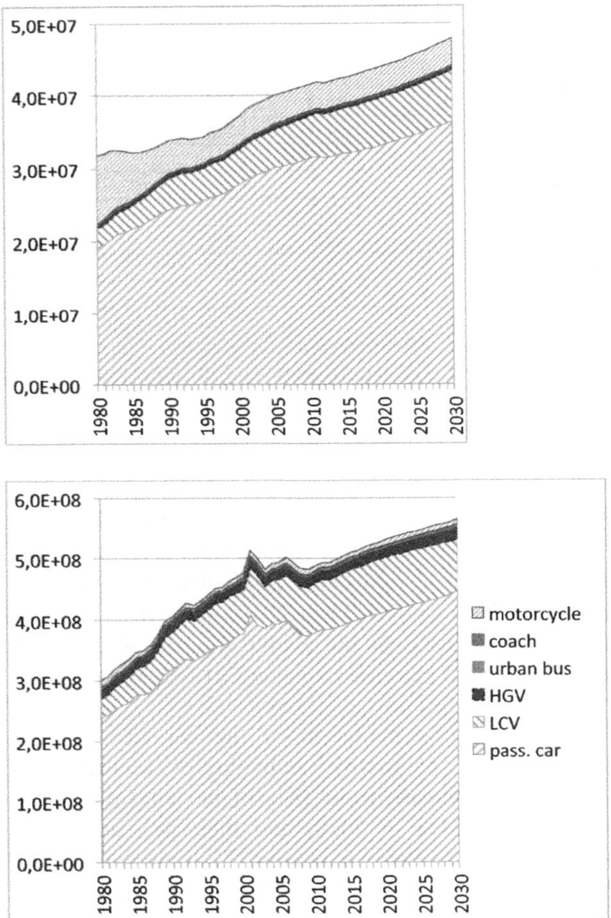

Figure 27.4. *Evolution of the fleet composition, in vehicle number (top) and traffic volume, 103 veh x km (bottom)*

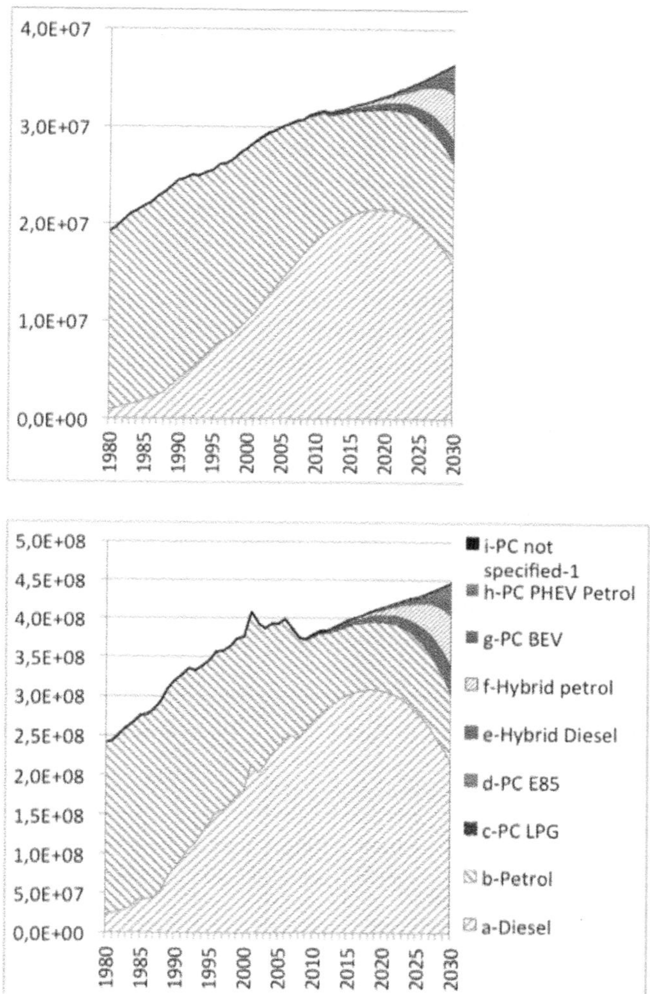

Figure 27.5. *Evolution of the car fleet composition regarding technologies, number a) and traffic volume, 103 veh x km b). For a color version of the figure, see www.iste.co.uk/jacob/energy.zip*

27.7. Local observations and vehicle fleet variability

As mentioned in the introduction to this chapter, most emission estimation studies concern local application (assessment of urban transport policies, etc.), but rely on national fleet estimation through lack of data. To improve the knowledge of local fleets, an experiment was conducted in the greater Paris area, using video to record the vehicle registration and get their technical characteristics through the vehicle registration database. Nine points were selected to cover the diversity of the areas, and to assess fleets depending on their socioeconomic characteristics. Depending on the sites, local (intra) traffic or vehicle-flows coming from a larger area (regional) were observed. These sites were monitored during 2 week-days or a full week. In all, 600,000 observations were recorded, and 375,000 vehicles were identified. Fleet compositions are described in Tables 27.3 and 27.4 and analyzed according to the area.

Compared to the national estimation, lower diesel rates were observed in the Ile de France area, and this rate is particularly low for the Hauts-de-Seine area (with higher living standards) where petrol and alternatives are more frequent.

When comparing the vehicle emission standards (and ages), the overall distribution is quite similar to the national estimation, but a higher share of Euro 5 is observed, meaning a higher renewal of the vehicles. On the other hand, we observe that Euro 4 and 5 represent 44–47% of the traffic in Seine-Saint-Denis, and between 58 and 68% in Hauts-de-Seine (and an intermediate rate in Paris). This high difference (around 18%) is compensated by older cars in the same proportion. These results demonstrate a high distortion of the fleet composition according to the area and also its high variability.

	Current works			Paris (75)		Hauts de seine (92)		Seine-St-Denis (93)	
	2012	*2013*	9 sites	Intra (4–5)	Ring (6)	Intra (7–8)	Regional (9)	Intra (1–2)	Regional (3)
Diesel	*73%*	*75%*	63	66	66	60	57	64	70
Petrol	*26%*	*25%*	35	31	32	37	41	34	28
Hybrid, electric, others	*0.4%*	*0.5%*	2.3	2.5	2	3.2	1.8	1.2	1.5

Table 27.3. *Comparison of the French car fleet (2012) with local video observation in the Paris areas*

	Current works		*Video obs.*	*Paris (75)*		*Hauts de seine (92)*		*Seine-St-Denis (93)*	
	2012	*2013*	9 sites	Intra (4–5)	Ring (6)	Intra (7–8)	Regional (9)	Intra (1–2)	Regional (3)
Pre-Euro	*2%*	*1%*	1	1	0.5	0.3	0.3	1	1
Euro-1	*4%*	*3%*	5	5	3	3	5	6	7
Euro-2	*10%*	*8%*	11	11	9	7	11	15	16
Euro-3	*28%*	*24%*	23	23	21	17	23	29	29
Euro-4	*38%*	*36%*	25	24	26	26	26	25	22
Euro-5	*18%*	*27%*	32	32	37	42	32	22	22
NR			3.5	4	3	4	3	3	3

Table 27.4. *Comparison of the French car fleet (2012 and 2013) regarding emission standards with local video observation in the Paris areas*

27.8. Other data required for emission calculation

Several other input data such as vehicle load and usage (trip lengths, parking conditions, etc.) are also required, for the emission calculation. These data are discussed in [AND 13]. Among them, the traffic conditions (speed, congestion, etc.) are important due to the sensitivity of the emission, and as they are often envisaged as a possible parameter to control the air pollution from transport. In the COPERT approach, average speed and mileage distribution according to speed conditions are considered. In HBEFA, traffic situations (combination of road configuration and traffic conditions) are proposed. In both cases, the main challenge is to elaborate the data/assumptions corresponding to a given case study. For local applications, speeds are generally estimated through (macroscopic) traffic models and the pertinence of the resulting figures is really questionable (underestimation of the congestion events, too high speed estimations, etc.). For aggregated estimation, average speeds are hardly derived from mobility/travel surveys. In these works, we have attempted to compile available data constituting each vehicle category: (1) traffic and average speed statistics and (2) traffic distribution regarding traffic situations. These works-in-progress, however, came up against several difficulties: (1) the lack of data (for most vehicles except PC) and (2) the heterogeneity, inconsistency and incomplete coverage of the speed and traffic data. Furthermore, it was observed that emission tools do not properly consider low speeds that are quite frequently recorded in urban situations (below 15 km/h or even 10 km/h).

27.9. Discussion and conclusions

The elaboration of fleet-related data is one of the main input data needed for pollutant emission estimation. It is of high complexity, as it relies on a variety of data and the setting-up of numerous crucial assumptions. These works and assumptions are, however, rarely made explicit through publications, and the resulting estimations can diverge due to unexplained reasons.

In a systematic approach, we have attempted to collect and compile the better available data for the elaboration of the fleet composition of the road transport in France. Several difficulties were identified to approach detailed technical characteristics of the vehicles. Survival and mileage functions are among the most crucial questions and require improvements. Prospective assumptions are also a key element to estimate emissions.

The comparison between different well-known estimations for France demonstrates their high variability and weakness and the need to conduct works to harmonize, improve and make explicit the underlying assumptions and calculation approaches. Such objectives are currently being searched for in France with a working group around that topic. The significant differences obtained between different fleet estimations also show the fragility of the related emission estimations.

Finally, as far as local application is concerned, video observations enabled the *in-situ* description of fleet. These data demonstrate that local fleets can diverge from national estimation, and significant contrasts can be observed according to the areas, demonstrating the need of a taking into account of local data for local applications.

The works presented in this frame concern the elaboration of the French data. However, it is likely that similar difficulties and weaknesses are encountered for the elaboration of fleet compositions from other European countries, also making questionable a correct assessment of the air pollution from transport and the effectiveness of traffic/transport-related measures and policies questionable. A systematic questioning and an experience feedback around these topics would be of high value.

27.10. Acknowledgments

These works were supported by the French Ademe and by the French department of Ecology and sustainable development MEDDE in the framework of PREDIT and PRIMEQUAL research projects.

27.11. Bibliography

[ADE 12] ADEME, Vision Ademe 2020-2030, Technologies/Carburants pour les véhicules particuliers neufs, Document de travail Ademe, p. 26, 2012.

[AND 13] ANDRÉ M., ROCHE A.L., BOURCIER L., Traffic statistics for the calculation of pollutant emissions from road transports in France, Statistiques de parcs et trafic pour le calcul des émissions de polluants des transports routiers pour la France, report, IFSTTAR-LTE, Bron (France), p. 132, 2013.

[GRE 96] GREENSPAN A., COHEN D., *Motor Vehicle Stocks, Scrappage, and Sales*, Washington D.C., Federal Reserve Board, 1996.

[HIR 07] HIROTA K., "Passenger car ownership estimation toward 2030 in Japan-BAU scenario with socio-economic factors-studies in regional science", *Japan Society of Regional Science*, vol. 37, no. 1, pp. 25–39, 2007.

[KOL 12] KOLLI Z., Car fleet dynamic of renewal and environmental impacts from exhaust emissions, PhD thesis, Univ. Paris, 2012.

[PAR 13] PARCAUTO, available at: http://www.tns-sofres.com/etudes-et-points-de-vue/parc-auto (24/09/2013), 2013.

[SOE 13] SOES, available at: http://www.statistiques.developpement-durable.gouv.fr/sources-methodes/theme-transports/1543.html, September 29, 2013.

[ZAC 95] ZACHARIADIS T., SAMARAS Z., ZIEROCK K.H., "Dynamic modeling of vehicle populations: an engineering approach for emissions calculations", *Technological Forecasting and Social Change*, no. 50, pp. 135–149, 1995.

PART 6

Noise and Vibration

Introduction to Part 6

Transports systems have been known for a long time to be responsible of significant noise (and vibration), and numerous traffic places in urban areas are exceeding noise limit levels, with well-identified annoyance- and health-effects for the local residents. Consequently, a large amount of research has already been conducted for assessing noise and health impacts, for developing mitigation and compensation measures, setting-up regulation, particularly for road traffic in dense areas, urban motorways, railways and high-speed train related noise and effects, as well as harmful effects due to airplanes and nuisance around airports.

The works reported here could appear as being far from these usual problems, but they certainly constitute an interesting set of works, enlarging and enriching to a certain extent the current ways of approaching acoustic impacts of transport.

Indeed, the first two chapters address shipping noise and their acoustic impacts on marine life. Methodology and indicators are addressed for assessing the underwater noise footprint on marine life due to shipping. A multidisciplinary and international team (AQUO project) has performed several experiments at sea on different ships, developed measurement procedures, and elaborated noise prediction models, while dedicated bio-acoustic studies were conducted on different representative marine species to derive noise acceptable limits. The need of coherent and agreed regulation and the importance of methodology to assess the efficiency of mitigation measures (technical or regulatory) were also pointed out.

The reduction of the noise emitted by vessels is considered as the most outstanding novelty and the most difficult challenge that the shipbuilding industry has ever faced, in a "confusing" context of regulation and scientific knowledge regarding the actual impacts. However, here the shipbuilding industry demonstrates

---

Introduction written by Michel ANDRÉ.

the feasibility of "silent vessels", – the "Ramón Margalef" and "Ángeles Alvariño" for the Oceanographic Spanish Institute – in compliance with high-level standards regarding underwater-radiated noise.

Two complementary chapters address noise and vibration impacts from railway trains. Vibrations from freight trains are a significant issue. The mitigation of low-frequency vibration – mainly due to track defects – is explored at the source, for which a long-lasting good geometry quality associated with wheel and rail maintenance is a key solution, whereas soil barrier systems could significantly reduce the vibration propagation.

In a context of reinforcement of noise emission limits for new and upgraded railway rolling stock, the certification procedure becomes an important issue. It is currently based on real tests such as pass-by noise measurements while there can be minor changes between different configurations. Works were thus conducted within the Acoutrain research project to set up the bases of a certification relying on a noise modeling platform, and that would enable the assessment while considering only the modified components. The proposed concepts are a first step toward a "virtual" certification.

Finally, the last chapter also offers an interesting contribution in the field of environmental acoustics, with the need to make electric or hybrid vehicles more noisy. Indeed, these silent vehicles can be seen an advantage in urban areas, but that makes them dangerous for pedestrians and visually impaired people. Through perception experiments, the eVADER research project studied the detectability, meaning and unpleasant characteristics of sounds, and proposed thus optimized additional sounds for these vehicles, with low levels of noise but high warning efficiency.

In all, this final part demonstrates the diversity of the environmental noise problem, and how it can be enriched by diverse approaches and different application cases.

28

AQUO European Collaborative Project – Development of Methods and Indicators for the Assessment of Shipping Noise Footprint on Underwater Environment and Impact on Marine Life

The goal of the EU project AQUO (www.aquo.eu) is to provide policy makers with practical guidelines, acceptable by shipyards and ship owners, in order to mitigate underwater noise footprint due to shipping, which causes disturbances to marine life. Thanks to a pluridisciplinary team of specialists from 8 different European countries, AQUO project results are expected to have significant impacts, contributing to and meeting the requirements of the MSFD. After a general presentation of the project, this chapter presents the methodology followed to assess an underwater noise footprint on marine life due to shipping, taking into account recent results from the project. In particular, focus will be given on the expression of the needs, and by proposed definitions of noise footprint and associated indicators. Special care will be given on the methodology to assess the efficiency of mitigation measures, which are of two types (technical or regulatory). The presentation will conclude with the ongoing research and its use to derive the expected final guidelines.

28.1. Introduction

Recent directives outline the need to mitigate the underwater noise footprint caused by shipping, to prevent negative consequences to marine life. The need is becoming

Chapter written by Christian AUDOLY, Céline ROUSSET, Thomas FOLEGOT, Michel ANDRÉ, Lanfranco BENEDETTI and Eric BAUDIN.

even more acute because of shipping traffic, which is steadily increasing. In that context, the final goal of AQUO project (Achieve QUieter Oceans by shipping noise footprint reduction – www.aquo.eu) is to provide policy makers with practical guidelines, acceptable by shipyards and ship owners. The list of solutions will be split into solutions regarding ship design (including propeller and cavitation noise), and solutions related to shipping control and regulation. This European Collaborative Project, which started in October 2012 for a duration of 3 years, is supported by the 7th Research Framework Program and coordinated with the topic "Oceans of Tomorrow". Thanks to a pluridisciplinary team of specialists and experts from different companies and institutions belonging to 8 different European countries, results from the AQUO project are expected to have significant impacts and contribute to meet the requirements of the MSFD (Marine Strategy Framework Directive).

After a general presentation of the project, this chapter presents in more detail the methodology followed to assess an underwater noise footprint on marine life due to shipping, taking into account the first results from the project. In particular, focus will be given on the following topics:

– expression of the needs, by a review of current regulations and standards, by identification and characterization of sensitive maritime areas: presence of fauna crossed with significant maritime traffic areas, and by the characterization of ship traffic and its expected evolution;

– proposed definitions for noise footprint, and associated indicators taking into account sensitivity of marine life, to be implemented in the noise footprint assessment tool developed in AQUO.

Special care will be given on the methodology to define and assess the noise footprint of shipping. The presentation will conclude with the ongoing research and its use to derive the expected final guidelines.

AQUO project results are available on its website www.aquo.eu.

28.2. Overview of AQUO project

AQUO is split into 5 technical work packages (WP), interconnected one to the other as shown in Figure 28.1.

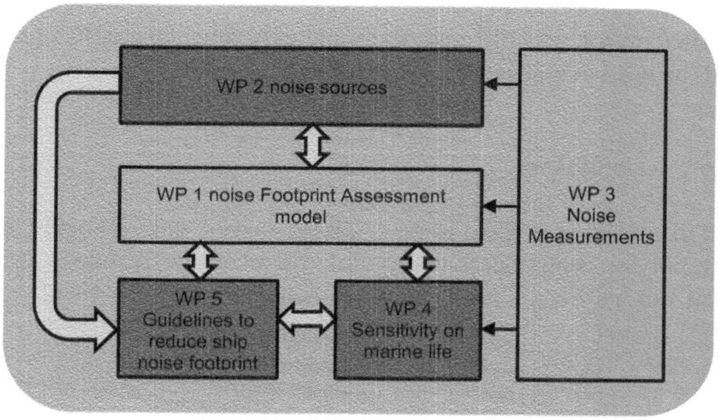

Figure 28.1. *AQUO project technical workpackages*

The project is supported by relevant methods and tools, which are used to assess the effectiveness of noise mitigation measures in order to select the most appropriate ones:

– WP1: a noise footprint assessment tool, derived from Quonops© [FOL 09], an existing operational underwater noise prediction system, connectable with Automatic Identification System (AIS) shipping data. The tool will be adapted to the problem considered and validated by comparison with *in situ* measurements at sea. The AQUO model will use the live stream of acoustic data provided by autonomous buoys deployed in selected areas and processed under the LIDO acoustic architecture [AND 11];

– WP2: computer methods will be developed and scale model experiments will be carried out to predict radiated noise from ship propellers, including cavitation effects and interaction with ship hull. These predictive techniques will be validated by comparison with measurements;

– WP3: to support the analysis, several experiments at sea will be conducted on different ships, including commercial vessels. Improved measurement procedures for ship underwater radiated noise are also proposed;

– WP4: dedicated bio-acoustic studies will be conducted on different marine species, representative of European maritime areas, aiming at deriving criteria regarding shipping underwater noise acceptable limits;

– WP5: in this synthesis work package, different mitigation measures for the underwater noise impact of shipping on marine life will be proposed. These can be either technical (reduction of the noise sources themselves, which are the ships), or related to traffic control. The different tools and experimental results obtained in the other WP will be used to assess their effectiveness.

28.3. Needs and policies

The topic of underwater noise impacts on the marine environment can logically be seen as one of the components that pollute the oceans. Therefore, the first level for addressing this particular problem is to include it in the general environment policy.

28.3.1. *Overview of the existing regulatory framework*

The observatory's location is the first point to be addressed in order to build a relevant and applicable tool for the legislator. Depending on its location, some laws, rules or other local initiatives established for the protection of the marine fauna can be applied:

– United Nations Convention on the Law of the Sea (UNCLOS): the convention defines the rights and responsibilities of nations in their use of the world's oceans. Different jurisdictional marine waters limits are defined in this convention (Territorial Seas, Contiguous Zone, Exclusive Economic Zone, Continental Shelf and High Seas);

– Marine Protected Areas (MPA): regions in which human activity has been placed under some restrictions in the interest of conserving the natural environment, its surrounding waters and the occupant ecosystems, and any cultural or historical resources that may require preservation or management. Typical restrictions in MPAs include ones on fisheries, oil and gas mining, and access for tourism. Other restrictions may include the use of ultrasonic devices like sonar, other development and construction. However there are also zones in which even navigation is at times regulated, sometimes as strictly as ban, either as a preventive measure or to avoid direct disturbance to certain species;

– European Framework: after the adoption of Directive 79/409/EEC on the conservation of wild birds and the Directive 92/43/EEC on the conservation of natural habitats of wild fauna and flora, an important step was made in 2008: the EU adopted Directive 2008/56/EC, establishing a framework for community action in the field of marine environmental policy. This new legal advance was called the Marine Strategy Framework Directive (MSFD). In Article 10, the Directive drew the assessment process on the basis of establishing a comprehensive set of environmental targets and associated indicators. The goal through this setting of criteria and methodological standards is to allow a consistent approach in evaluating whether the Good Environmental Status has been achieved. Two indicators were retained regarding “underwater noise and other forms of energy”: low and mid-frequency impulsive sounds from 10 Hz to 10 kHz (descriptor 11.1), and continuous low frequency sound third octave bands centered at 63 and 125 Hz (descriptor 11.2), this latter directly addressing the issue of shipping noise.

28.3.2. *Actions conducted by organizations involved in the maritime sector*

IMO

The International Maritime Organization (IMO) is the United Nations specialized agency with responsibility for the safety and security of shipping and the prevention of marine pollution by ships. In 2012, the IMO (Design and Equipment DE57 Committee) issued a report, which is a first step toward the reduction of noise from commercial shipping and its adverse impacts on marine life [IMO 12]. The Non-Mandatory Guidelines, applying to Commercial Shipping, are intended to provide general advice about reduction of underwater noise to designers, shipbuilders and ship operators. The core of the Guidelines is articulated along three main aspects:

– hydrodynamic noise (propeller, hull, appendages, etc.);

– on board machinery noise with a special attention to the propulsion machinery, HVAC and pipes;

– operational and maintenance considerations focusing on propeller and hull cleaning mainly.

ISO

The role of ISO is standardization at international level. ISO is organized in Technical Committees (TC). The scope of ISO/TC43/SC 3 “Underwater acoustics” is: Standardization in the field of underwater acoustics (including natural, biological, and anthropogenic sound), including methods of measurement and assessment of the generation, propagation and reception of underwater sound and its reflection and scattering in the underwater environment including the seabed, sea surface and biological organisms, and also including all aspects of the effects of underwater sound on the underwater environment, humans and aquatic life. ISO/TC8/SC2 “Ship and marine technology – Marine environment protection” is also contributing significantly to this effort.

Classification societies

Their purpose is to provide classification, statutory services and assistance to the maritime industry and regulatory bodies as regards maritime safety and pollution prevention, based on the accumulation of maritime knowledge and technology.

Currently, acoustics issues linked to health and comfort on board is the main matter of concern for class societies. Underwater Radiated Noise (URN) services and notation are not yet widely used by them. DNV offers a notation concerning URN and Germanischer Lloyd (GL) provides some services (calculation or consultancy).

28.3.3. *Ship traffic and its evolution*

Global maritime commercial trade has exploded during the last 50 years. The European seas catch a large part of this maritime traffic. This evolution is due to the increase of the number of vessels on one side (for tankers and bulk carriers) and the size of others vessels (for container ship). Finally, it has resulted in the increase of noise emitted by the maritime traffic.

A way to assess the maritime traffic is to monitor AIS data (Figure 28.2(a)). Since the beginning of 2004, a regulation requires AIS to be fitted aboard all ships of 300 gross tonnages and upward engaged on international voyages, cargo ships of 500 gross tonnages and upward not engaged on international voyages and all passenger ships irrespective of size.

Growth and dynamics of the global economy are, to a large extent, driven by changes in the world's demography, productivity, relative welfare, demand and availability of energy, of raw materials and of food. As the population increases, the international waterborne trade is likely to also increase which will result in a noisier ocean, as shown by several studies by IMO and others (Figure 28.2(b)).

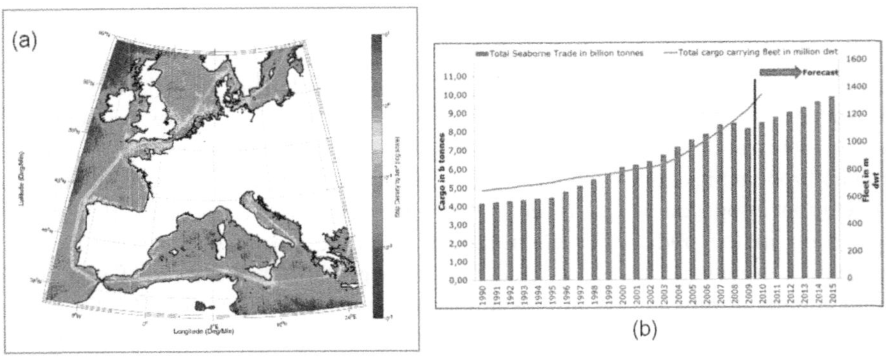

Figure 28.2. *a) Density map of ship traffic based on AIS data from mid-December 2012 to mid-January 2013. The color scale is logarithmic and represents the average number of ships present in a 1km2 cell at any time (source: Quiet-Oceans) based on Marine Traffic website; b) Seaborne Trade projection for 2010–2015 from 1990 to 2009 data (CESA, 2012). For a color version of the figure, see www.iste.co.uk/jacob/energy.zip*

28.3.4. *Spatial distribution of marine species*

The Habitat Directive requires, for each species listed into its Annex IV (Species of Community Interest) that each Member State reports on: maps of presence and estimate of conservation status. It is possible to find the presence probability maps of some marine species in EU waters. It is particularly useful to superimpose the maritime routes and the spatial distribution of protected marine species to define sensitive areas to noise. This allows identifying qualitatively areas where:

– there is a significant anthropogenic pressure from shipping, together with a known and reported habitat;

– significant anthropogenic pressure associated to shipping is juxtaposed to a known and reported habitat;

– for instance, Figure 28.3 shows the superposition of ship traffic density and known habitat for two mammal species. These maps show clearly the interaction between shipping routes and reported habitat for these species, in particular the west coast of Portugal, the north of the Bay of Biscay, and the English Channel.

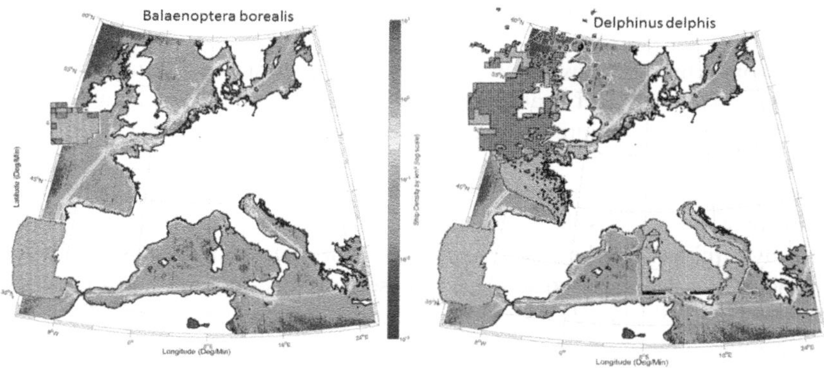

Figure 28.3. *Superimposition of density map of ship traffic and reported habitat of two marine mammal species (balaenoptera borealis and delphinus delphis). For a color version of the figure, see www.iste.co.uk/jacob/energy.zip*

28.4. Noise footprint characterization

Some metrics or indicators are being implemented in the "noise footprint assessment model" to be used to determine the efficiency of different noise mitigation measures under consideration. As the idea of underwater noise footprint is emergent, and several options are available, a specific effort has been conducted in the AQUO project to clarify the definitions.

28.4.1. *Result from a bibliographic analysis*

A literature review has been carried out, focusing on papers referring to the underwater noise footprint and related topics. It appears that the terminology "underwater noise footprint" appears increasingly in literature. However the definition is generally vague and differs strongly between one author and another. For example, the underwater noise footprint is defined either as:

– the area where underwater noise produced by one particular ship exceeds a given limit;

– the percentage of maritime area of interest where the average noise cumulated over a long period, exceeds a given limit, following in that sense "continuous low frequency descriptor" defined in the scope of the European Framework (see section 3.1);

– the total noise power of some activity (e.g. shipping) above the ambient noise, including other human activity.

We can note also that some authors take into account the sensitivity of marine life to noise in the definition of the noise footprint.

It is worthwhile to draw some parallels with airborne environmental acoustics issues. As a matter of fact, transport noise (either air, rail or road traffic), and its impact on human population, has been studied for several decades.

Regarding air traffic, the main issue is noise in the vicinity of airports. For that reason, some regulations have been enforced, requiring the reduction of noise emitted by airplanes by their design, and by some specific measures, e.g. adapting flight routes or schedules. Noise maps representing the "footprint" associated to air traffic are often displayed, taking into account the frequency sensitivity of the human ear and some averaging procedures.

28.4.2. *Bio-acoustic criteria*

A recent report from the 7th Framework Program project, SILENV [SIL 10], has given a synthesis of literature on the effects of anthropogenic noise pollution on cetaceans. Regarding ship noise, as the signal can be assumed to be stationary, the two main effects are masking and habitat displacement. For higher underwater noise levels, more severe effects are, increasingly, TTS (Temporary Threshold Shift), PTS (Permanent Threshold Shift), injury and death often associated with standings (Figure 28.4(a)). For lower noise levels, behavior changes can also be observed, but are more difficult to assess. For the present study, the most relevant issues were identified to be masking and population displacement effects, which are the consequences of an increased background noise level and could affect some vital functions in mammals, fish and invertebrates (communication, target detection, environment sensing, etc.) (see Figure 28.4(b)). Clearly, some parallels can be drawn with the sonar equation used for the assessment of detection ranges of sonar systems, used mainly in underwater naval warfare.

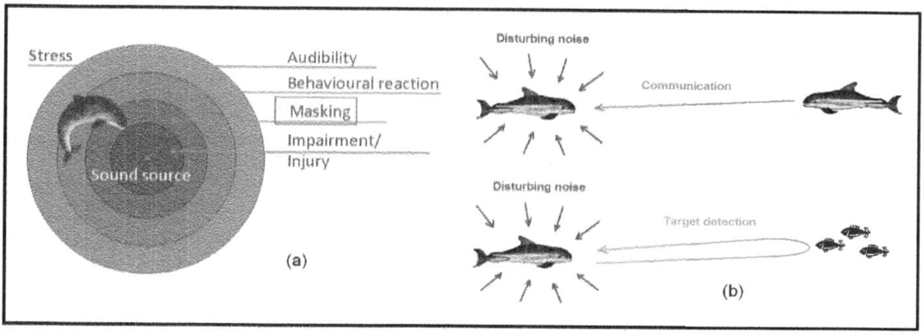

Figure 28.4. *a) The different effects of noise on a marine mammal; b) schematic representation of the masking effect*

Some data from the literature can be found regarding the hearing thresholds and sensitivity curves to underwater noise of different marine species. In particular, M-weighing curves, which can be considered as the transposition of the dB(A) level to marine mammals, have been defined and can be useful for this study.

However, bioacoustics applied to marine species is a hot scientific topic which is evolving quickly. Contrarily to noise limits associated to high impulsive noise, it is difficult to achieve consensus among the scientific community on limit criteria for continuous noise, and these are due to evolve in the near future. This is the reason why one part of the AQUO project, the WP4, is dedicated to specific bioacoustics studies on different marine species (fish, marine mammals and cephalopods).

28.4.3. *Proposed definitions*

There should not be confusion between:

– the radiated noise level of a ship, which is a physical, measurable quantity, used to characterize a ship as a source of underwater noise;

– the underwater noise map due to shipping, which is a physical quantity used to determine underwater noise due to shipping level in a maritime area. It can be spatially and/or temporally averaged;

– the underwater noise impact on marine life, which is the main issue in the present study. The assessment of the noise footprint must obviously take into account the receiver(s), through the sensitivity of different marine species to underwater noise.

This results in three levels, schematically represented on Figure 28.5.

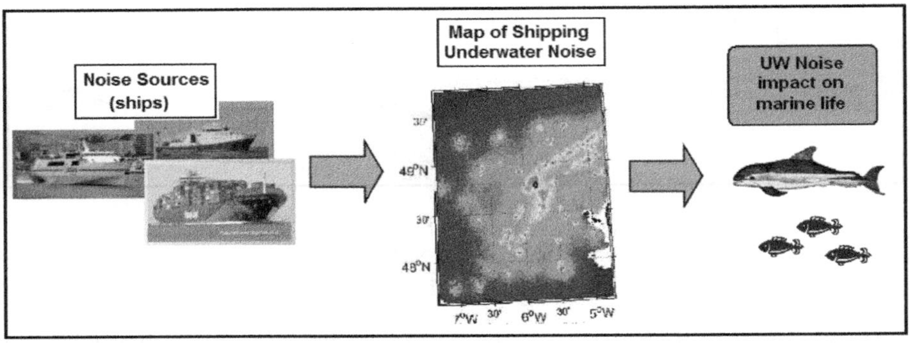

Figure 28.5. *Three levels of characterization of underwater noise footprint due to shipping. For a color version of the figure, see www.iste.co.uk/jacob/energy.zip*

After analyzing the different elements available, the definitions proposed by the AQUO team are the following:

– signature or underwater radiated noise source level: power spectral density of a sound source back-propagated to an arbitrary distance of 1 meter, possibly as a function of azimuth, bearing. It is a physical, measurable quantity, which can be used to characterize a ship as a source of underwater noise;

– noise map: spatial distribution of underwater sound in a maritime area of interest. A noise map, which represents a physical quantity, is obtained either from measurements, either by simulation by superimposing one or more noise sources in the area of interest. It can be either (but not limited to):

- determined in a given frequency band (for example a 1/3 octave band),

- integrated in a wider band domain, with weighing coefficients if needed (for example M-weighing coefficients for different functional mammal groups),

- in the form of a spectral level at a given frequency,

- deterministic or statistical.

– noise footprint: the representation of the noise level arising from maritime activities that affects a portion of the sea. It includes the description of the noise sources and the propagation of the sound in the ocean environment that can be represented as a noise map, without frequency weighing. It can be used to assess the effect or impact of anthropogenic sound on marine life, using suitable indicators;

– indicator: combination of metrics that allows the partial or exhaustive evaluation of a descriptor. In the present study, the indicators must be relevant to both the noise anthropogenic pressure and the biological effect of underwater noise on marine life.

Obviously, for the AQUO project, noise maps are not sufficient. Indicators of the impact of noise on marine life, obtained after post-processing of noise maps, are needed to quantitatively compare the efficiency of one noise mitigation measure to the others. In the present case, some relevant indicators can be defined for example as the percentage of a maritime area of interest where time-averaged underwater noise due to shipping exceeds some limit. Different noise limits can be defined, according to the current knowledge, referring either to low frequency sound, either to masking effect.

28.4.4. *Test case to illustrate the idea of noise footprints*

A test case has been implemented for a physical noise footprint in order to illustrate the potential use of this powerful concept. The scenario is taking place in the Baltic Sea based on real maritime traffic data. Shipping generates broadband continuous sound emissions covering infrasonic, audible and ultrasonic frequency bands, from a few tens of Hertz (Hz) to several tens of kHz. A description of shipping activity in the Baltic Sea has been compiled in order to account for sources of anthropogenic noise associated with vessel movements in the study area. The primary source of data used is the Automatic Identification System (AIS). Individual underwater noise signatures were assigned to each vessel class (given by the AIS message), defined in terms of sound pressure and frequency. Vessel noise signatures were assigned according to reference values obtained from the literature.

The maximum received level has been modeled using Quonops©, Quiet-Oceans' operational ocean noise prediction system at a given day and time. The distribution of ships and types of ship were taken from a "frozen" situation given by the AIS data the 17th May 2013 at 1400 UTC. The resulting noise field in the 125 Hz third-octave (to be relevant in the framework of the MSFD) is shown on Figure 28.6(a). Figure 28.6(b) represents the noise footprint of the cargo-related activities in the Baltic Sea, defined as the emergence of the noise generated by all tanker vessels above the noise generated by all the other shipping activity (fishing, tanker, tug, pleasure, etc.). For that particular situation, we can see that tanker traffic is the main source of underwater noise in some areas.

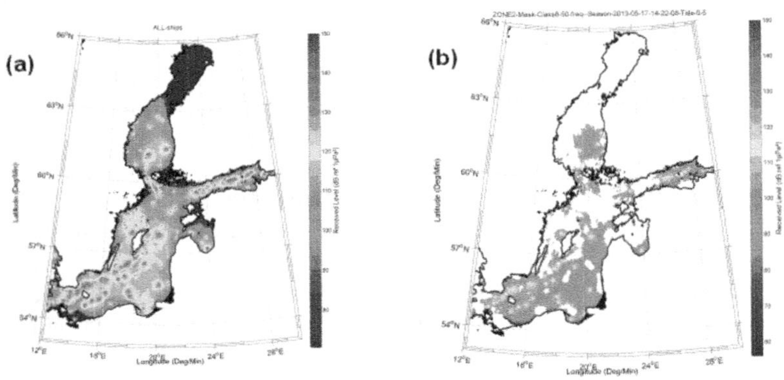

Figure 28.6. *a) Maximum of noise at a given time in the Baltic Sea, resulting from the contribution of all types of ship; b) noise of the tanker-related activity in the Baltic Sea (emergence with respects to other ship types). For a color version of the figure, see www.iste.co.uk/jacob/energy.zip*

28.5. Methodology for the derivation of noise mitigation guidelines

At the end of the project, the main deliverable will be practical guidelines documents providing support to policy makers, to meet the requirements of the MSFD. The purpose is to protect marine life against excessive noise produced by shipping. The design recommendations for the reduction of underwater radiated noise of ships and measures for the mitigation of impacts of shipping noise on the marine environment must be practical and economically feasible (for instance, without affecting fuel efficiency). If not, the different stakeholders of the maritime activity sector would not support this initiative. Indeed, it would not be suitable to impose a general underwater noise limit for all vessels, in all areas, in all cases. The issue is much more complex, and a more global approach must be followed. The approach followed in the AQUO project to build consistent practical guidelines is represented schematically in Figure 28.7.

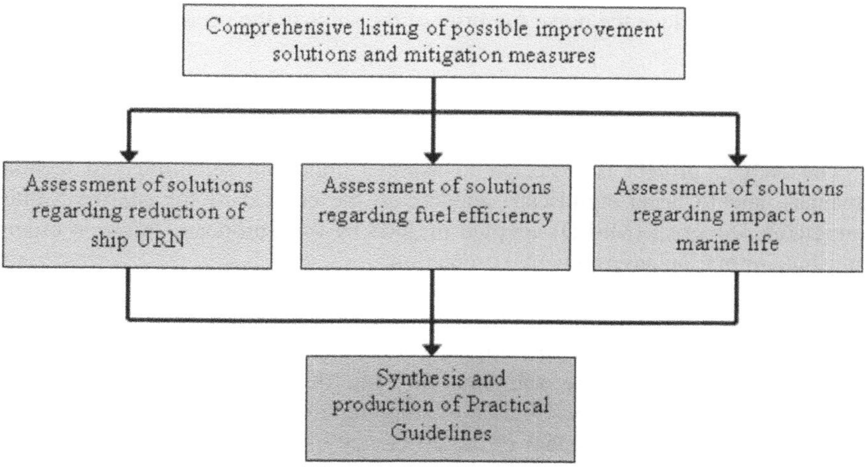

Figure 28.7. *Methodology for noise mitigation measures assessment*

Based on the experience of the different partners in the AQUO consortium, a list of solutions will be first established, split into:

– solutions regarding ship design, including propeller and cavitation effects. They would consist of design options for new ships (to be considered at the earliest stages of the design phase) and possibly improvement of existing ships;

– solutions related to shipping control and regulations (for example speed reduction).

In a second stage, the efficiency of these solutions will be assessed using different predicting tools and results from experiments at sea regarding three aspects: reduction of ship underwater radiated noise, effective fuel consumption and impact on marine life. As far as possible, attention will be paid to the economical aspect of the solutions and their implementation. The noise footprint model introduced in WP1 will be used to assess the efficiency of solutions regarding reducing the noise impact on marine life.

A crossed synthesis of the 3 previous matrices will issue underwater radiated noise control guidelines. The "practical guide" will provide not only the list of solutions technically and experimentally supported but also the right and consistent "methodology" to be applied from the early stage of the design to operational best practices in order to develop ships with low underwater noise signatures and consequently to reduce the noise footprint of the ships in the oceans.

28.6. Conclusions and ongoing actions

The AQUO project is expected to achieve important scientific and technical advances, and to bring significant impacts, in the context of the Marine Strategy Framework Directive (MSFD) adopted in 2008 by the European Union, in order to contribute to marine environmental protection.

The ultimate goal of the AQUO project is to provide policy makers with practical guidelines in order to mitigate underwater noise impacts of shipping on marine life. It will propose solutions for ship design, including propeller and cavitation noise, and solutions related to shipping control and regulation. As a first step to determine the assessment of the efficiency of the solutions, a proposition of the noise footprint has been provided in this chapter.

Main ongoing actions are the following:

– to develop and validate a shipping underwater noise footprint assessment tool related to AIS (Automatic Information System) data;

– to improve and/or validate models and methods to predict radiated noise from ship propellers, taking into account cavitation phenomena and interaction with hull effects;

– to propose a standard for ship underwater radiated noise measurement method;

– to develop new vibro-acoustic measurements tools and methods applied to real experiments at sea;

– to establish criteria and good practices for noise protection on marine life, based on analysis of available data and specific bio-acoustic experiments on representative marine species;

– to establish a list of design improvement solutions to reduce ship underwater radiated noise, without reducing fuel efficiency of the ship;

– to establish practical guidelines to reduce shipping noise for a quieter ocean, providing responses to the needs of policy makers.

Most technical results will be available to general public and to the scientific community through the publication of reports (available on the AQUO website at www.aquo.eu).

28.7. Acknowledgments

The AQUO addresses to SST.2012.1.1-1. "Assessment and mitigation of noise impacts of the maritime transport on the marine environment" of the call FP7-SST-2012-RTD-1 of the TRANSPORT theme in the Sustainable Surface Transport, coordinated topic "Oceans of Tomorrow", grant agreement no. 314227. The content of this chapter does not reflect the official opinion of the European Union. Responsibility for the information and views expressed in the chapter lies entirely with the authors.

28.8. Bibliography

[AND 11] ANDRÉ M., VAN DER SCHAAR M., ZAUGG S. *et al.*, "Listening to the deep: live monitoring of ocean noise and cetacean acoustic signals", *Marine Pollution Bulletin*, vol. 63, pp. 18–26, 2011.

[CES 12] CESA, Market Monitoring Report Q4 2012, Community of European Shipyards Associations, 2012.

[FOL 09] FOLEGOT T., Method for monitoring, predicting and reducing the level of acoustic energy of a plurality of sources in an aquatic environment, and method for monitoring, predicting and reducing the risk of noise annoyance for marine species, European Patent Register, WO2010FR52017, 2009.

[IMO 12] IMO, Provisions for reduction of noise from commercial shipping and its adverse impacts on marine life, International Maritime Organization, Sub-committee on ship design and equipment DE 57/7, 2012.

[SIL 10] SILENV, Underwater radiated noise, SILENV Collaborative Project FP7 no. 234182, Deliverable D1.3, 2010.

29

Reduction of the Underwater Radiated Noise by Ships: New Shipbuilding Challenge. The Vessels "Ramón Margalef" and "Ángeles Alvariño" as Technological References of How to Build Silent Vessels

The abatement of the noise radiated by all kind of vessels, but especially in terms of underwater radiated noise, because of its impact on marine life, has become the most outstanding novelty and the most difficult challenge that the shipbuilding industry has ever faced. The industry is hence looking for new solutions in order to comply with the new directives and requirements that have been recently introduced and promoted by the EU, marine institutions and the scientific community. This new topic, already studied by navies because of their strategic requirements, is now in the limelight and all the different aspects are being studied and discussed between the different aforementioned actors. In this new trend, the appearance of some class notations with "different limits", assertions such as that the "propeller" is the major noise source and the absence of consensus among the biological community about what the limits makes be, it difficult to define the technological steps that should be followed. In this "confusing scenario" this paper is a clear example of how the Spanish shipbuilding industry is moving ahead to reduce the impact of the new ships. The dynamic and acoustic design, developed by the authors, of the "Ramón Margalef" and "Ángeles Alvariño" fishing research vessels, for the Oceanographic Spanish Institute and the experimental results obtained, in full compliance with ICES No. 209 Underwater Radiated Noise requirements, makes them reference vessels for the construction of "silent vessels".

Chapter written by Publio BELTRÁN PALOMO, Raul SALINAS MULLOR and Alfonso MORENO RODRÍGUEZ.

29.1. Introduction

Within the framework of the European Commission's (EC) environmental policies and other maritime institutions such as the International Maritime Organization (IMO), all the targets defined are focused on combating and reducing the current "environmental impact" of all type of vessels that make up the current European fleet. To fulfil this main target, special EC R&D projects have been promoted and launched to reduce the emissions as well as the noise & vibration (N&V) of vessels and hence their environmental impacts.

A common conclusion brought out from these projects[1] is that the N&V-Full Signature (N&V-FS) of the vessel is required to properly assess environmental impact. This indicator is made up by the N&V on board, Noise Radiated to Harbour (NRH) and Underwater Radiated Noise (URN), the URN and NRH being the most outstanding, and in the case of URN, the newest and greatest challenge to be faced.

In fact very little data from "civilian vessels" regarding this aspect is available apart from some interesting results of URN measurements in the publications [ADV 00, KIP 02, KIP 04a, KIP 04b]. These publications are very interesting because several URN signatures of various commercial vessels (mainly cruise ships) are presented. These URN levels come with detailed descriptions of the measurement conditions and of the ships' characteristics, which allows for the analysis of the data. This lack of URN data is due to there being no URN requirements in the vessel contract specification. The only exceptions are the fishing research vessels due to their performance requirements (they must not disturb fish species during their studies).

This undesirable trend, related to the absence of URN experimental data within the current fleet, has been totally confirmed by the analysis carried out by the SILENV project [BEL 09]. In the SILENV-N&V database generated from a total of existing 151 vessels, only of them (3.1% of the total sample) have included specific URN requirements in their contract specifications. To enrich this poor scenario, dedicated on-site measurements activities were carried out to obtain the URN signature of 10 additional vessels (of different types), and therefore more available experimental data was collected.

Another important point to consider is the current measurement procedures. In a first attempt to standardize these, ANSI-ASA launched its standard "Quantities and Procedures for Description and Measurements of Underwater Sound from Ships" [ANS 09]. This URN measurement standard is thought to carry out measurements of underwater sound pressure in the far field and deep waters (to avoid reflections).

1 Breakthrough in European Ships and Shipbuilding Technologies (BESST) and Ships Oriented Innovative Solutions to Reduce Noise & Vibration (SILENV), projects funded by the EC 7th Framework Project. Information available at www.transport-research.info.

The procedure selected has no inherent limitation with regard to a vessel's size. Usually the results include the noise pressure levels in third-octave in decibels referenced to 1μPa at 1 m from both hull sides. Band levels are associated with the mid-frequency of the third-octave band. For many purposes, it is convenient to use the power spectral density per unit bandwidth (spectrum level).

An important parameter is transmission loss (TL), used to calculate the noise radiated at 1 m from the ship. The "real" transmission loss depends on many parameters (sound frequency, salinity, sea state, sea bottom, water depth, local sound velocity, etc.) and to obtain a real and precise value would be very difficult. That is the reason why, in measurement standards, simplified methods are used to assess TL. Usually the formula TL = 20Log D_{CPA} (D_{CPA} being Distance of Closest Point of Approach) is applied for all the frequency bands (ANSI-ASA standard), which produce considerable uncertainties in the results of the measurement, especially for low frequencies. Finally, ANSI-ASA does not define targeted limits.

Recently, DNV published a new standard for the measurement of the underwater noise emitted by ships [DNV 10]. This "SILENT" standard specifies targeted limits and associated measurement procedures for shallow waters. The method used to define these limits is not known. The different targeted limits specified in this new standard are different, depending on the type of vessel considered. According to this class notation, the Society, based on an operational speed profile for typical hydro-acoustics operations submitted to it, will decide the "speed(s) at which the vessel shall be tested". It is important to keep in mind that based on the practical expertise of the authors, this parameter becomes essential for the suitable dynamic-acoustic design of silent vessels, and it should be defined and agreed at the beginning of the project. Likewise, in this class notation, the TL is calculated according to the following formula: TL= 18.Log D_{CPA} – 5dB.

Finally, since the 1990s, the URN target limits defined by the International Council Exploitation of Seas (ICES) no. 209 Regulation [ICE 95] became a widespread standard used within fishing research vessels, especially in the more modern ones. This standard specifies limits for 1/3 octave spectrum levels, at a speed of 11 knots, as it can be seen in Figure 29.1.

Within the current regulation framework, several operative experiences have been compiled. The first are related to the URN on-site measurements data processing and data presentation. In Figure 29.2 the usual format for the presentation of URN spectra: pressure (ref 1 μPa) @ 1 m- 1/3 octave band (1 Hz) against frequency (Hz) is reported.

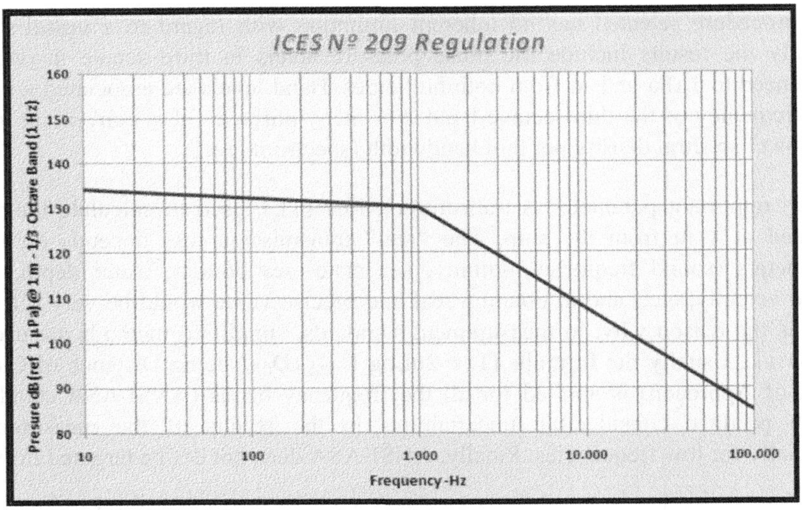

Figure 29.1. *ICES No. 209 Regulation. URN target limits at 11 knots*

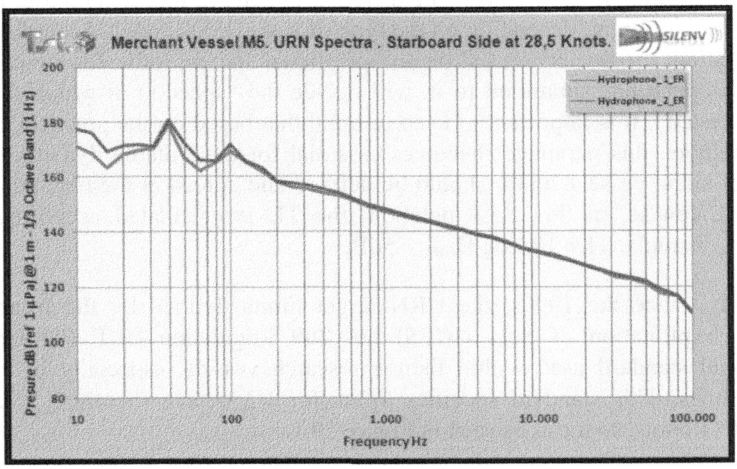

Figure 29.2. *Typical URN 1/3 octave band spectra for merchant vessel at 28.5 knots. For a color version of the figure, see www.iste.co.uk/jacob/energy.zip*

In parallel, an intense debate about which the main URN sources affecting the URN signatures of the vessels is ongoing. Traditionally, the propeller is considered to be the principal cause of the URN signature of the vessels with no experimental support. These kind of assertions are made based on "old figures" and outdated information, where the propeller, of course, appeared as one of the main root causes of vibration on board. Based on the authors' point of view, with 36 years of experience as N&V consultants in the marine sector, with these "assertions" the shipbuilding industry is getting confused and cannot reach an understanding on what has to be done to move ahead on the improvement of the current situation.

To shed light on this matter, in the framework of the SILENV project, on-site measurements activities were carried out to obtain the URN signature with sufficient frequency accuracy to identify the main URN sources (machinery and propeller) in the spectrum (Figure 29.3).

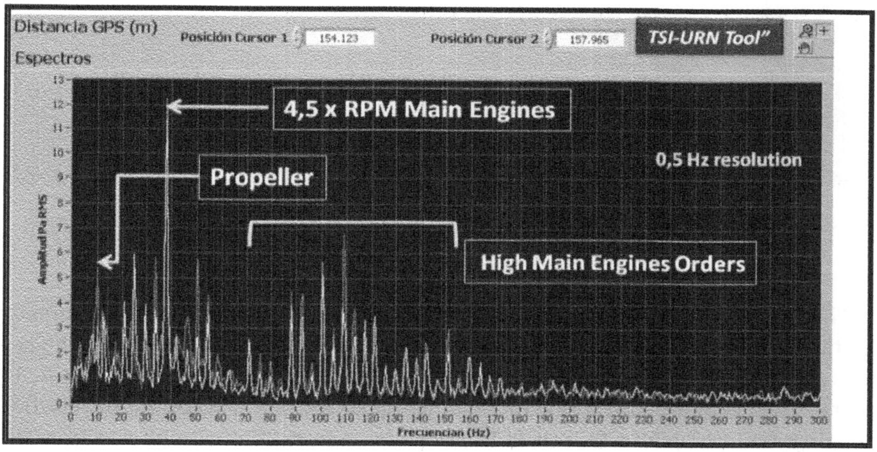

Figure 29.3. *Narrow band analysis for accurate identification of URN contributors. For a color version of the figure, see www.iste.co.uk/jacob/energy.zip*

These experimental results have identified the machinery and the propeller (Figure 29.4), in this order, as the main contributors to the URN signature in a significant sample of vessels of the current European fleet. This approximation is sound enough with the experiences that were documented at Glacier Bay, mainly related to passenger vessels [ADV 00, KIP 02, KIP 04a, KIP 04]. Currently, and based on these experimental data, the message to the shipbuilding industry is clear: the abatement of the URN signature of the "new vessels" will require a preventive control of the sources and actions focused on reducing the vibration energy transmission from the main mechanical sources, from the main machinery to the hull. The intensity of the excitation induced by the propeller must be controlled as well as its cavitation.

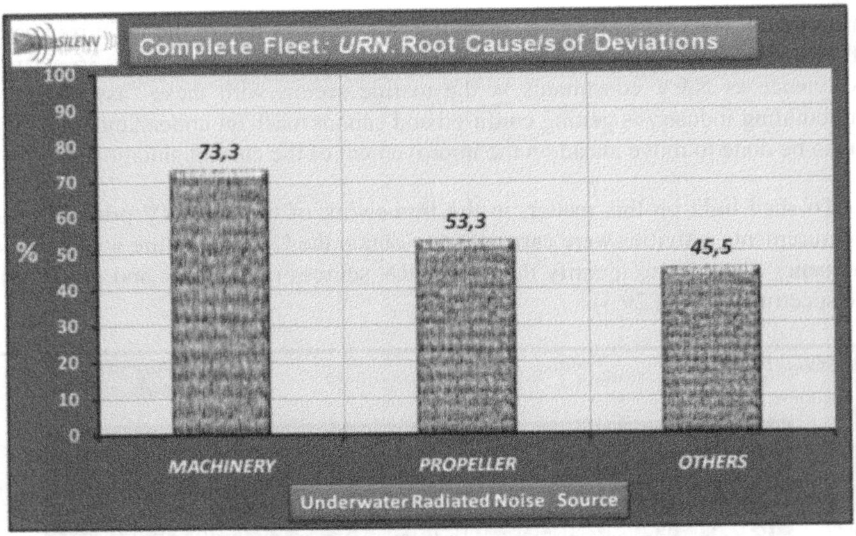

Figure 29.4. *Statistical assessment: identification of URN root cause(s)*

Therefore, as a summary we can conclude that, first of all, it is unquestionable that the current debate amongst the scientific and technical marine communities as far as the impact of the underwater footprint of commercial vessels in the marine life is concerned, is in the limelight. On the other hand, we find that for the current fleet, with very few exceptions (e.g. fishing research vessels), there are no specific URN requirements included in their contract specifications. Consequently, the availability of URN data is limited, hence the impossibility to assess, with enough accuracy, the real impact of each family of vessels in the oceans and in the marine life. To make up for this lack of URN data, special mention must be given to the experimental database that was generated within the framework of the BESST and SILENV projects (see footnote on page 456).

Another reason for the lack of regulation is the novelty that this topic means for the civilian Shipbuilding Industry and, from the authors' point of view, "the general confusion" that there is in relation to the harmonization of the measurement procedures and mainly concerning the limits. The limits suggested in [ICE 95] are, for some scientists, "still quite permissive" but excessive for the shipbuilders: the conflict is still unresolved. As will be detailed in our conclusions, the harmonization, the improvement of the measurement procedures and a "preliminary agreement" as far as the targeted limits are concerned between the scientific community and the marine sector is vital to define the technological strategy that should be followed for improving the current situation.

Based on the authors' expertise in building "silent ships", fishing research vessels especially the most modern ones, have become a point of reference and a clear example of the technological strategy that must be applied for minimizing the impact of the vessels in the oceans. They are a "technological reference" that make it possible to look at what type of dedicated solutions have been applied for reducing the URN signature. Of course, it must be kept in mind that other factors such as cost/benefit ratios, fuel efficiency and technical viability should be considered and properly weighted. They will also be a "strategic reference" because it will be possible to discover why fishing research vessels, contrarily to the other types, are fighting and moving ahead in the abatement of their underwater footprints. This will enable the sector to increase the level of sensitization of the rest of the fleet with regard to this new topic.

29.2. Case study: "Ramón Margalef" and "Ángeles Alvariño"

29.2.1. *Introduction*

At the end of August 2011 and in the summer of 2012, the "Ángeles Alvariño" and "Ramón Margalef" fishing research vessels were delivered to the IEO Spanish Oceanographic Institute. As a result of sea trial tests, the following main aspects can be highlighted:

– for the "Ramón Margalef" the maximum vibration level on board: *0.7 mm/s-rms*; Noise on board at Cabins: between *37 dB(A) and 57 dB(A);* and *87 dB(A)* at the Engine Room; Noise levels radiated to Harbour (NRH) in compliance with Directive 2006/87 EC (lower than 65 dB(A) at 25 m of both hull sides); and Noise Radiated to the Water (URN) in compliance with the ICES No. 209 Regulation;

– for the "Ángeles Alvariño" the maximum vibration level on board *0.7 mm/s-rms*; Noise on board at Cabins: between *50 dB(A) and 55 dB(A);* and Noise Radiated to the Water (URN) practically in compliance with the ICES Nº 209 Regulation.

These experimental results, respecting the most strict N&V comfort and environmental requirements, made the *Ramón Margalef* and *Ángeles Alvariño* two of the most silent vessels of all the European fleet and perhaps of all the civilian fleet in the world. Thus these vessels became a clear milestone and indicators that the European shipbuilding industry and so, the Spanish one too, are capable of meeting the current need to abate the N&V environmental impact of vessels.

These results are not a question of luck, but are based on solid pillars: the first one is the high level of sensitization shown by the IEO, who defined a technical specification that included specific requirements to guarantee the comfort and good working conditions of the crew and scientists, the reduction of the environmental impact of the vessel and, of course, the high efficiency of the scientific echo-sounders on board. Indeed, SIMRAD/KONSBERG, in another fishing research vessel in which the same methodology described below has been applied by the authors, has confirmed that the “low background noise of the vessel” has enhanced the efficiency of the electronic equipment by more than 40%.

On the other hand, the excellent experimental results obtained enable us to validate and consolidate the N&V integrated management methodology that the authors have been applying for over 36 years.

The aim of this chapter is to present a brief description of this methodology applied in the *Ramón Margalef* and *Ángeles Alvariño* vessels. It is widely detailed in [BEL 01, BEL 06, BEL 07, BEL 08, BEL 09, BEL 10]. A summary of the experimental results obtained during sea trial tests will also be presented.

29.2.2. *Description and Main Particulars of the vessels*

Because of its size and capacity, *Ramón Margalef* is classified as a regional fishing research vessel. It has autonomy for 12 days and can provide accommodation for 11 researchers and technicians in addition to a crew of 12 members. *Ramón Margalef* has diesel-electric propulsion made up of three gen-sets driven by GUASCOR F480TA-SG diesel engines with an output of 868 kW at 1.500 rpm, elastically mounted according to the criteria defined by the consultant (the authors). This plant produces all the power required to drive the two INGETEAM-INDAR KN-800-5-b-c DC Electric Motors, as well as all the electric power for the ship’s onboard services and machinery. The vessel have two shafts each driven a by single fixed pitch propeller. The main particulars and an overview of the vessel are reported in Figure 29.5. The *Ángeles Alvariño* has the same characteristics as the *Ramón Margalef* except that the latter only has one shaft driven by a single fixed pitch propeller.

29.2.3. *The methodology: N&V integrated management*

Vessels are elastic systems under periodic forces that come from different sources. Consequently they are sensitive to vibration. The vibration levels obtained in the system (the vessel) depend, mainly, on three parameters: 1) the intensity or magnitude of the excitation forces; 2) how stiff the structure is; 3) the dynamic amplification at different frequencies due to resonance phenomena: local and global. As a consequence, the possible actions that could be implemented if we want to keep vibration levels under certain limits are: 1) minimize the excitation forces of the system; 2) avoid flexible structures from a dynamic point of view; 3) avoid resonance phenomena (coincidence of structural frequencies and excitation frequencies).

Overall Length	46.70 m
Maximum Breadth	10.5 m
Design Draft	4.00 m
Depth to Deck Nº 3	7.10 m
Gross Tonnage	988 t
Dead Weight	230 T
Main Propulsion	2x 900 kW
Maximum Speed	13 knots
Classification	BUREAU VERITAS

Figure 29.5. *Vessel's overview and main particulars*

Similarly, and from an acoustic point of view, the vessel has built-in sonorous sources: main engine and auxiliary ones, propellers, hydraulic systems, HVAC, etc., that are either are airborne and structure-borne noise generators. The noise is transmitted or spread through all the structure of the vessel (path) and reaches all the different locals and spaces (receptors) and is radiated to the water. In the same way, the possible actions that could be implemented in order to maintain the noise levels under the predefined limits are: 1) minimize the sound power and vibration power at the noise sources; 2) reduce or diminish their transmission to the paths; 3) isolate the receivers.

From an elastic-acoustic point of view, the hull becomes a wave radiator to the water due to the acceleration induced on it by the action of the main mechanical excitation sources and to the propeller. In this case, preventive actions and source control can be applied to reduce the vibrational energy transmitted to the hull by the main machinery and by the propeller. So, dedicated Factory Acceptance Tests (FAT) (for the main machinery) pressure pulses induced by the propeller and cavitation tests should be carried out to avoid this phenomenon.

For the practical application of the basic principles of *Noise and Vibration Integrated Management* (N&VCM) as a control mechanism, dynamic and acoustic specific requirements have been incorporated in the purchase specifications of the different supplies as well as FAT procedures to control and verify their compliance before installing the most critical supplies on board. This first activity level has focused on the fact that the shipyard can control the main N&V and URN sources that are usually under the scope of the suppliers. The intention with this "control mechanism" is to achieve and guarantee the so-called "contractual sensitivity" of the suppliers toward the dynamic and acoustic objectives of the project. If this easy task is not performed, the shipyard will have to apply the solutions or countermeasures later on that can be more expensive. The wide expertise of the suppliers, involved in similar projects, has permitted, for the case of the *Ramón Margalef*, for an easy and efficient control methodology.

From the beginning of the project, the shipyard has assumed that "dynamic and acoustic vessel design" is mandatory. Thus, once the N&V main sources are under control, a "second module or block of activities" considers all the aspects of the competence of the shipyard, such as the supplying of a vessel structure with an appropriate dynamic-acoustic design with no resonances and with enough acoustic isolation to guarantee the compliance with the N&V contract requirements. Focused on achieving these N&V targets, a whole dedicated package of simulation tools and dynamic tests, such as N&V prediction calculations (Figure 29.6) and radiated noise (outdoor and underwater, prediction calculations) were put in play for the achievement of the optimal design.

To verify the fulfilment with the "design criteria" related to the intensity of the excitations induced by the main machinery and by the propeller, as well as the theoretical results provided by the simulation tools applied, a whole package of model tests and on-site dynamic tests (FAT-Factory Acceptance Tests- and Inertance Tests, Figure 29.7), have been applied at different stages of the project.

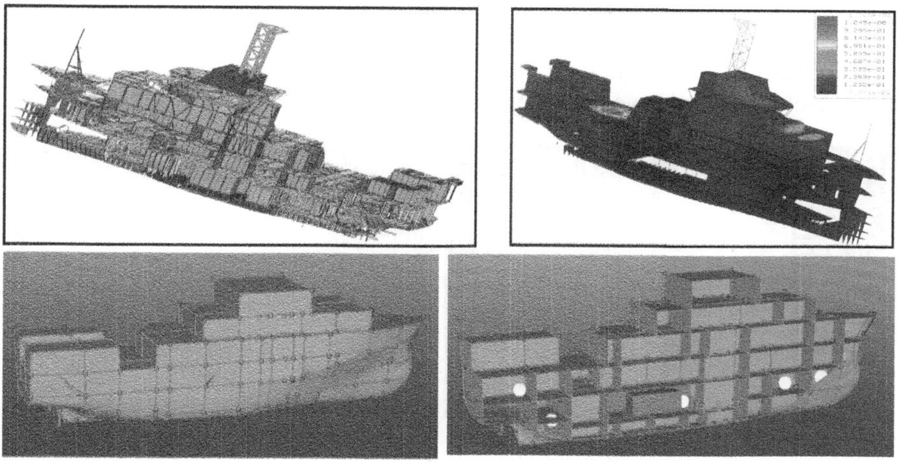

Figure 29.6. *Simulation tools applied for the dynamic-acoustic design of the "Ramón Margalef". For a color version of the figure, see www.iste.co.uk/jacob/energy.zip*

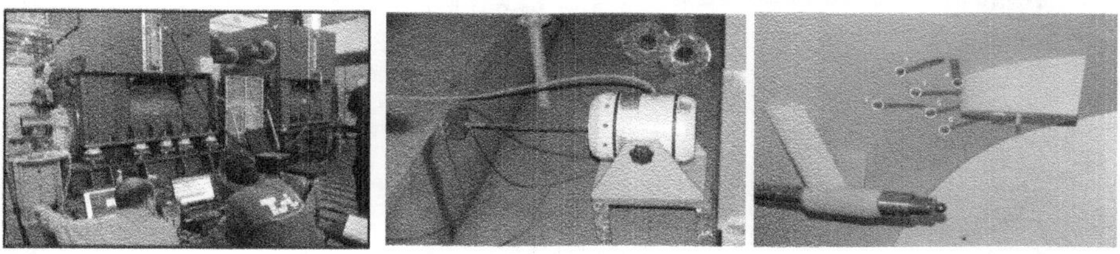

Figure 29.7. *FAT, inertance and model tests, developed for the design of the "Ramón Margalef"*

After this entire analytical-experimental process, the vessel is ready to pass the most important test: the sea trial tests.

29.2.4. *Sea trial tests: results*

All the sea trial measurements were carried out at Vigo Bay (Spain) according to the protocols that were previously approved by the IEO, the shipyard and the BUREAU VERITAS. Some of the results are summarized in the following figures.

29.2.4.1. *Noise onboard*

For the Accommodation/Laboratory-Public Engine Room/Work spaces, the noise results obtained have been reported in the diagrams in Figures 29.8 and 29.9.

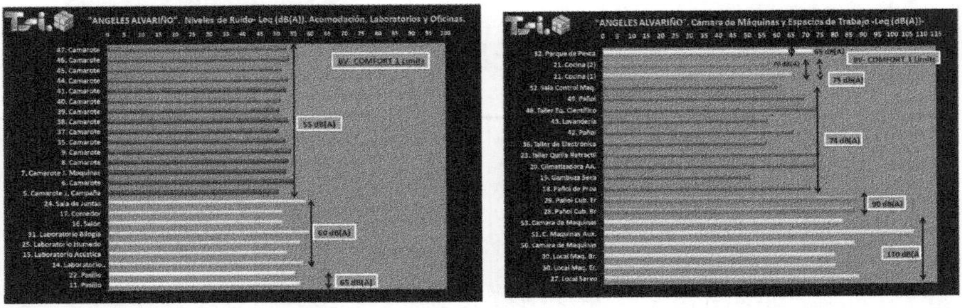

Figure 29.8. *"Ramon Margalef" noise onboard: Accommodation and Engine Room spaces. For a color version of the figure, see www.iste.co.uk/jacob/energy.zip*

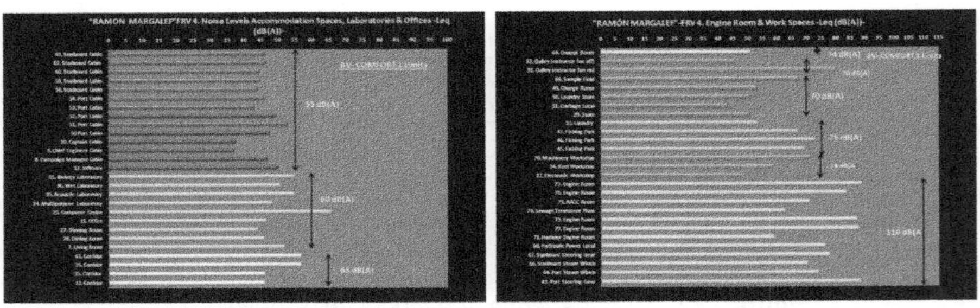

Figure 29.9. *"Ángeles Alvariño" noise onboard: Accommodation and Engine Room spaces. For a color version of the figure, see www.iste.co.uk/jacob/energy.zip*

The numerical results reveal that:

– "Ramón de Margalef": For Accommodation/Laboratory/Public spaces the average noise onboard values are *-9.6 dB(A)* and *-10.1 dB(A)*, respectively, *lower* than the corresponding BV-COMF VIB 1 limits (55, 60 and 65 dB(A), respectively). A special mention must be done for the noise levels achieved in the Engine Room and in the Machinery spaces with an average value of *-32.6 dB(A) lower* than the BV's limit (110 dB(A)) and than the new IMO amendment limits, in progress.

– "Angeles Alvariño": For Accommodation/Laboratory/Public spaces the average noise onboard values are *-2.7 dB(A)* and *-6.2 dB(A)*, respectively, *lower* than the corresponding BV-COMF VIB 1 limits.

29.2.4.2. *Vibration onboard*

For the Accommodation/Laboratory/Public/Engine Room/Work spaces, the vibration results obtained have been reported in the diagrams in Figures 29.10 and 29.11.

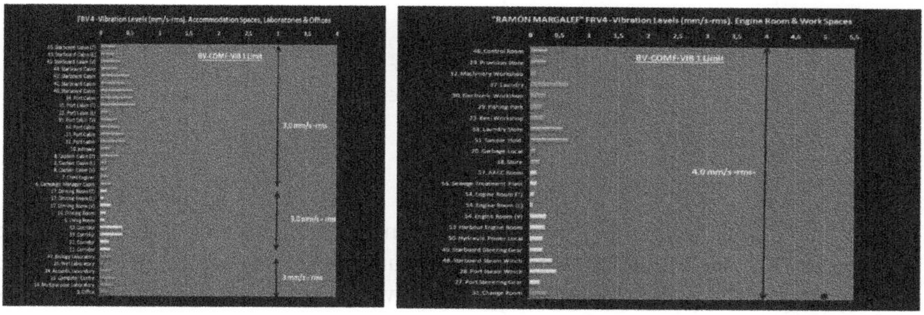

Figure 29.10. *"Ramón Margalef" vibration onboard: Accommodation and Engine Room spaces. For a color version of the figure, see www.iste.co.uk/jacob/energy.zip*

The numerical results reveal that:

– "Ramón de Margalef": For a total of 36 locations at the Accommodation/ Laboratory/Public spaces average vibration onboard achieves a significant value of *-2.7 mm/s-rms, lower* than the corresponding BV-COMF VIB 1 limit (3 mm/s-rms). A special mention must be done again for the vibration levels at 23 locations in the Engine Room and in the Machinery spaces with an average deviation of *-3.7 mm/s-rms*, when compared with the corresponding BV-COMF VIB 1 (4 mm/s-rms).

– "Ángeles Alvariño": As with the previous vessel for a total of 36 locations at the Accommodation/Laboratory/Public spaces the average vibration onboard achieves a significant value of *-2.7 mm/s-rms, lower* than the corresponding BV-COMF VIB 1 limit (3 mm/s-rms), and in the Engine Room and Machinery spaces with an average deviation of *-3.6 mm/s-rms*, when compared with the corresponding BV-COMF VIB 1 (4 mm/s-rms). All these values are a clear indicator of the low vibrational energy distribution achieved along the vessel's structure due to the preventive actions adopted.

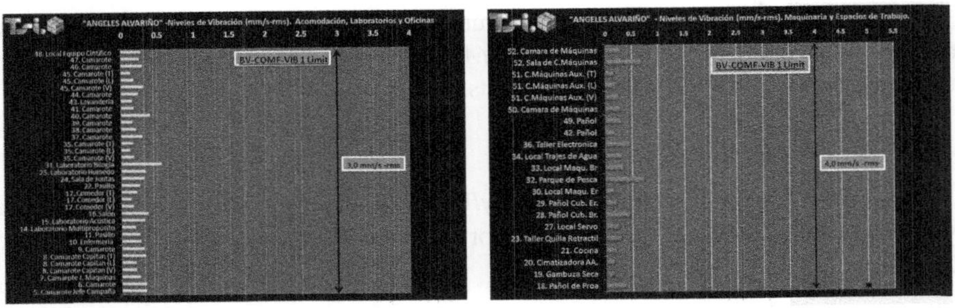

Figure 29.11. *"Ángeles Alvariño" vibration onboard: Accommodation and Engine Room spaces. For a color version of the figure, see www.iste.co.uk/jacob/energy.zip*

29.2.4.3. *Noise Radiated to Harbour (NRH)*

For the *Ramón Margalef*, the corresponding results related to this topic are reported in Figure 29.12 and were carried out in the framework of the on-site measurement activities of the SILENV project (which is why *Ángeles Alvariño* does not have the NRH signature).

As can be seen, the outdoor noise levels at 20 m of the starboard hull-side along all the length of the *Ramón Margalef* are below 65 dB (A), according to the preliminary SILENV limits adopted as well as the limits established by Directive 2006/87 EC and by ISO 2922/200 Acoustic [ISO 00].

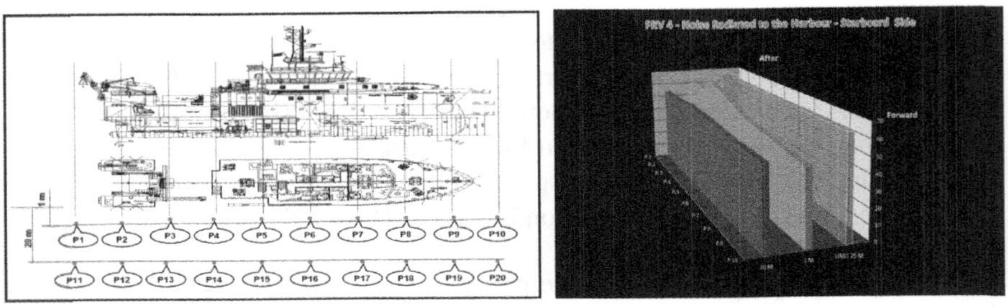

Figure 29.12. *"Ramón Margalef" starboard hull side outdoor noise levels. For a color version of the figure, see www.iste.co.uk/jacob/energy.zip*

29.2.4.4. *Underwater Radiated Noise (URN)*

A sequence of the URN measurement tests and the results obtained in terms of URN pressure (1 μPa) @ 1 m 1/3 octave band spectra (1 Hz) at an operational speed of 11 knots, for both hull sides, are reported in Figures 29.13 and 29.15:

– “Ramón Margalef”: An examination of Figure 29.13 enables us to state that the *Ramón Margalef* complies with the ICES No. 209 Regulation [ICE 95]. The narrow band analysis has permitted us to prove that the small deviations observed at low and high frequencies are not related to URN sources but to the disturbance that the measuring device suffered because of the sea and because of some interference that came from a meteorological sound-boy operating in the area.

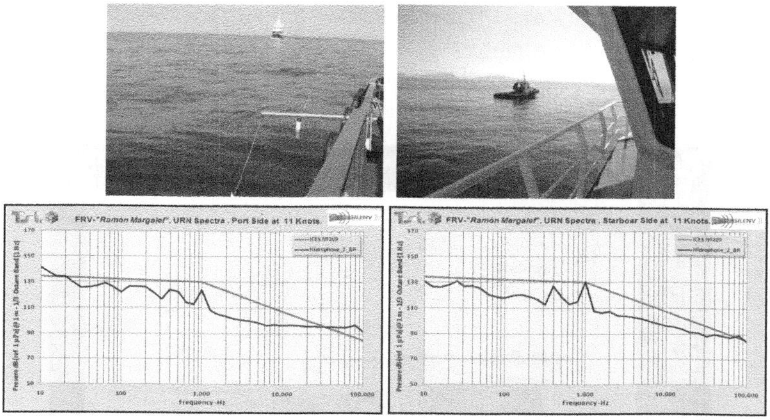

Figure 29.13. *“Ramón Margalef” URN measurements & URN 1/3 octave band spectra @ 11 knots. For a color version of the figure, see www.iste.co.uk/jacob/energy.zip*

Finally, as a complementary “external indicator” of the low underwater footprint of the *Ramón Margalef*, Figure 29.14 shows the “high resolution and spectacular definition” of the data obtained by the IEO (courtesy of SIMRAD/KONSBERG) with the onboard echo-sounders, during the geological research carried out by the vessel in its first investigation about the sudden appearance of a new volcano in the coast of “El Hierro Island” (Canary Islands, Spain).

– “Angeles Alvariño”: An examination of Figure 29.15 enables us to state that this fishing research vessle almost complies with the ICES N° 209 Regulation [ICE 95]. The narrow band analysis and the comparison of these results with the ones corresponding to the *Ramón Margalef* has permitted us to prove and confirm that the small deviations observed are due to the existence of a single shaft driven by a fixed pitch propeller which produces flow problems at the stern of the vessel because the inflow to the propeller is not as good as in the *Ramón Margalef*.

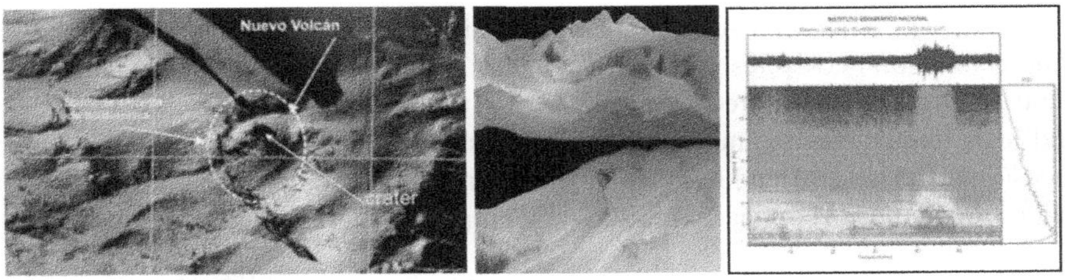

Figure 29.14. *"Ramón Margalef" geological data appearance of new volcano (SIMRAD). For a color version of the figure, see www.iste.co.uk/jacob/energy.zip*

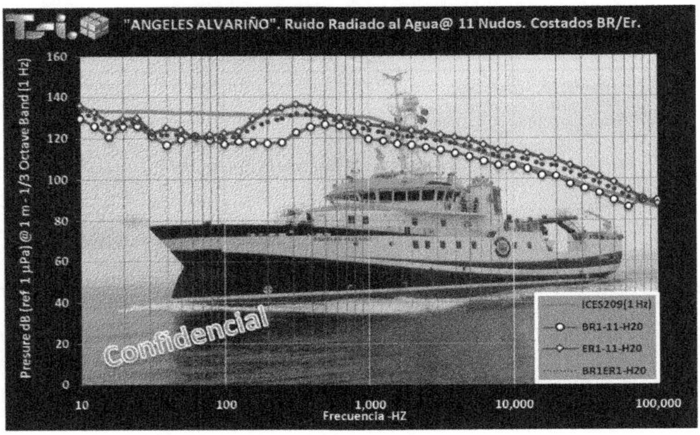

Figure 29.15. *"Ángeles Alvariño" URN Measurements & URN 1/3 octave band spectra @ 11 knots. For a color version of the figure, see www.iste.co.uk/jacob/energy.zip*

29.3. Conclusions

Some of the most interesting lessons learnt, according to our judgment, are detailed in the following paragraphs:

– The experimental results support the *Ramon Margalef* as a civilian technological reference within the field of "silent vessels" showing that the European, and in particular, the Spanish shipbuilding industry is ready to meet current and future environmental requirements in the framework of the new EC Green Policy.

– The practical cases of the *Ramón Margalef* and *Angeles Alvariño* have shown the validity of the N&V Integrated Management methodology to be applied as a practical guide in the design and building of "silent vessels". This methodology can also be applied for commercial vessels considering and properly weighting the cost/benefit ratio, fuel efficiency and technical viability.

– Once these two essential pillars, i.e. shipbuilding capabilities and the availability of successful "practical guides", have been consolidated, the battle against the high noise in the oceans requires, from the authors' point of view, the establishment of clear, well-founded and coordinated policies based on the following points:

- *technological*: harmonization of the measurement procedures and improvements in the measurements devices and in the acquisition and post-processing process, are essential. The precise identification of the URN sources; in order to properly attack them, will not be submitted to "unsupported assertions" but based on consistent experimental results;

- *preliminary limits definition*: the target is "to reduce the underwater noise for protecting the marine fauna". Two key questions arise and need to be answered: *how much must we reduce the noise* (unfeasible for the current fleet), and *how much will it cost for a new vessel?* A first consensus between the scientific community and the shipbuilding industry is essential if some preliminary URN limits want to be defined. Perhaps these preliminary limits are not as strict as many biologists would like, but they will oblige the shipbuilding industry, to move ahead and to assess the impact that the achievement of these limits has on the costs, and so their technical-economic viability;

- *regulatory framework*: the complete absence of URN requirements in the contract specifications has been confirmed as the main cause of URN data unavailability. The premise "no one does nothing if they are not forced to" is well stated. This status needs to be changed, and for changing the following actions must be carried out: *firstly*, there should be a thorough dissemination of the EC policy within the Marine Sector (currently immerse in a "hard crisis") focused on not forcing but convincing about the advantages of fighting against the URN impact. *Secondly*, the absence of URN data for most of the current fleet made impossible the suitable "URN Management" and the identification of which vessels should not enter protected areas. The URN signatures from a representative sample of vessels are therefore needed, independently of the compliance with whichever the URN limits are. The same criteria should be applied for the new constructions.

– Finally the retrofitting of old vessels in order to reduce the current underwater noise levels is technically and economically unfeasible. Fleet renewal must be based on the full compliance with the "Environmental Requirements: Emissions, Fuel consumption reduction and N&V".

29.4. Acknowledgments

The IEO (Spanish Oceanographic Institute) deserves a special mention for the initiative and the motivation of all the actors: shipyards, suppliers and consultants, on achieving the objectives of the project. Some of the results presented have been obtained in the framework of the BESST and SILENV project. We would like to thank all the partners that have participated.

Finally, it is important to note that the building of *Ramón Margalef* benefitted from EU Regional Structural Funds under Grant FICTS-2011-03-01.

29.5. Bibliography

[ADV 00] ADVERSON P., VENDITTIS D., "Radiated noise characteristics of a modern cargo ship", *Acous. Soc. Am.*, vol. 107, pp. 118–129, 2000.

[ANS 09] ANSI/ASA, Part 1: Quantities and Procedures for Description and Measurement of Underwater Sound from Ship – Part 1: General Requirements, S12.64-2009, 2009.

[BEL 01] BELTRÁN P., "Noise vibration on ship prediction: Basic engineering tool to fulfil the current requirements of comfort and quality", *Naval Engineering Magazine*, 2001.

[BEL 06] BELTRÁN P. *et al.*, "Silent Fishing Vessels, A milestone for small and big Spanish shipyards. Lessons learnt", *Naval Engineering Magzine*, 2006.

[BEL 07] BELTRÁN P., "The Fishing Research Vessel Miguel Oliver", *Naval Engineering Magazine*, 2007.

[BEL 08] BELTRÁN P., "Oceanographic Ship 'Miguel Oliver' The excellence on Noise and Vibration on board fulfilling ICES-Nª209", *47th Congress of Naval Engineering and Maritime Industry*, 2008.

[BEL 09] BELTRÁN P., SALINAS R. *et al.*, Collection exiting and available N&V data, SILENV Project D 2.2.1, 2009–2011.

[BEL 10] BELTRÁN P., TESORERO M.A., "Spanish shipbuilding ready to meet the new and demanding environmental requirements: new challenges and business opportunities", *49th Congress of Naval Engineering and Maritime Industry*, Bilbao, Spain, 2010.

[DNV 10] DNV, Silent Class Notation, available at: https://rules.dnvgl.com/docs/pdf/DNV/rulesship/2011-07/ts624.pdf, January 2010.

[ICE 95] ICES, Underwater Noise of Research Vessels, Review and Recommendations, Research Report no. 209, May 1995.

[ISO 00] ISO 2922/2000, Acoustics – Measurement of airborne sound emitted by vessels on inland waterways and harbours, 2000.

[KIP 02] KIPPLE, Southeast Alaska Cruise Ship Underwater Acoustic Noise, Naval Surface Warfare Center-Detachment Bremerton Technical report NSW CCD-71-2002-574, 2002.

[KIP 04a] KIPPLE, NVSSCCD Report: Coral Princess Underwater Acoustic Levels, 2004.

[KIP 04b] KIPPLE, NVSSCCD Report: Volendam Underwater Acoustic Levels, 2004.

30

Mitigation of Ground Vibrations from Freight Trains

This chapter reports on the results of the EC research project CARGOVIBES. The difference in vibration emission between freight trains and conventional trains is discussed. We also discuss methods to design vibration mitigation solutions for freight trains, and propose solutions at the source and in the propagation path. Deterioration of track quality (a concern for ballasted tracks) leads to increased ground borne vibrations at low frequencies which are directly related to the speed of the train and the defect wavelengths of the track. These low frequency vibration problems cannot be solved by conventional vibration control measures such as under sleeper pads and ballast mats. The use of special sleeper track on ballast to obtain a sustainable good track geometry quality is discussed as a solution at the source. A mitigation measure in the vibration propagation path is the use of specific soil barrier systems, which do not impact the track system itself and significantly reduce the transmitted low-frequency vibrations.

30.1. Introduction

In an urban area, railway traffic-induced low-frequency vibrations may damage buildings, cause annoyance to people and cause malfunctioning of sensitive equipment inside nearby buildings. This is especially the case for low-frequency vibrations as generated by freight trains. The study of ground-borne vibrations caused by road or railway traffic and their effects on nearby structures has received considerable attention and concern during the last two decades [LOM 09, PEP 07, FRA 07].

Chapter written by Patrick VANHONACKER and Hamid MASOUMI.

30.2. Vibrations from freight trains

As Figure 30.1 illustrates, in comparison with passenger trains, results of field measurements show that freight trains exhibit higher dynamic forces (+20 dB) in the lower frequency ranges below 20 Hz [VAN 13]. This is mainly due to the differences in bogie design and especially the lack of a secondary suspension system on the freight bogies.

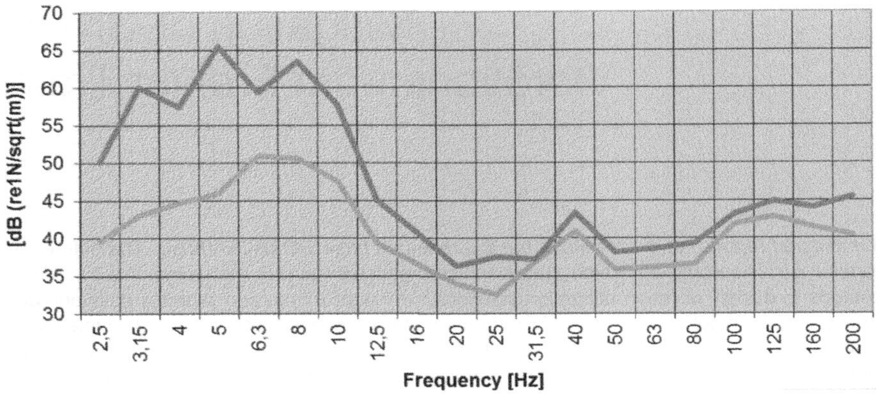

Figure 30.1. *Average force density spectra of passenger (red line) and freight (green line) trains in the speed range of 70–80 km/h obtained during field measurements in Heverlee (Belgium) [VAN13]. For a color version of the figure, see www.iste.co.uk/jacob/energy.zip*

Vibrations from freight trains in the lower frequency band below 25 Hz are generated by wheel and track defects such as wheel out of roundness, wheel flats, rail discontinuities and track irregularities. These vibrations show little attenuation as a function of distance and have a high probability of coinciding with building resonance frequencies, especially floors. Vibrations in the higher frequency band (between 25 and 100 Hz) are generated by wheel/rail support resonances, with a higher attenuation as a function of distance and are responsible for the generation of ground-borne noise. Unlike vibrations below 25 Hz, these higher frequencies can be controlled by conventional solutions such as ballast mat (for ballasted track), floating slab (for slab track) and resilient rail fasteners. The only track solution available to mitigate the lower frequency vibrations is maintaining a good track geometry.

Good track geometry in the 3–50 m wavelength range is very important. Indeed, a train passing at 72 km/h (20 m/s) over a track geometry defect with a wavelength of 2 m generates vibrations at 10 Hz. Good track geometry is easier to obtain and maintain with one of the four following different track concepts: ladder track, half-space sleepers, wide sleepers or conventional sleepers with their spacing reduced to 0.5 m. All sleepers must be equipped with under sleeper pads. If the track is not upgraded to one of these configurations, then regular tamping is a valid alternative. These four track systems and regular tamping produce a good track geometry and rail surface quality as a function of time and passing tonnage. The benefits of tamping are further explored in this chapter.

30.3. Decomposing the vibration problem

The problem of train-induced vibrations and the transmission of these vibrations to a building is termed as a "dynamic soil-structure interaction problem" combined with a wave transmission mechanism. The main problem can be decomposed into four subproblems:

1) the *dynamic vehicle–track interaction* that generates a dynamic contact force on the railhead, and which depends highly on the vehicle and the track receptance, the track geometry condition and the wheel/rail surface quality;

2) the *track–ground interaction* refers to how the track reaction is transferred to the ground, and depends on the track system and the soil conditions under the track;

3) the *transmissibility of the ground* refers to how the stress waves are transmitted through the ground between the source (the track) and the receiver (the building), and depends on the dynamic characteristics of the soil and its stratification. Existing public utilities such as cables, pipes and sewers in the propagation path can also significantly influence the transmissibility of the ground;

4) the *ground–building interaction* refers to how the ground motion is transferred to the building and depends mainly on the foundation type and the soil conditions around the foundation. The structural response of the building elements depends on the dimensions of floors and walls, their material properties and the frequency content of the incoming vibrations.

According to the above decomposition of the problem and focusing on the low frequency-induced vibrations, two innovative vibration mitigation techniques at different levels, (1) at the source and (2) in the propagation path, are presented.

At the source, the low-frequency vibration problem cannot be solved by conventional vibration mitigation measures such as under sleeper pads, floating slabs and ballast mats. Here, the main focus is on reducing the contact force by removing the irregularities on the wheel and the rail surfaces. Therefore, it was investigated how improvement in the maintenance intervention levels for track irregularities such as rail corrugation, track geometry errors and wheel out-of-roundness may reduce the intensity of the train-induced vibrations in the free field. The rail and the wheel irregularities are the source of the train-induced vibrations [THO 10, RAI 06, RAI 98]. The dynamic vehicle–track interaction of a train running over a track generates a contact force on the rail contact surface that depends strongly on the train speed, the wheel-rail surface quality, the track and the wheel receptance [THO 10].

In this chapter, improving wheel and rail maintenance to reduce train-induced vibrations is termed as a source-related mitigation measure. Furthermore, the benefit of mitigation from optimizing the track structure was studied for two track systems: (1) conventional sleeper track and (2) ladder track. The dynamic behavior of the track structure has been investigated by several authors [BLA 11, SU 05, ASA 04]. The main objectives of these studies were related to the impact of the track systems on safety, economy, maintenance as well as environmental issues, e.g. the induced vibrations in an urban area.

The use of soil barriers to control the transmitted vibrations in the propagation path has been studied by several authors [MAS 13a, ADA 05, CEL 09]. The presence of an isolating barrier between the vibration source and the receivers can reduce the vibration level by reflecting and scattering the incoming waves, as well as by affecting the wavelength of the incoming waves by wave modulation of the barrier layers (in the case of a multi-layer barrier). The efficiency of the barriers was verified using a small-scale laboratory test of the problem in a test bench before installation on site. In addition, the numerical prediction model is validated by the experimental measurements in the test bench [MAS 13A].

30.4. Numerical modeling

The procedure adopted in this simulation follows two main steps: (1) dynamic source modeling: the force exerted by a running freight train on the railway track is calculated with a multi-body dynamics model, and (2) the numerical simulation of the vibration transmission from the railway infrastructure through the soil using the finite element method (FEM).

30.5. Mitigation measure at the source

The dynamic vehicle–track interaction of a train running over a track generates a contact force on the rail contact surface that depends strongly on the train speed, the wheel-rail surface quality, the track and the wheel receptance, but is less dependent on the vehicle suspension system. However, it will be shown that freight trains with a nonlinear primary suspension system generate vibrations in a way that is different from that of conventional passenger trains [MAS 13c]. The quality of the wheel and rail surfaces is very important for vibration control. Errors in the track geometry (wavelengths between 1 and 30 m) lead to increased ground-borne vibrations at frequencies, which are directly related to the speed of the train and the wavelength of the irregularities. The long wavelength of the track geometry errors results in low-frequency vibration problems (below 20 Hz) for trains at the medium-to-high speed (above 100 km/h).

Track geometry errors or irregularities such as track alignment, track level, track gauge and cant are categorized into three levels: *corrugation* (very short wavelength), with wavelengths of 30–80 mm and with amplitude of a few hundreds of a millimetres; *short waves* with wavelengths of 150–300 mm and with amplitudes up to 1 mm; and *long waves*, with wavelengths up to 2 m [RAI 06].

30.5.1. *Track geometry correction by tamping*

The effect of the track geometry correction was studied by introducing a geometrical deviation on the track with a wavelength range from 1–36 m. The effect of tamping was modeled by defining two vertical deviations on the track: for the case "before" 6 mm (the maximum allowed deviation according to Track Standards Manual GC/RT5017) [RAI 98] and for the case "after" 1 mm. Then, using the presented numerical methodology, the vibration level on a sleeper is computed for the passage of a Y25 freight bogie over a tangent track at a speed of 80 km/h.

Figure 30.2 shows the insertion loss obtained by the difference between the vertical vibration velocity in the sleeper before and after the tamping. The results show a reduction of about 20 dB below 20 Hz. Considering the train speed of 80 km/h and the wavelength of the track deviation (1–36 m), the results are valid in the frequency range below 22 Hz (~v/λ). It should be noted that the continuous tamping of the track is not a realistic option [LAK 11].

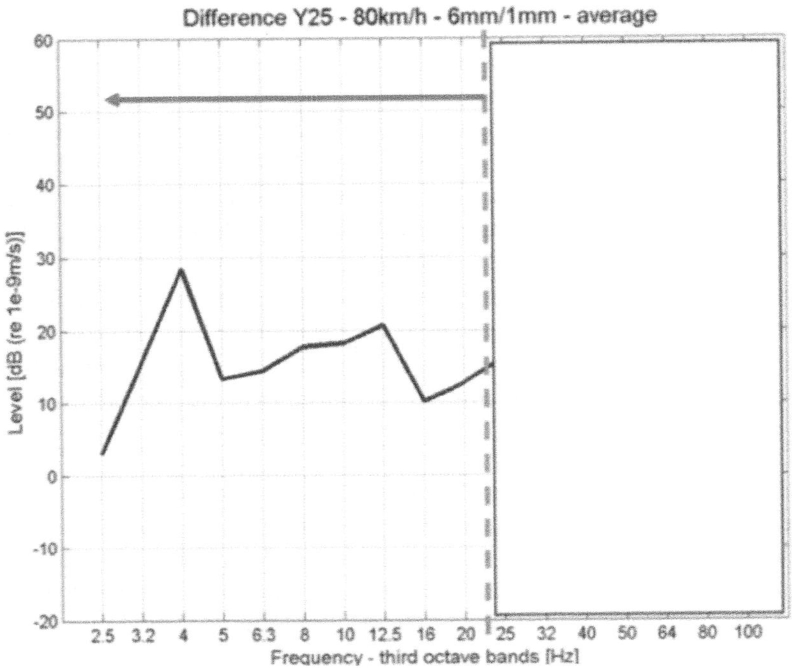

Figure 30.2. *Computed insertion loss on the sleeper*

30.5.2. *Rail roughness and grinding*

The most important mechanical cause of noise and vibration from a passing train is the wheel and the rail roughness in the contact area. In the following section, it is shown how removing the rail roughness by grinding results in a reduction in the low frequency freight-induced vibrations. A train passing over a rail with high roughness results in a dynamic excitation and consequently, ground-borne vibrations near the railway.

Simulations were performed for the passage of Y25 freight train at speeds of 10 and 80 km/h on tangent track with a measured rail roughness profile. Therefore, a vehicle with perfect wheels (no wheel roughness) and only an experimentally obtained rail roughness excitation (with max. wavelength of 50 cm) is considered. The considered rail roughness spectra and their related excitation frequencies are shown in Figures 30.3(a) and (b), for, respectively, the cases before and after grinding.

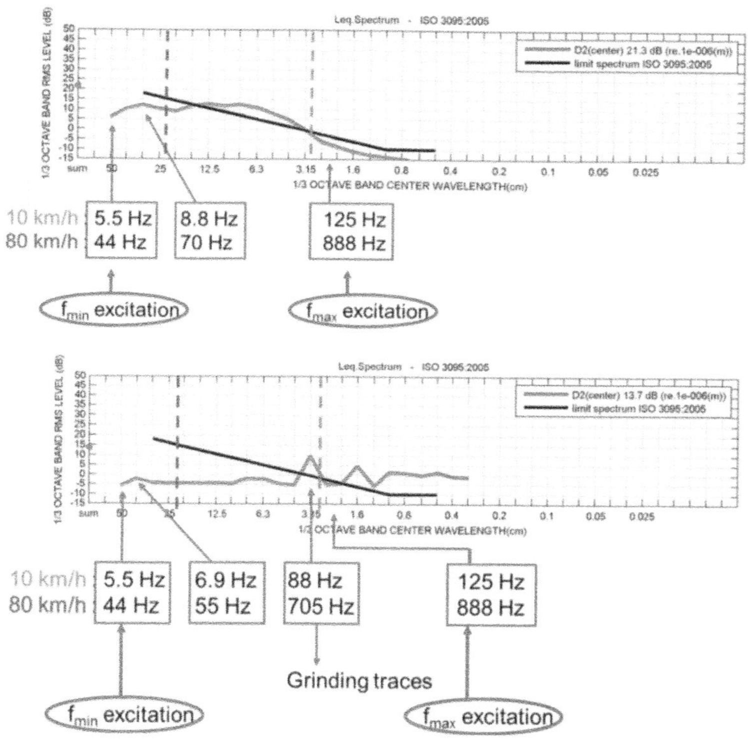

Figure 30.3. *Considered rail roughness spectra a) before and b) after grinding. For a color version of the figure, see www.iste.co.uk/jacob/energy.zip*

Figures 30.4(a) and(b) show the insertion loss obtained for the vertical vibration velocity on the sleeper before and after grinding for train speeds of 10 and 80 km/h. The results show a reduction of about 20 dB below 20 Hz.

30.5.3. *Track structure system*

In the following, the intervention regarding the mitigation in the track structure is studied. Two track systems are discussed: (1) the sleeper track system and (2) the ladder track system, and their vibration effects under different excitation conditions are studied.

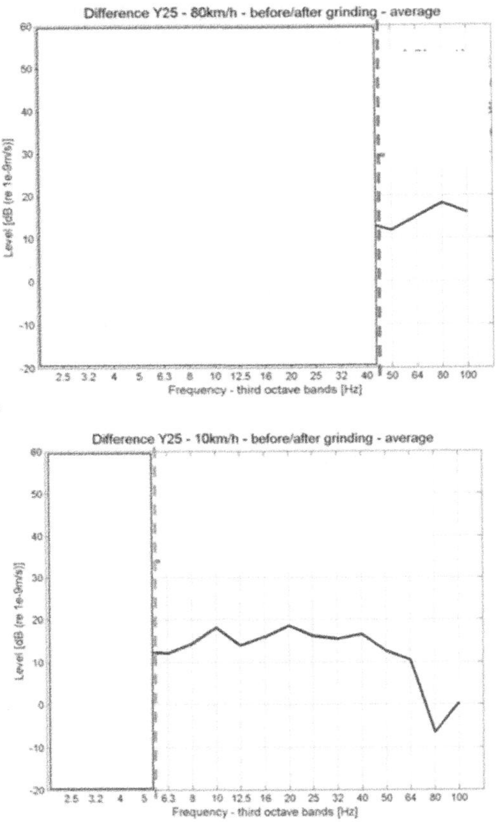

Figure 30.4. *Insertion loss on the sleeper after grinding obtained for different train speeds at a) 80 and b) 10 km/h*

In this project, it was anticipated that the use of ladder track on ballast would significantly reduce the track settlement, helping to maintain a good geometrical track quality as a function of tonnage on the line, and thus preventing the increase in ground vibration levels over time. The ladder track is a ladder-shaped structure consisting of two longitudinal prestressed concrete beams connected by transversal steel pipes and concrete beams. The Structural Engineering Group of Japan Railway Technical Research Institute has developed the ladder track system to improve track performance and reduce the maintenance cost. The main idea behind the ladder track system is to combine a longitudinal concrete sleeper with the rail. It results in higher rigidity and stability of the track against the lateral and the longitudinal train loads [ASA 04].

The MNF files generated for the two track systems are transferred to the multi-body software VI-Rail, and a dynamic simulation is performed for each model for a freight bogie Y25 passing at 80 km/h over a perfect rail/wheel contact surface track: a quasi-static excitation. The results show that there is no significant difference between the total trainload generated by running a freight train with a perfect wheel-rail contact surface over the sleeper track or the ladder track (Figure 30.5).

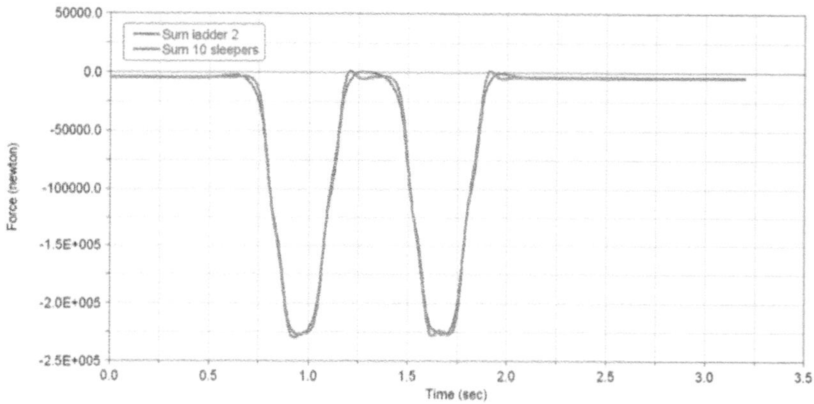

Figure 30.5. *Comparison between the sum of attachment forces over 10 sleepers (green line) and that obtained over two ladders (red line). For a color version of the figure, see www.iste.co.uk/jacob/energy.zip*

An FE model is used to determine the free field vibrations due to passage of a freight bogie Y25 at a speed of 80 km/h. Two models are prepared for each track type. Figure 30.6 shows the vibration levels in different points on the soil surface. The results show that in the near field, a freight train over a ladder track generates higher vibrations above 10 Hz than when running over a sleeper track. However, the difference between the vibration levels decreases with increasing distance from the source, such that at 34 m from the track, this difference is no longer observed.

The same results have been reported by Lak *et al.* [LAK 11], when they compared the soil response due to an impact force applied on different foundation types: a continuous beam, a finite plate and a small disk. They have shown that in the near field and at high frequencies, the results for different cases show a significant discrepancy. However, at low frequency and in the far field, these differences become negligible and the soil response is almost identical [LAK 11].

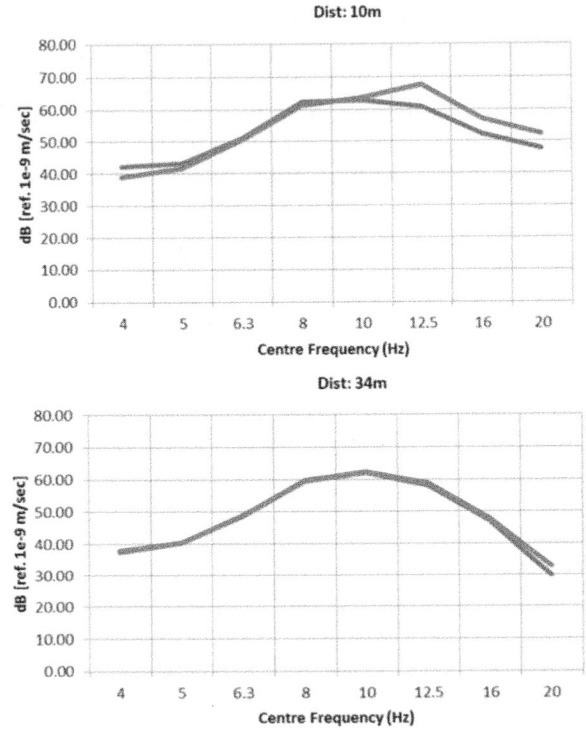

Figure 30.6. *Vertical velocity computed by FE analysis at different distances of a) 10 m and b) 34 m from the track. Results are obtained for a freight train with a perfect wheel–rail contact surface (quasi-static excitation) passing at a speed of 80 km/h over the two track systems: the sleeper track (blue line) and the ladder track (red line). For a color version of the figure, see www.iste.co.uk/jacob/energy.zip*

Although the initial vibration levels of the ladder track are the same as those of the conventional sleeper track, it remains an interesting concept from a technical point of view, as it reduces ballast loading, increases vertical, longitudinal and transversal track stability and maintains a stable track geometry as a function of tonnage (reduced track maintenance), and thus ground vibration emissions remain constant over time. However, the negative points, high investment and technical installation difficulties, outweigh these benefits [PIN 04]. Therefore, TTCI[1] has studied some alternative track systems.

The first alternative is a half-frame sleeper with higher vertical and transversal contact surface with ballast in comparison with the standard sleeper. It comes at a

1 Transportation Technology Center, Inc. – Pueblo, Colorado (US) – www.aar.com.

far lower cost, is easier to install and track maintenance (tamping) remains possible with standard machines. The half-frames are equipped under sleeper pads. The disadvantages of the half-frames as evidenced by use under heavy axle loads were the development of cracks in the concrete and broken rail fasteners [LOP 11, LOP 12]. The second alternative consists of standard sleepers with a spacing of 50 cm instead of 60 cm. This alternative has all the advantages of the half-frame sleeper without its disadvantages [LOP 01].

The measurements at TTCI show that the initial track geometry of the half-frame sleeper track is not as good as that of new standard sleeper track. Half-frame sleepers are more difficult to align and level. The half-frame track remains stable in terms of track geometry. Even after deterioration, the standard sleeper track still shows a slightly better geometry than the initial half-frame sleeper track. In order to perform better in terms of track geometry and hence in terms of low-frequency vibration emission, the half-frame track (and ladder track) must be installed with at least the same track geometric quality as the standard sleeper track. This is the real challenge.

The use of standard sleepers with 50 cm spacing or the use of single sleepers with increased surface contact with ballast seems to be the best options to obtain the advantages of ladder track without its disadvantages. Within CARGOVIBES, wide SSL sleepers (type HDS) equipped with CDM undersleeper pads were chosen as a solution to obtain and maintain good track stability in curves, a necessary condition for reduced low-frequency ground-borne vibrations from freight trains. It is anticipated that less tamping will be required with these larger sleepers, equipped with under sleeper pads.

30.6. Mitigation measure in propagation path

Besides several mechanisms that contribute to the performance of soil barriers, the two main mechanisms are (1) the wave reflection due to the impedance jump on the soil–barrier interface, and (2) the vibration scattering due to the bending waves of the barrier as an infinitely long beam. The latter effect depends strongly on the relative stiffness difference between the barrier and the soil. In addition, this mechanism clearly shows how the barrier performance can be improved by increasing its depth.

The wavelength of the incoming waves is the key parameter in the design of an isolating barrier. Based on a data sheet catalog for the barrier predimensioning prepared within the frame of the CARGOVIBES project, an efficient barrier must have a depth of at least 1.2 times the longest incoming wavelength, and a width of at least 0.06 times the longest incoming wavelength [MAS 13b].

Hereafter, two types of barriers are studied, using FE modeling (Figure 30.7): (a) a pure concrete barrier and (b) a concrete barrier covered with a layer of resilient material (CDM-RR43).

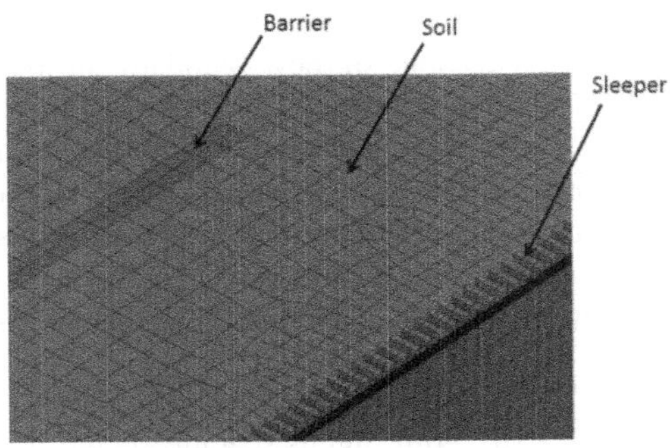

Figure 30.7. *FE model of the track, the soil and the barrier*

The ground consists of a soil layer overlaying the bedrock at 25 m depth. The soil has a Young's modulus E = 123 Mpa, a Poisson's ratio $\nu = 0.3$, a density $\rho = 1800$ kg/m^3 and a material damping of 5%. The shear and compression wave velocities of the soil are 160 and 320 m/s, respectively.

The concrete barrier has a depth of 16 m (around two times the wavelength at 20 Hz), a width of 1.0 m and a length of 80 m. It is located at 11 m from the track centreline. The concrete has a Young's modulus E = 30,000 Mpa, a Poisson's ratio $\nu = 0.25$ and a density of $\rho = 2500$ kg/m^3.

In the second barrier type, the concrete barrier is covered by 12 cm of the resilient material (CDM-RR43). According to the datasheet of the manufacturer (CDM Resin Bonded Rubber), this material has a Young's modulus E = 5.2 Mpa, a Poisson's ratio $\nu = 0.05$, a density $\rho = 700$ kg/m^3 and a material damping of 10%.

Figures 30.8(a) and (b) show the insertion loss obtained for different isolating barriers at points located behind the barrier at distances of (a) 6 m and (b) 26 m from the barrier.

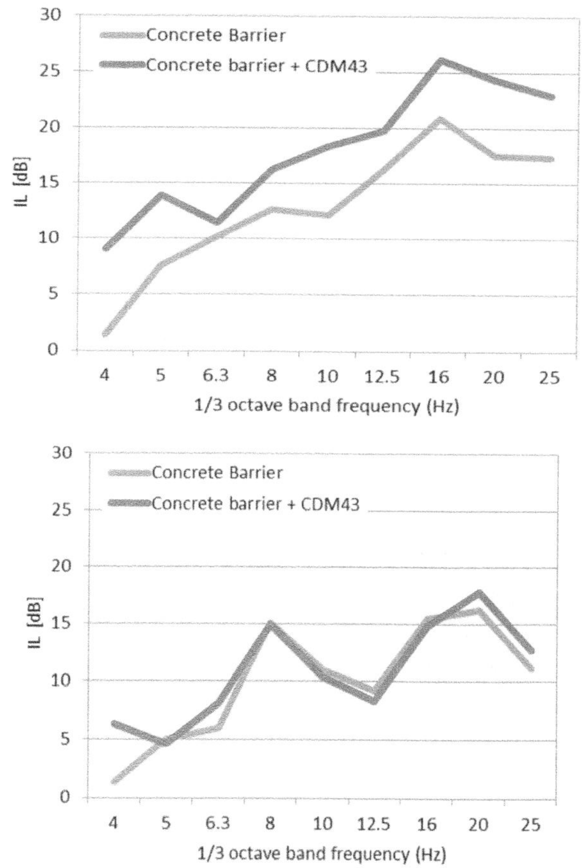

Figure 30.8. *Insertion loss obtained for different barriers at a point located behind the barrier at distances of a) 6 and b) 26 m from the barrier. For a color version of the figure, see www.iste.co.uk/jacob/energy.zip*

At 10 Hz, the dominant frequency of the quasi-static excitation, an average insertion loss of 10 dB is obtained with the concrete barrier, and 15 dB with the concrete barrier covered with resilient material. In addition, the results show that a higher efficiency is obtained at points closer to the barrier; and that the concrete barrier covered with resilient material is more efficient in the near field than the pure concrete barrier. In the far field, however, the same efficiency is obtained. It was also found that a lower insertion loss is obtained for lower frequencies than for higher frequencies.

These results are in agreement with those reported in the literature [ADA 05, CEL 09]. In addition, the results of the numerical modeling show a reasonable agreement with the experimental measurements performed in a reduced-scale test bench [MAS 0].

Recently, within the frame of the SOILVIBES project (a EUROSTARS project), a reduced-scale test bench was fabricated to evaluate the efficiency of the isolating barriers. The test bench consists of a soil container (4 m × 2 m × 2 m) and an isolating screen.

The container was filled with a very fine sand, using the pluviation technique to guarantee the uniformity of the soil and the repeatability of the test. A small foundation excited by a shaker over different frequency ranges is used as the vibration source. The soil responses are measured by accelerometers placed on the soil surface at different distances from the source (Figure 30.9). The results of the reduced-scale tests were used for the validation of the numerical modeling [MAS 10].

Figure 30.9. *Measurement configuration for the concrete barrier test [MAS 13a)*

30.7. Conclusions

Low frequency ground-borne vibrations caused by freight trains can be controlled at different levels by some innovative solutions. This study shows that:

1) At the source: (a) a track geometry correction by tamping can result in a vibration reduction of 20 dB for frequencies below 20 Hz, (b) improving the wheel maintenance quality can result in a vibration reduction of 10–15 dB.

2) In the propagation path, a well-designed concrete barrier can control the induced low-frequency vibrations by at least 6–10 dB. In addition, the experimental and numerical results show that the performance of a pure concrete barrier can be improved by adding a layer of resilient material.

30.8. Acknowledgements

The solutions presented in this chapter are part of the results obtained within the frame of the following EC research programs: SOILVIBES, CITYHUSH and CARGOVIBES. These projects are funded by IWT-Vlaanderen and European Commission FP7 programs. Their financial supports are gratefully acknowledged.

30.9. Bibliography

[ADA 05] ADAM M., VON ESTORFF O., Reduction of train-induced vibrations by using open and filled trenches, *Computers and Structures*, vol. 83, pp. 11–24, 2005.

[ASA 04] ASANUMA K., Ladder track structure and performance, *Railway technology avalanche*, no. 6, 2004.

[BLA 11] BLANCO-LORENZO J., SANTAMARIA J., VADILLO E.G. *et al.*, Dynamic comparison of different types of slab tracks and ballasted track using a flexible track model, *Proc. IMechE Part F: Journal of Rail and Rapid Transit*, vol. 225, pp. 574–592, 2011.

[CEL 09] CELEBI E., FIRAT S., BEYHAN G. *et al.*, Field experiments on wave propagation and vibration isolation by using wave barriers, *Soil Dynamics and Earthquake Engineering*, vol. 29, pp. 824–833, 2009.

[FRA 07] FRANÇOIS S., PYL L., MASOUMI H., The influence of dynamic soil-structure interaction on traffic induced vibrations in buildings, Soil *Dynamics and Earthquake Engineering*, vol. 27, no. 7, pp. 655–674, 2007.

[LAK 11] LAK M.A., DEGRANDE G., LOMBAERT G., The effect of road unevenness on the dynamic vehicle response and ground-borne vibrations due to road traffic, *Soil Dynamics and Earthquake Engineering*, vol. 31, no. 10, pp. 1357–1377, 2011.

[LOM 09] LOMBAERT G., DEGRANDE G., Ground-borne vibration due to static and dynamic axle loads of InterCity and high speed trains, *Journal of Sound and Vibration*, vol. 319, nos. 3–5, pp. 1036–1066, 2009.

[LOP 11] LOPRESTI J.A., KALAY S., Evaluation of track performance under 39-ton axle loads, *Proceeding of the AREMA 2011 Annual Conference* Minneapoliz, September 2011.

[LOP 01] LOPRESTI J.A., Evaluation of track performance under 39-ton axle loads, TTCI, 2001.

[LOP 12] LOPRESTI J.A., High strength tie/fastener systems, TTCI/AAR, 2012.

[MAS 10] MASOUMI H.R., EUROSTARS SOILVIBES Project, Deliverable 3.1: Railway vibration mitigation in transmission path (soil): Test bench and physical modeling of soil barriers, EUREKA and the European Community, 2010.

[MAS 13a] MASOUMI H.R., VANHONACKER P., Physical modeling of the vibration mitigation by an isolating screen, *Proceedings of the 18th International Conference on Soil Mechanics and Geotechnical Engineering*, Paris, 2013.

[MAS 13b] MASOUMI H.R., VANHONACKER P., EU FP7 CARGOVIBES Project, Deliverable 2.6. Attenuation of ground-borne vibration affecting residents near railway lines: Mitigation performance data sheets, European Commission FP7, 2013.

[PEP 07] PEPLOW A.T., KAYNIA A.M., Prediction and validation of traffic vibration reduction due to cement column stabilization, *Soil Dynamics and Earthquake Engineering*, vol. 27, pp. 793–802, 2007.

[PIN 04] PINNEY C., Ladder Sleeper Economic Analysis, TTCI report, 2004.

[RAI 06] RAIL DEFECTS HANDBOOK, some rail defects, their characteristics, causes and control, Engineering Practices Manual, Australian Rail Track Corporation Ltd, 2006.

[RAI 98] RAILWAY GROUP STANDARD-GC\RT5017, Track Standards Manual, Section 8: Track Geometry, Railtrack PLC, 1998.

[SU 05] SU B., Effects of railroad track structural components and subgrade on damping and dissipation of train induced vibration, PhD Thesis, University of Kentucky, 2005.

[THO 10] THOMPSON D.J., *Railway Noise and Vibration: Mechanisms, Modeling and Means of Control*, Elsevier Ltd, 2010.

[VAN 13] VANHONACKER P., EU FP7 CARGOVIBES Project, Deliverable 2.2. Attenuation of ground-borne vibration affecting residents near railway lines: Engineering tools to assess the efficiency of rolling stock and infrastructure based mitigation measures for freight train vibrations, European Commission FP7, 2013.

31

ACOUTRAIN: Virtual Certification of Acoustic Performance for Freight and Passenger Trains

In 2006, the Noise Technical Specification for Interoperability (TSI) came into force, introducing limiting values in noise emission for the certification of new or upgraded rolling stock. The Noise TSI certification process today is based on real tests such as pass-by noise measurements. This procedure is often seen as very restricting considering that several new rolling stocks are widely based on existing designs and minor changes therefore have to be taken into account in terms of noise emission. Acoutrain is a European research project within the 7th framework programme co-funded by the European Commission, with the goal of simplifying and improving the acoustic certification process of new rolling stock. This chapter gives an overview of the Acoutrain results that should improve the acoustic certification process of new rolling stock.

31.1. Introduction

In recent years, awareness has grown among European politicians and policy makers regarding the negative health effects caused by environmental noise. During the same period, it has also become obvious that the emission of greenhouse gases has to be reduced considerably in order to limit climate change. One strategy to reach this target is a significant expansion of the European railway sector and growth in rail traffic for the next decades. However, the already strained noise situation for residents living close to strategic railway lines could impede this important development.

Chapter written by Nicolas FURIO, Maria STARNBERG, Estelle BONGINI, David THOMPSON, Ulf ORRENIUS and Nathalie CUNY.

One way to create the conditions for railway traffic to grow, without risking human health because of increased noise levels, is to control and limit the noise emission of the railway vehicles. The Technical Specification for Interoperability (TSI)–Noise for conventional rail came into force in 2006 and sets limiting values for noise emission of new or upgraded railway rolling stock. It puts pressure on manufacturers and suppliers to develop quieter products.

Although the Noise TSI is strategically important to ensure that the railway remains an attractive means of transport in the future and to allow the sector to grow, it also means restrictions and increased costs for the individual stakeholders.

31.1.1. *Current noise conformity assessment*

The conformity assessment required in the current Noise TSI is mainly based on field tests. This can be a very expensive and time-consuming process which includes measurements of stationary noise, starting noise, pass-by noise as well as interior noise in the driver's cab. One further complication consists of the necessity of performing pass-by noise measurements at a specific low-noise track. The relatively high demands on track quality and maintenance cause a limited access to such reference tracks.

Modern rolling stock development is often based on platform design that results in a family of similar vehicles. Sometimes, these variations only have a small influence on the noise levels that are relevant for the conformity assessment. After a couple of years of experience of the Noise TSI certification procedure, the question was raised whether it was really necessary to perform full-scale measurements in cases where only minimal changes are expected in comparison to already tested vehicles, for instance of the same family.

This issue was addressed in the limited revision of the Noise TSI in 2011 (2011/229/EU) that now offers more flexibility in the conformity assessment procedure by permitting a simplified evaluation method.

31.1.2. *The simplified evaluation method*

The simplified evaluation method can be applied to avoid some or all of the tests by "acoustically comparing the type under assessment to an existing type with documented noise characteristics compliant with the noise TSI". This approach consists of "providing evidence to show that the acoustically relevant systems and characteristics are either identical to those of the reference type, or such that they will not result in higher noise emission of the unit under assessment". It is further stated in the TSI Noise that the evidence could include a calculation, simplified measurement or a combination of both.

The introduction of the simplified evaluation method in the Noise TSI was the first step to making the conformity assessment of new railway rolling stock less complex and expensive. However, this method still has its limitations. One is that the new rolling stock can only be verified through this method if it can be proved that the noise level is lower than that of the reference. This is required even though the levels of the reference might be far below the Noise TSI limiting values. Furthermore, it has not been clarified what kind of evidence is needed. This has partly resulted in a reluctance to use this method.

One of the reasons for these limitations is the lack of commonly agreed procedures for acoustically assessing modifications with calculations and simplified measurements. Another reason could be the fact that there are no tools today that have been recognized to reliably demonstrate the effect of the modifications on the total noise of the vehicle.

In the European research project Acoutrain, work has been done to clarify the simplified evaluation procedure by providing flow charts for common cases. The next step in the direction of developing a less complex and time-consuming conformity assessment procedure is to address the issues mentioned above by developing procedures for so-called virtual testing. This is one of the main objectives of the Acoutrain project. This chapter gives a general introduction to the project.

31.2. Acoutrain: a European research project

Acoutrain is a project which is co-funded by the European Commission within the 7th framework programme for research. It is coordinated by UNIFE, the European Rail Industry Association, and the consortium consists of 15 partners representing railway operators, rolling stock manufacturers, universities and research institutes. The work package leaders are representatives from DB Systemtechnik (on behalf of DB), SNCF, Bombardier Transportation, Alstom Transport and the University of Southampton. The Acoutrain project term started in October 2011.

The main objective of Acoutrain was to reduce time and costs of the TSI Noise conformity assessment procedure by developing procedures for acoustic virtual testing. It should also contribute to a harmonization of the noise conformity assessment across Europe by providing standard procedures.

31.2.1. *The simplified evaluation method in Acoutrain*

The last TSI Noise revision (2011/229/EU) already introduced some flexibility in the certification process in specific cases:

– *vehicle families*: very often variants of a vehicle type are produced to satisfy different requirements of end-users; this results in several similar vehicle types belonging to the same vehicle family;

– *different formation of multiple units*: often multiple units are produced with different combinations of powered and unpowered vehicles;

– *retrofitted vehicles*: it should be demonstrated that noise levels are not increased if a component is replaced on a vehicle (for instance, an HVAC unit).

A so-called "simplified evaluation method" is allowed for these cases, but further clarifications on the process to be applied were needed: Acoutrain has proposed some procedures to be applied in this framework.

A number of different application cases have been selected, corresponding to the most common cases where a simplified method will be useful:

– modification of number of axles;

– modification of the maximum speed;

– modification of the wheels;

– modification of the brake system;

– different formations of multiple single cars (add or remove a single car);

– selection of the worst case within a family of vehicles;

– modification of equipment (traction equipment or auxiliary system) or its installation.

These cases do not represent an exhaustive selection of what could be encountered when applying the Noise TSI; the type of rolling stock to which the method is applicable has to be specified case-by-case.

The procedures that could be used as a simplified method have been described with dedicated flowcharts in a document by Acoutrain. The flowcharts define the limits of application of a simplified approach in each of the cases mentioned above and give a procedure to check if noise levels (for stationary and/or pass-by running condition) are lower or equivalent to the noise levels of a reference vehicle (already certified as Noise TSI compliant). For example, for the case of a modification of an equipment or its installation, the following flowchart (Figure 31.1) proposes a simplified approach to check that there would be no effect (no change in the noise level) or a positive effect (decrease in the noise level) on the stationary, pass-by or starting noise so that the train equipped with the new equipment is automatically certified.

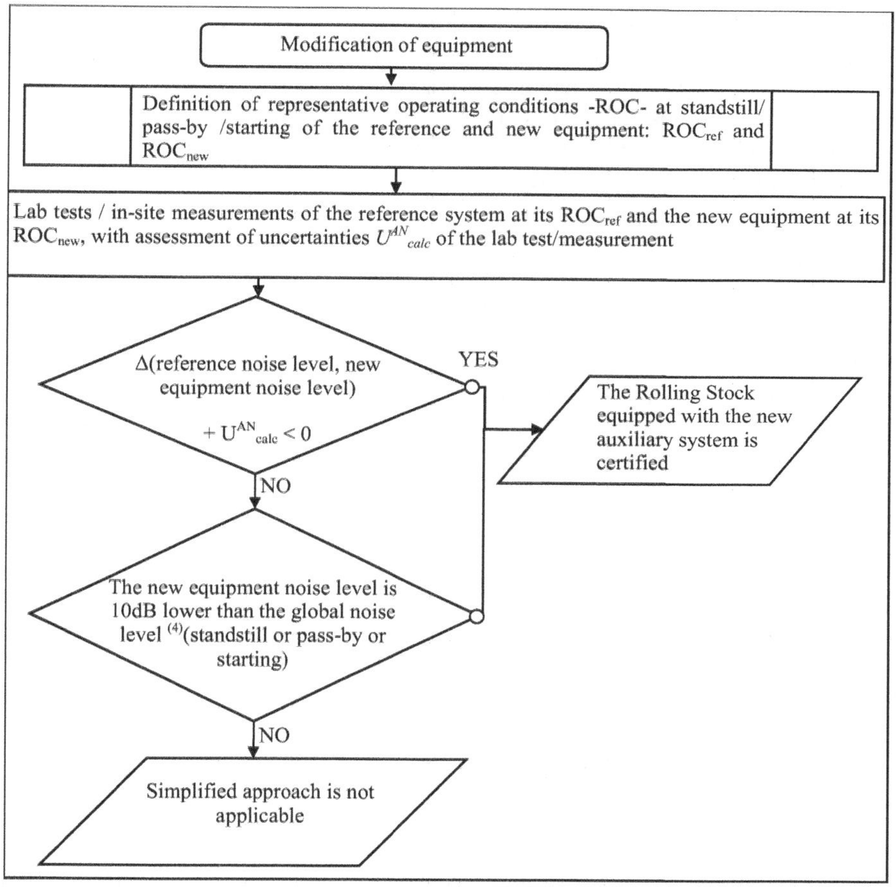

Figure 31.1. *Example of flowchart*

For all the flow charts, it is considered that the proposed methodologies can only manage one modification at the time. Of course, positive effects (reduction of noise level) can be considered as cumulative and in this case several modifications can be handled. In addition, indications are given to justify their reliability, considering either previous experience or analytical justifications.

31.2.2. *Virtual testing concept in Acoutrain*

The virtual testing (VT) procedures that are being developed in Acoutrain should provide an alternative to the real testing procedures (measurements) that are required today by the Noise TSI. The indicators and the limiting values that are defined in the TSI should, however, remain and be consistent for both real and virtual procedures. Consequently, the general concept of virtual testing as proposed in Acoutrain is to simulate the measurement situation and set up with corresponding computations. This means that, for instance, a pass-by noise measurement is simulated considering parameters such as operating conditions, speed and receiver positions as found in the description of the measurement method in the current Noise TSI.

Generally, to perform virtual testing, a specific simulation tool is required. By defining the geometry of the vehicle in question and loading sound power levels of the individual noise sources in the tool, a virtual vehicle is created. The characteristics of the individual sources can be assessed with measurements, numerical models or a combination of both. The simulation tool takes physical effects such as wave propagation, ground reflection and the Doppler effect into account. Such simulations have often been done already by the manufacturers in advance of the certification measurements, to manage the development of the vehicle in terms of acoustic aspects.

In Acoutrain virtual testing, procedures are mainly developed for stationary and pass-by noise. A simulation tool has been developed that includes the basic functions that are required for virtual testing for these two cases. It is useful for demonstrating the new procedure but also made available at the end of the project. An example from the Acoutrain simulation tool can be seen in Figure 31.2. Other tools that are used in Acoutrain to apply the virtual testing procedure are internal tools developed by the partners of the consortium. All these tools are solely developed to simulate exterior noise and are therefore not suited for simulation of noise in the driver's cab.

There are two work packages in Acoutrain that are dedicated to the typical sound sources of railway rolling stock and the development of methods to characterize these to create the input for a simulation tool. Sources that are investigated are, for instance, wheels and tracks (rolling noise), cooling and ventilation systems and electric traction systems. Figure 31.3 shows the measurements that were made in Acoutrain on a converter mock-up in the acoustic laboratory of project partner KTH.

Some important aspects for these work packages are to evaluate the possibility of characterizing individual noise sources in an acoustic laboratory while at the same time simulating realistic operating conditions, source integration effects, accounting for source directivity and the modeling of wheels and track.

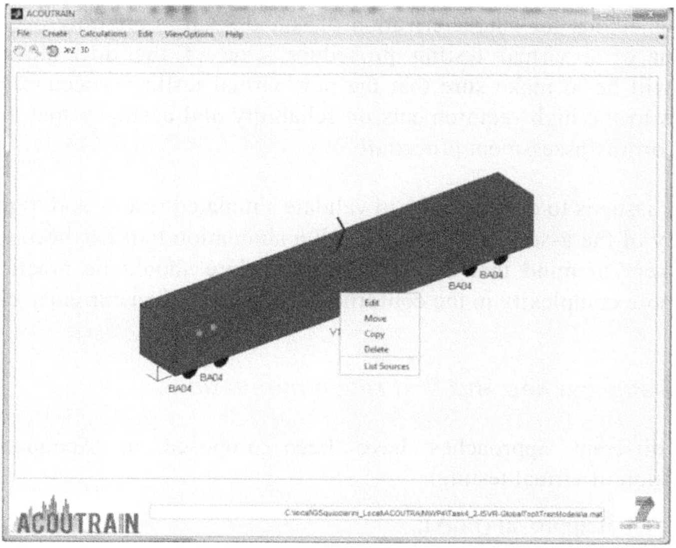

Figure 31.2. *Screenshot from the Acoutrain simulation tool showing the definition of the virtual vehicle. The green points represent noise sources*

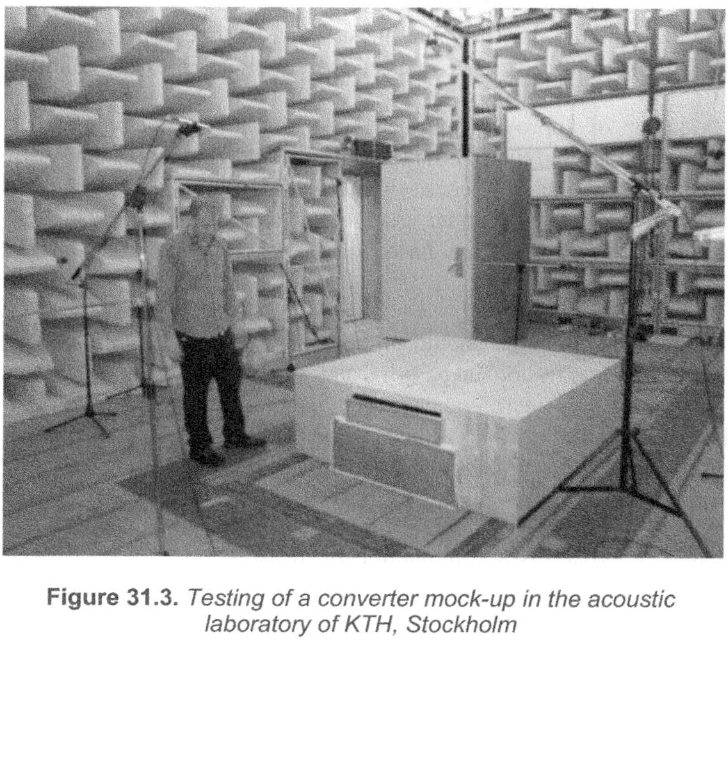

Figure 31.3. *Testing of a converter mock-up in the acoustic laboratory of KTH, Stockholm*

But the methods for characterization of sound sources are only one part of the development of a virtual testing procedure. One of the difficulties following Acoutrain will be to make sure that the new virtual testing procedure as a whole corresponds to the high requirements on reliability and accuracy that are expected from a conformity assessment procedure.

One key issue is to define a way to validate simulated results and to maintain the transparency of the assessment even though a simulation tool has been used. It also has to be kept in mind that the new VT-procedure should be practical and not introduce more complexity in the conformity assessment than currently exists.

31.2.3. *VT-approaches and first recommendations*

Three different approaches have been proposed in Acoutrain for the implementation of virtual testing:

– extension of approval (EoA);

– hybrid testing;

– full virtual testing.

Extension of approval and hybrid testing are the two approaches that at this point are considered to be the most reliable ones, since the assessment is based on a mix of well-known measurement procedures and relatively simple computations.

The extension of approval approach is to create a virtual vehicle of the type under assessment in the simulation tool by altering the inputs and settings of a preexisting and validated reference virtual vehicle model. The reference vehicle should be similar to the type under assessment; ideally the two vehicles belong to the same vehicle family and are based on the same design with only a limited number of modifications.

A measurement campaign should be performed with the reference vehicle. The results of the noise emission in pass-by and stationary situations will be used to validate the accuracy of the reference virtual vehicle. If it is approved, it can be used for the homologation of further vehicles. For these succeeding vehicles, it is only necessary to assess the modifications that would affect the noise emission in comparison to the reference vehicle.

The hybrid approach is a combination of measurements and computations, but unlike the extension of approval approach it does not rely on the comparison with a reference vehicle. The hybrid approach could, for instance, be useful to transpose pass-by measurements results from a normal track to correspond to a TSI compliant reference track.

The full virtual testing approach is when the type testing is performed solely by computations in a simulation tool. Theoretically, for this approach, no real assembled vehicle would be required for the certification, only the inputs to create a virtual vehicle. This can consist, for instance, of measurements of individual sound sources. The full virtual approach can be seen as a long-term objective that could be reached after more experience with virtual testing and the tools have been gained. It remains to be examined how the results from this approach could be validated without having access to full- scale measurements.

These three approaches will be validated in Acoutrain before the proposal a new virtual certification procedure. A number of application cases have been defined that represent scenarios that are common in the field of rolling stock certification and for which a virtual testing procedure would be profitable. These will be demonstrated and evaluated by comparing simulation outputs to measurement results. The focus will be on validating the accuracy of the new procedure, taking variability into account as well as to assess its practicability.

31.3. Conclusion

The main objective of the European research project Acoutrain was to develop a virtual testing procedure that could be introduced in the next revision of the Noise TSI as an alternative to the field tests that are required for the current conformity assessment. The aim was to reduce costs but at the same time maintain a high level of reliability and practicability. A basic concept for virtual testing has been developed, including the three different VT-approaches described above. The next step will be to validate the proposed procedure by using the methods and tools that have been developed so far in Acoutrain and comparing the results to experimental data. Both measurement and calculation uncertainties as well as possible sources for errors will be evaluated during this process.

More information regarding Acoutrain can be found on the project home page: www.acoutrain.eu

31.4. Acknowledgments

We would like to thank all contributing partners of the Acoutrain project.

31.5. Bibliography

[BON 12] BONGINI ESTELLE, Clarification of the simplified method in the partial revision of the TSI, Acoutrain research project Deliverable 1.1, 2012.

[COM 11] COMMISSION DECISION of concerning the technical specifications of interoperability relating to the subsystem 'rolling stock – noise' of the Trans-European conventional rail system, 4 April 2011.

[EUR 11] EUROPEAN COMMISSION, Roadmap to a single European Transport Area – Towards a competitive and resource efficient transport system, White Paper, 2011.

32

Additional Efficient Warning Sounds for Electric and Hybrid Vehicles

Electric and hybrid vehicles are very silent, which represents a major advantage for the reduction of noise annoyance in urban areas. However, this makes them dangerous for pedestrians, especially vulnerable ones such as visually impaired people. Current solutions consist of using warning sounds so that the exterior noise of an electric vehicle is as high as for a conventional one. This may cancel the benefit of electric vehicles for the reduction of sound annoyance. The eVADER project (funded by the European Commission) aims at proposing a prototype car which combines a high safety and a low noise level. Part of the work program consists of perceptual studies. The goal of these studies was to evaluate the influence of various timbre parameters on the detectability and the unpleasantness of a warning sound. Results show that it is possible to make an electric vehicle easily detected while keeping its sound level much lower than the one of a conventional car.

32.1. Introduction

Electric and hybrid vehicles are very quiet at low speeds (typically below 30 km/h). The level of the noise emitted by such vehicles is 5–7 dB(A) lower than the one of conventional diesel cars. This is very beneficial to people living in an urban area, because transportation noise is a major source of annoyance in cities. On the other hand, this represents a hazard for pedestrian, who may not hear an approaching car. Vulnerable people, such as the visually impaired, have a very strong concern about such cars and manufacturers use additional warning sounds in order to prevent this risk. Though several studies have already been conducted in this field (e.g. [YAM 11, EME 12]), the easiest way to solve this safety issue is to use loud warning sounds, which cancel the noise reduction advantage of electric

Chapter written by Etienne PARIZET, Ryan ROBART, Perceval PONDROM, Jean-Christophe CHAMARD, Guillaume BAUDET, David QUINN, Karl JANSSENS and Manfred HAIDER.

cars. Regulations, either already decided (NHTSA) or under preparation (in Europe), are headed this way. As an example, the regulation currently prepared in Europe states that "the (warning) sound level may not exceed the sound level of a similar internal combustion engine vehicle". It should be possible to use more efficient warning sounds.

The eVADER project brings together partners from universities, research centers, car manufacturers and suppliers. Last but not least, the European Blind Union is a partner, so that endusers are represented in the consortium. The goal of this project is to develop a prototype vehicle combining a high safety for pedestrian and a low noise annoyance for city residents. Several technical solutions are developed:

– an automatic pedestrian detection device;

– a set of loudspeakers focusing the sound in the direction of the detected pedestrian;

– a warning sound designed so that it can be detected in an urban environment at a low level.

This chapter will describe studies focusing on this third objective. It will present three listening test experiments: two of them aimed at evaluating the influence of some timbre parameters on the detectability and annoyance of warning sounds, and a third one intended to determine whether some information about the speed of the car can be evoked by the warning sound, without any training of the listener.

These three experiments were simultaneously conducted by various partners of the project, which allowed us to use large subject samples.

32.2. Detectability

The first part of the study was related to the detectability of warning sounds. Given the background traffic noise of an urban environment, is it possible to make a warning sound easily detectable in spite of a low level? It was decided to limit the study to multi-tone sounds in a middle frequency range (300–1,500 Hz). The lowest frequency was due to technical limitation of the loudspeakers to be used in the prototype (for a high radiation efficiency at low frequencies, very large loudspeakers would have been necessary). And the high frequency limit was selected because people suffering from presbyacousis have high hearing thresholds at higher frequencies.

32.2.1. *Procedure*

In this experiment, three timbre factors were investigated: the number of tones, the frequency variation and the temporal variation. Each factor could have three levels: as an example, the number of tones could be 3, 6 or 9. A fractional factorial design was used, so that nine combinations were used (instead of 27 in a full factorial design). More details about the stimuli definition can be found in Robart *et al.* [ROB 13].

Therefore, nine stimuli were synthesized; they all had the same A-weighted level. Then, they were modified in order to represent a moving source, passing in front of a listener at the speed of 20 km/h. Finally, each modified stimulus was added to the recording of an electric vehicle recorded at 20 km/h by a dummy head located close to the road. This way, it was possible to simulate the situation of a pedestrian facing the road, waiting to cross this road and paying attention to any approaching car (Figure 32.1).

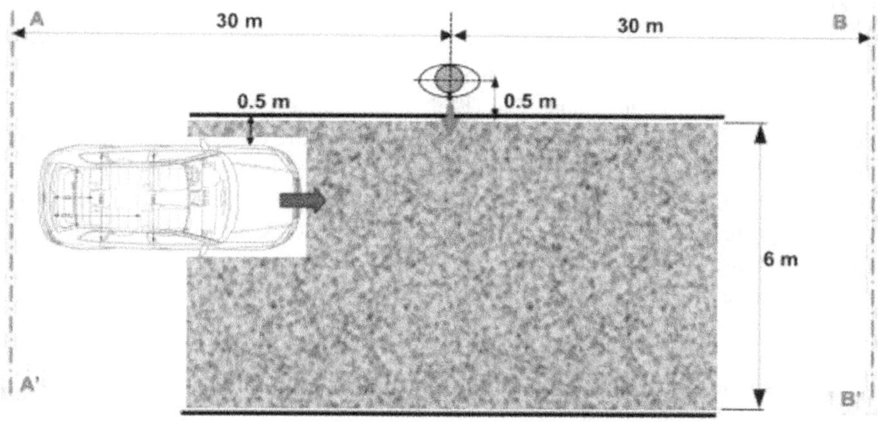

Figure 32.1. *"Waiting to cross" scenario*

The recording of the electric vehicle alone as well as the one of a similar diesel car was added to this set of stimuli. The level of warning sounds was adjusted so that they increase the level of the electric vehicle very slightly (less than 2 dB(A), see Figure 32.2). The level of the diesel car was higher by more than 5 dB(A).

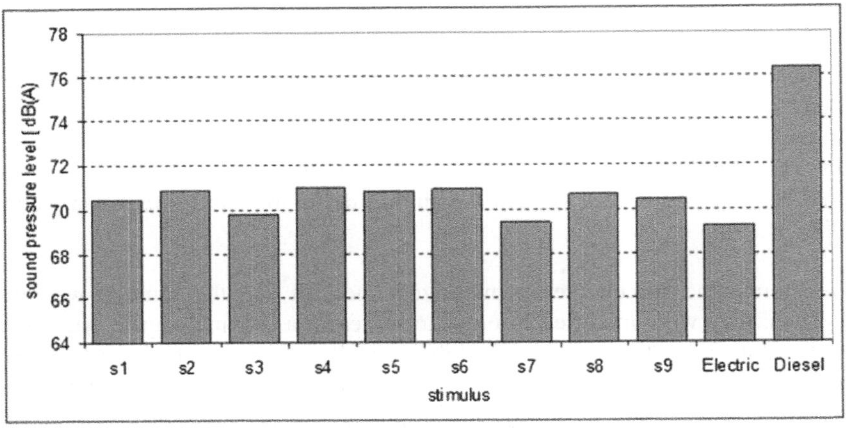

Figure 32.2. *Peak level (A-weighted SPL) of each stimulus used in experiment 1*

During the experiment, the listener was hearing a background traffic noise through headphones. Rain noise was added, as this represents a very difficult situation for blind pedestrians. The level of this noise was 69 dB(A). At randomly selected times, one of the car stimuli was added to the noise. This car could arrive from the left or from the right of the listener. The task of the listener was to detect the approaching car as soon as possible and its direction. He gave his answer by pressing a key of a computer keyboard. Two keys were used: the <Enter> key in the case of a "right" answer and the <Space> bar for the "left" answer. The response time (from the starting of the stimulus) was measured and stored by the computer.

Each sound was presented eight times (four times from each direction) so that a listener was presented 88 stimuli (in a random order).

One hundred and ten subjects participated to the experiment; among them, 33 were visually impaired people.

32.2.2. *Results*

The averaged response time was converted into distance from the listener at detection. These distances are shown in Figure 32.3.

In this figure, the "risk area" is represented by a hashed section. If the pedestrian starts crossing the road while the car is closer than 5 m away, he may be hit by the car, given the averaged reaction time needed by the driver to start breaking [KER 08]. The electric vehicle is detected in this area, which confirms that EVs can be dangerous for pedestrians [GLA 12].

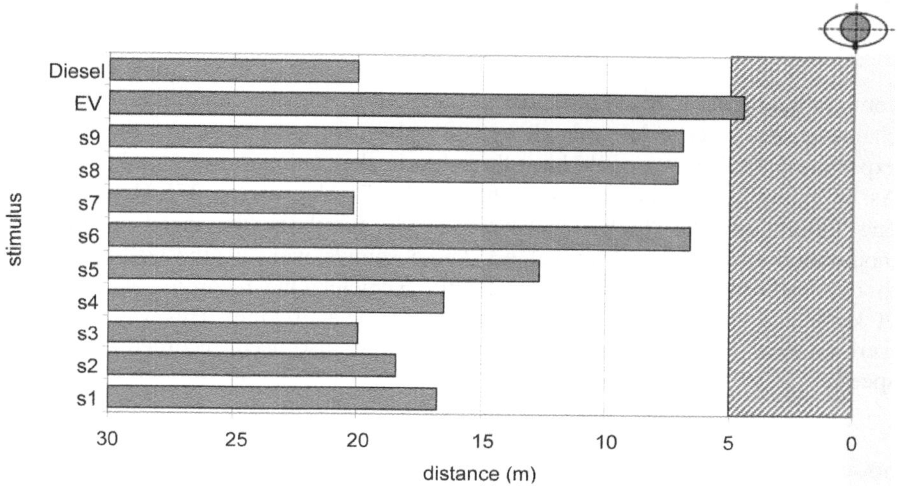

Figure 32.3. *Averaged distance from the listener at detection. The hashed area represents the "risk area" (see text)*

Warning sounds had quite different efficiencies, depending on their timbre. Some warning sounds are nearly inefficient (s6, s8 and s9). Some other ones make the EV as easy to detect as the diesel car (s3 and s7). Figure 32.2 clearly shows that such differences are not related to the level of these sounds – timbre features are the only reason for such differences. Further analysis showed that some controlled features were favorable to detection:

– a low number of tones. The overall sound level was kept constant, so that the difference between each tone level and its detection threshold was greater when the number of tones was low;

– amplitude modulation. Fluctuations in the signal amplitude help the listener to detect it in the background of traffic noise.

32.3. Sound meaning

In a second experiment, the question was raised whether it is possible to design the warning sound so as to provide the pedestrian with some information about the distance and the speed of the car. The hypothesis was that a higher pitch and a quicker temporal variation of the warning sound would evoke a faster car.

32.3.1. *Procedure*

To test this hypothesis, a full experimental design was used. Sounds had three components and a pseudo-random fluctuation of the amplitude, as these characteristics were determined as favorable for detection in the previous experiment. Each factor could have three levels, denoted as "low", "mid" or "high". As an example, one factor was pitch; the "low" value means 275 Hz, the "mid" one 300 Hz and the "high" level was 325 Hz. Similarly, the speed of amplitude modulation could have three values. These sounds were synthesized and added to two recordings of a passing-by electric vehicle. One recording was made at 20 km/h as in the first experiment and the other one at 30 km/h. This way, 18 stimuli were created. Two recordings of a diesel car were also used, recorded at the same speed.

These sounds were presented to listeners in a background noise. The level of this noise was lowered as compared to the first experiment (64 dB(A) instead of 69 dB(A)) so that each approaching car could be easily detected. The task of the listener was to imagine that he was going to cross the road in spite of this approaching car. He had to indicate when he thought that it was no longer safe to do it, given his appreciation of the speed of the car. Once again, he gave his answer by pressing a key and the response time was stored.

Each sound was repeated four times, giving a number of 80 responses from each listener. One hundred and sixteen people participated to the experiment (77 with normal vision and 39 visually impaired).

32.3.2. *Results*

The effect of each factor at each level was computed as the averaged response time for this level subtracted by the grand mean of all data. These effects are shown in Figure 32.4.

The speed effect is clear and natural: people feel unsafe sooner when the car is approaching at 30 km/h compared to 20 km/h. It should be noted that this 10 km/h difference is greater than the velocity threshold as mentioned in a previous study from Lutfi and Wang [LUT 99], who measured a threshold of 8 km/h at 36 km/h. But the response time difference between the two speeds is small: 0.4 s. If the response times are converted into distances from the car, averaged values are 14 m at 20 km/h and 10 m at 30 km/h. This indicates that listeners clearly detected speeder cars, but underestimated the speed difference.

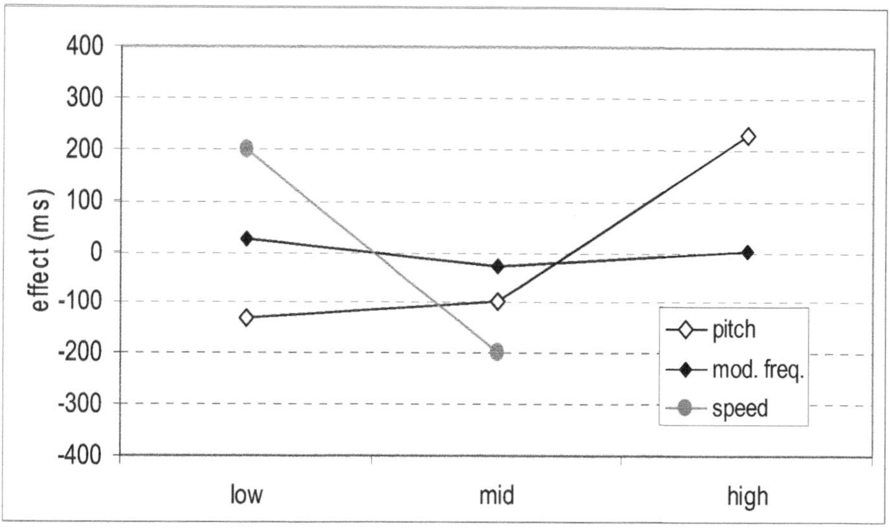

Figure 32.4. *Factor effects*

Results also show that varying the speed of the temporal fluctuation of the additional sound gives no information to listeners: the effect of this factor is very close to zero, for any of the three levels. And, very surprisingly, the effect of the pitch is that the highest value gives a later response time. This was totally unexpected.

The main conclusion of this experiment is that it seems very unlikely to use the warning sound for conveying some information about the speed of the car which can be understood without any training. The speed of the car is certainly evaluated from the temporal increase in loudness as the car is approaching. Any warning sound will participate to this temporal loudness variation, whatever its characteristics. However, varying the pitch of this sound with the speed of the car may indicate an accelerating (or decelerating car); this was not evaluated by this experiment, as the speed was constant for each stimulus.

32.4. Unpleasantness

32.4.1. *Procedure*

The third experiment was devoted to the evaluation of the unpleasantness of all warning sounds used in the previous parts of the study. Twenty sounds were presented to listeners: the 11 stimuli of experiment 1 and the nine representing an EV with a warning sound at 20 km/h in experiment 2. The experiment was conducted in two conditions: some listeners were presented the stimuli without any background noise and for some other ones, a low-level traffic noise (57 dBA) was added to each stimulus. In both cases, the task of the listener was to evaluate the unpleasantness of the sound. He gave his answer by moving a cursor on a continuous scale, labeled from "not at all unpleasant" to "extremely unpleasant". The position of the cursor was stored as a number between 0 (for "not at all unpleasant") to 1,000 ("extremely unpleasant"). Each sound was presented twice, the order of presentation being randomly selected. 145 people participated to this last experiment, which was conducted in four laboratories.

32.4.2. *Results*

The repeatability of each listener was evaluated by computing a mean squared difference between the two values he gave for each sound, namely:

$$C = \sqrt{\frac{1}{20}\sum_{n=1}^{20}\left(x_{n1} - x_{n2}\right)^2} \qquad [32.1]$$

where x_{n1} and x_{n2} represent the two evaluations of sound n.

Individual coefficients range between 50 and 450 (mean value: 179, standard deviation: 65). For 23 subjects, this coefficient is higher than 250, which represents a full category of the scale: such subjects can be considered as inconsistent. Such a high number of inconsistent subjects means that the task was difficult. So, it was decided to select the most reliable subjects for further analysis. The maximum value for C was fixed to 150, which allowed to select 56 people. Twenty-six of them did the experiment with the background traffic noise and 30 without this noise.

An analysis of variance was realized (repeated measures, background noise condition as an interindividual factor and stimuli as intraindividual ones). The only influent factor was stimuli [$F(19, 988) = 48.5$, $p < 0.0001$]. The unpleasantness of each sound was averaged over this subpanel and results are shown in Figure 32.5.

Homogeneous groups of sounds are represented by thick horizontal lines (these groups have been determined using Scheffe's technique).

As it can be seen in Figure 32.5, most warning signals strongly increase the unpleasantness of the sound. This can be due to the fact that people felt very unfamiliar with such warning sounds. Three of them can be considered as equally unpleasant as the diesel car. The particularity of these sounds is that no temporal fluctuation was applied to their amplitude. Amplitude fluctuation increased unpleasantness; the first experiment had shown that it also increased the detectability of sounds. The relation between efficiency and unpleasantness had already been proved for other kinds of warning sounds [TAN 95, RIC 10, FAG 11].

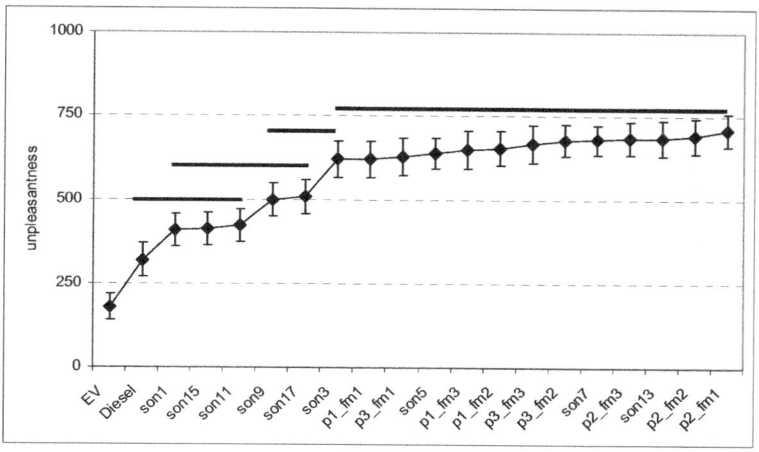

Figure 32.5. *Unpleasantness of each sound*

32.5. Conclusion

This chapter presented three experiments aiming at evaluating the influence of basic timbre parameters of a warning signal on its detectability and annoyance. It has been shown that some warning sounds can make an electric vehicle as detectable as a diesel car, for a much lower sound level. Nevertheless, people reported these signals to increase the unpleasantness of the car sound. It is hypothesized that this was due to the unnaturalness of such signals and to their novelty. Further studies during which subjects could get used to such sounds would be useful. Nevertheless, some signals seem to provide a correct compromise between the two objectives.

Finally, it should be recalled that the goal of these studies was not to define a warning signal for a typical application, but to investigate the influence of timbre parameters. This way, it is expected that car manufacturers will have some guidelines when defining their own signal, which should also fulfill some brand image requirements.

32.6. Acknowledgments

The eVADER project is funded by the European Commission in the frame of the 7th Framework Program.

And the first author is a member of the Labex CeLyA (Lyon Acoustics Center), funded by the French National Research Agency (ANR-10-LABX-60).

32.7. Bibliography

[EME 12] EMERSON R., NAGHSHINEH K., KIM D. *et al.*, "Blind pedestrians and quieter vehicles: how adding artificial sounds impacts travel decision", *Proceedings of Internoise Conference*, New York, 2012.

[FAG 10] FAGERLÖNN J., Urgent alarms in trucks: effects on annoyance and subsequent driving performance, IET Intelligent Transport Systems, available at: http://pure.ltu.se/portal/files/34271457/Johan_Fagerl_nn.pdf, 2010.

[GLA 12] GLAESER K.P., MARX T., SCHMIDT E., "Sound dectection of electric vehicles by blind or visually impaired persons", *Proceedings of Internoise Conference*, New York, 2012.

[KER 08] KERBER S., FASTL H., "Prediction of perceptibility of vehicle exterior noise in background noise", *Proceedings of DAGA conference, Dresden*, 2008.

[LUT 99] LUTFI R., WANG W., "Correlational analysis of acoustic cues for the discrimination of auditory motion", *Journal of the Acoustical Society of America 106*, pp. 919–928, 1999.

[RIC 11] RICHIR B.A., Effects of rhythm on the perception of urgency and irritation in auditory signal, MSc Thesis, Utrecht University, available at: http://igitur-archive.library.uu.nl/student-theses/2010-0907-202354/finalthesis–basalleinricher 0819.pdf, 2011.

[ROB 113] ROBART R., PARIZET E., CHAMARD J.C. *et al.*, "eVADER: a perceptual approach to finding minimum warning sound requirements for quiet cars", *Proceedings of AIA-DAGA conference*, Merano, 2013.

[TAN 95] TAN A.K., LERNER N.D., Multiple attribute evaluation of auditory warning signals for in-vehicle crash avoidance warning systems, Report no. DOT HS 808 535, National Highway Traffic Safety Administration, 1995.

[YAM 11] YAMAUCHI K., MENZEL D., FASTL H. *et al.*, "Cross-cultural study on feasible sound levels of possible warning sounds for quiet vehicles", *Proceedings of Internoise conference*, Osaka, 2011.

List of Authors

Maysam ABBOD
Centre for Electronics Systems Research
Brunel University
Middlesex
UK

Bijan ADL-ZARRABI
Chalmers University of Technology
Gothenburg
Sweden

Lars AKKERMANS
Transport & Mobility Leuven
Belgium

Michel ANDRÉ
Laboratory Transport and
Environment (LTE)
IFSTTAR
Bron
France

Michel ANDRÉ
Universitat Politecnica de Catalunya
Barcelona
Spain

Christian AUDOLY
DCNS
Toulon
France

Heidi AUVINEN
Technical Research Centre of Finland
(VTT)
Espoo
Finland

Wolfgang BACKHAUS
Rupprecht Consult Forschung und
Beratung GmbH
Cologne
Germany

Wamadeva BALACHANDRAN
Centre for Electronics Systems Research
Brunel University
Middlesex
UK

Eric BAUDIN
Bureau Veritas
Paris
France

Heike BARLAG
Siemens AG
Fuerth
Germany

Guillame BAUDET
Renault
Guyancourt
France

Radu BELECA
Centre for Electronics Systems Research
Brunel University
Middlesex
UK

Lanfranco BENEDETTI
SEA Europe
Brussels
Belgium

Wojciech BLAZEJEWSKI
Wroclaw University of Technology (WRUT)
Wroclaw
Poland

Simon BOLWIG
Technical University of Denmark
Denmark

Estelle BONGINI
SNCF
Innovation and Research Department
Paris
France

Antje BURMEISTER
Production Systems, Logistics, Transport Organization and Work (SPLOTT)
IFSTTAR
France

Kristof CARLIER
Transport & Mobility Leuven
Belgium

Marion CARTERET
Laboratore de Chimie Moléculaire en Environnement
University of Savoy
France

Massimo CERAOLO
Department of Engineering of Energy, Systems, Territory and Construction
University of Pisa
Italy

Jean-Christophe CHAMARD
PSA Peugeot Citroën
Vélizy-Villacoublay
France

Alberto COLORNI
Poliedra and Design – Politecnico di Milano
Milan
Italy

Mario CONTE
Italian National Agency
for New Technologies
Energy and the Sustainable Economic Development (ENEA)
Rome
Italy

Cecilia CRUZ
Economic and Social Dynamics of Transport Laboratory (DEST)
IFSTTAR
Marne la Vallée
France

Nathalie CUNY
ALSTOM Transport
La Rochelle
France

Peter DAEMON
Magna Steyr
Graz
Austria

Igor DAVYDENKO
TNO Sustainable Transport and Logistics
Delft
The Netherlands

Mauro DELL'AMICO
Interuniversity Consortium for Optimization and Operations Research
Reggio Emilia
Italy

Muriel DESAEGER
Toyota Motor Europe
Brussels
Belgium

Guido DI PASQUALE
Pluservice SRL
Senigallia
Italy

Annele EEROLA
Technical Research Centre of Finland (VTT)
Espoo
Finland

Verena EHRLER
Institute of Transport Research German Aerospace Center (DLR)
Germany

Thomas FOLEGOT
Quiet-Oceans
Plouzane
France

Nicolas FURIO
UNIFE
Brussels
Belgium

Lionel GANIPPA
Centre for Advanced Powertrain and Fuels Research
Brunel University
Middlesex
UK

Pawel GASIOR
Wroclaw University of Technology (WRUT)
Wroclaw
Poland

Savas GEIVANIDIS
Laboratory of Applied Thermodynamics
Aristotle University
Thessaloniki
Greece

Gabriele GIUSTINIANI
Centro di ricerca per il Trasporto e la Logistica
La Sapienza – Università di Roma
Italy

Christian GREGOR
Bundesanstaltfür Materialforschung und –prüfung (BAM)
Berlin
Germany

Richard GRIMAL
Economic and Social Dynamics of Transport Laboratory (DEST)
IFSTTAR
Marne-la-Vallée
France

Paul GRZESCHIK
Institute for Combustion Engines (VKA)
RWTH Aachen University
Germany

Leandro GUIDOTTI
Università degli studi di Modena e Reggio Emilia
Italy

Manfred HAIDER
Austrian Institute of Technology
Vienna
Austria

Heather HAMJE
CONCAWE
Brussels
Belgium

Bastian HOLDERBAUM
FEV GmbH
Aachen
Germany

Mohammad HOSEINI
Norwegian Public Roads Administration (NPRA)
Oslo
Norway

Arno HUSS
AVL List GmbH
Graz
Austria

Bernard JACOB
IFSTTAR
Marne-la-Vallée
France

Karl JANSSENS
LMS International
Leuven
Belgium

Stephen JONES
AVL List GmbH
Graz
Austria

Jerzy KALETA
Wroclaw University of Technology (WRUT)
Wroclaw
Poland

Dimitris KATSAOUNIS
Laboratory of Applied Thermodynamics
Aristotle University
Thessaloniki
Greece

Petros KATSIS
Emisia S.A
Thessaloniki
Greece

Jan KIEL
Panteia
Zoetermeer
The Netherlands

Bernhard KIENER
Ventrex
Graz
Austria

Michael KLESCHINSKI
xperion Energy & Environment GmbH
Kassel
Germany

Thomas KÖRFER
FEV GmbH
Aachen
Germany

Jean-Bernard KOVARIK
Ministry of Infrastructure, Transport and the Sea
Paris La Défense
France

Michael KRAIL
Fraunhofer-ISI
Karlsruhe
Germany

Emre KURAL
AVL List GmbH
Graz
Austria

Nikolaos KYRTATOS
HERCULES Coordinator
National Technical University of Athens
Greece

Matthieu DE LAPPARENT
Economic and Social Dynamics of Transport Laboratory (DEST)
IFSTTAR
Marne-la-Vallée
France

Hermann-Josef LAUMEN
FEV GmbH
Aachen
Germany

Anna LEINONEN
Technical Research Centre of Finland (VTT)
Espoo
Finland

Alan LEWIS
Transport & Travel Research Ltd (TTR)
Nottingham
UK

Heinrich LIENKAMP
Infraserv GmbH & Co.
Frankfurt
Germany

Andreas LISCHKE
Institute of Transport Research
German Aerospace Center (DLR)
Germany

Alessandro LUÈ
Poliedra and Design –
Politecnico di Milano
Milan
Italy

Sven MAERIVOET
Transport & Mobility Leuven
Leuven
Belgium

Georg MAIR
Bundesanstaltfür Materialforschung
und –prüfung (BAM)
Berlin
Germany

Nadarajah MANIVANNAN
Centre for Electronics Systems Research,
Brunel University
Middlesex
UK

Giorgio MANTOVANI
Altra S.p.A
Genova
Italy

Pietro MASCOLO
Interuniversity Consortium for
Optimization and Operations Research
Reggio Emilia
Italy

Hamid MASOUMI
APT
Brussels
Belgium

Alexander MASSONER
AVL List GmbH
Graz
Austria

Hedi MAURER
Panteia
Zoetermeer
The Netherlands

Giorgio MENZATO
Centro Ricerche FIAT (CRF)
Turin
Italy

Alfonso MORENO RODRÍGUEZ
Técnicas y Servicios de Ingeniería S.L
Madrid
Spain

Edoardo MORRA
AVL List GmbH
Graz
Austria

David MORRIS
Coventry University Enterprises
Coventry
UK

Christian NIKLAS
AIT-Austrian Institute of Technology
Vienna
Austria

Roberto NOCERINO
Poliedra – Politecnico di Milano
Milan
Italy

Leonidas NTZIACHRISTOS
Laboratory of Applied Thermodynamics
Aristotle University
Thessaloniki
Greece

Ulf ORRENIUS
Bombardier Transportation
Västeras
Sweden

York OSTERMEYER
Chalmers University of Technology
Gothenburg
Sweden

Publio BELTRÁN PALOMO
Técnicas y Servicios de Ingeniería S.L
Madrid
Spain

Thomas PAPAGEORGIOU
Emisia S.A
Thessaloniki
Greece

Etienne PARIZET
Laboratoire Vibrations Acoustique
INSA-Lyon
Villeurbanne
France

Anaïs PASQUIER
Laboratory Transport and Environment (LTE)
IFSTTAR
Bron
France

David PEIRCE
Centre for Advanced Powertrain and Fuels Research
Brunel University
Middlesex
UK

Luca PERSIA
Centro di ricerca per il Trasporto e la Logistica
La Sapienza – Università di Roma
Italy

Perceval PONDROM
Technical University Darmstadt
Germany

David QUINN
Nissan
Sunderland
UK

Ashok RASTOGI
Infraserv GmbH & Co.
Frankfurt
Germany

Christophe RIZET
Economic and Social Dynamics of Transport Laboratory (DEST)
IFSTTAR
Marne-la-Vallée
France

Ryan ROBART
Laboratoire Vibrations Acoustique
INSA-Lyon
Villeurbanne
France

Kenneth D. ROSE
CONCAWE
Brussels
Belgium

Céline ROUSSET
DCNS
Toulon
France

Raul SALINAS MULLOR
Técnicas y Servicios de Ingeniería S.L
Madrid
Spain

Christos SAMARAS
Laboratory of Applied Thermodynamics
Aristotle University
Thessaloniki
Greece

Zissis SAMARAS
Laboratory of Applied Thermodynamics
Aristotle University
Thessaloniki
Greece

Wolfgang SCHADE
M-Five GmbH Mobility, Futures, Innovation, Economics
Karlsruhe
Germany

Udo SCHLEMMER KELLING
FEV GmbH
Aachen
Germany

Saskia SEIDEL
Institute of Transport Research German Aerospace Center (DLR)
Germany

Christa SIMON
AVL List GmbH
Graz
Austria

Maria STARNBERG
DB Systemtechnik GmbH
Munich
Germany

David Mark STORER
Centro Ricerche FIAT (CRF)
Turin
Italy

Volker STRUBEL
xperion Energy & Environment GmbH
Kassel
Germany

Reinhard TATSCHL
AVL List GmbH
Graz
Austria

David THOMPSON
ISVR
University of Southampton
UK

Sophie TINTIGNAC
Volvo Group Trucks Technology
Advanced Technology and Research
Gothenburg
Sweden

Anu TUOMINEN
Technical Research Centre of Finland (VTT)
Espoo
Finland

Giancarlo TROTTI
DimacRed
Milan
Italy

Patrick VANHONACKER
APT
Brussels
Belgium

Francesco VELLUCCI
Italian National Agency for New Technologies Energy and Sustainable Economic Development (ENEA)
Rome
Italy

Norbert VIERHEILIG
Siemens AG
Fuerth
Germany

Inge VIERTH
Swedish Road and Transport Research Institute VTI
Sweden

Christian VOCK
AVL List GmbH
Graz
Austria

Ilias VOUITSIS
Laboratory of Applied Thermodynamics
Aristotle University
Thessaloniki
Greece

Martine VROMANT
Centre of Expertise and Engineering on Risks, Environment, Mobility and Town Planning (CEREMA)
DTer Nord-Picardie
Lille
France

Holger WALLBAUM
Chalmers University of Technology
Gothenburg
Sweden

Nina WESSBERG
Technical Research Centre of Finland (VTT)
Espoo
Finland

Thomas WITTKA
VKA
RWTH Aachen University
Germany

Index

G, H, I

M, N, O

P, R, S

T, U, V, W

Printed and bound by CPI Group (UK) Ltd, Croydon, CR0 4YY

13/09/2026

14961035-0008